Springer Theses

Recognizing Outstanding Ph.D. Research

Aims and Scope

The series "Springer Theses" brings together a selection of the very best Ph.D. theses from around the world and across the physical sciences. Each volume has been selected for its scientific excellence by an internationally renowned institute or department. For greater accessibility to non-specialists, the published versions include an extended introduction, as well as a foreword by the student's supervisor explaining the special relevance of the work for the field. The series as a whole provides a valuable resource both for newcomers to the research fields, and for other scientists seeking detailed background information on special questions. Finally, it serves as an accredited documentation of the valuable contributions made by today's younger generation of scientists.

How to Apply

Theses prepared at internationally leading universities and institutes may be nominated for publication by supervisors with the endorsement of the head of department. Each such department is entitled to nominate one thesis per year. Enquiries and requests for a nomination form should be sent, together with the thesis, to Ms. Nathalie Berg (see link on right).

Eligibility Criteria

To qualify for consideration, theses should fulfil the following criteria:

- They must be written in good English.
- The topic should fall within the confines of Chemistry, Physics, Earth Sciences, Engineering and related interdisciplinary fields such as Materials, Nanoscience, Chemical Engineering, Complex Systems and Biophysics.
- The work reported in the thesis must represent a significant scientific advance.
- A maximum of 30% of the text may be a verbatim reproduction from the author's previous publications.
- They must have been examined and passed during the 12 months prior to nomination.
- The theses should have a clearly defined structure including an introduction accessible to new PhD students and scientists not expert in the relevant field.
- After acceptance, theses should be supplemented by a "Supervisor's Foreword" outlining the wider scientific relevance of the work

Indexed by zbMATH.

Timothée Pascal

Beyond the Standard Model at the Large Hadron Collider: From Simplified Model Constraints to Artificial Proto-Modelling for Dispersed Signals

Springer

Timothée Pascal
Laboratory of Subatomic Physics
and Cosmology
Grenoble, France

ISSN 2190-5053 ISSN 2190-5061 (electronic)
Springer Theses
ISBN 978-3-032-09390-5 ISBN 978-3-032-09391-2 (eBook)
https://doi.org/10.1007/978-3-032-09391-2

This Springer imprint is published by the registered company Springer Nature Switzerland AG
The registered company address is: Gewerbestrasse 11, 6330 Cham, Switzerland

Supervisor's Foreword

At the early stage of Run 3 of the LHC, the quest for new physics is at a crossroads. Yet again, one might say: It has also been at a crossroads about a decade ago, at the end of Run 1, when a new Higgs-like boson had been found and there was hope that this discovery might provide a window to new physics. Back then, another of my Ph.D. students, Beranger Dumont, received a Springer Thesis Award for his work on interpreting the first Higgs measurements and on early attempts toward the re-interpretation of searches for new physics.

We have come a long way since then. The 125 GeV Higgs boson turned out perfectly compatible with Standard Model expectations, and limits on the masses of new particles in Beyond-the-Standard Model (BSM) theories have been pushed higher and higher, at least in simplified scenarios. The experimental analyses at the LHC, especially at the ATLAS and CMS experiments, are, however, sensitive to a far greater set of theories and parameter combinations than considered in the experiments' publications. A large effort has thus emerged in the BSM community regarding the (re)interpretation of the experimental results in all kinds of theories of new physics, including the development of new methods and public software tools for this purpose. As one referee of the thesis presented here put it, "Re-interpretation of analysis outcomes is one of the big challenges and work directions of contemporary LHC physics."

So why are we at a crossroads today? The point is that a change of paradigm is taking place not only in the theory-experiment interaction (reflected, e.g., in the level of detailed material provided for public reuse) but also in the approach to new physics. While experimental searches are still being pursued final state by final state, targeting specific (yet often unrealistic) simplified-model scenarios, global interpretations of the wealth of experimental results have become crucial in order to understand which scenarios are really excluded and where new physics may still be hiding. This is particularly true for theories like supersymmetry, whose phenomenology is extremely rich and realistic scenarios are expected to give signals in several different channels. A global approach is then required to elucidate regions of parameter space yet uncovered. The question is a burning one at this stage of LHC Run 3, as new analysis strategies may be in order to cover existing holes.

The thesis of Timothée Pascal presented in this volume is part of this change of paradigm on several levels: First, it presents major improvements to SModelS, a public tool for the fast reinterpretation of LHC searches for new physics on the basis of simplified-model results, and its database of experimental results. This puts the reinterpretation of LHC results with SModelS on a new level of sophistication, opening the way for global likelihood analyses. Second, this thesis contains the first global analysis of LHC constraints on electroweak-inos in the context of the Minimal Supersymmetric Standard Model. Combining up to 16 ATLAS and CMS searches for which electroweak-ino efficiency maps are available in SModelS, it studies which combinations maximize the sensitivity in different regions of the parameter space, how fluctuations in the data in individual analyses influence the global likelihood, and what is the resulting exclusion power of the combination compared to the analysis-by-analysis approach.

Finally, switching from exclusion to discovery mode, the thesis presents a model-agnostic bottom-up approach to figuring out where new physics might be hiding in the LHC data. This approach, dubbed artificial proto-modeling, looks for small, mutually compatible excesses in the data, fitting candidate BSM states with generic properties to them. While the idea has similarities to anomaly detection, a big difference is that it aims at characterizing potential dispersed signals guided by a global likelihood. A dedicated chapter explains the concept together with the improvements and novelties that Timothée incorporated in its concrete realization.

All this is embedded in a thorough physics discussion, from the Standard Model and the rationale for exploring new physics theories to the construction of supersymmetric models, specifically the Minimal Supersymmetric Standard Model, without forgetting non-supersymmetric models. Also given are a comprehensive review of the experimental status after Run 2, and technical details regarding statistical inference. Quoting another referee, the thesis is "exceptionally well-written and organised, making the topic of LHC result reinterpretation accessible through comprehensive explanations." Moreover, all computer codes and generated data are publicly available, making this thesis a showcase example of adherence to open-access and FAIR principles.

At an early stage of his Ph.D., Timothée told me he wanted to do a thesis to be proud of. To my mind, and I think the reader will agree, this goal was achieved—and achieved quite well!

Grenoble, France
June 2025

Dr. Sabine Kraml

Preface

Our current understanding of the physical laws of our Universe relies on four fundamental interactions: gravitation, electromagnetism, the strong interaction, and the weak interaction. They dictate how matter (or energy) should behave when subject to them. Gravitation describes the space-time dynamics. It couples to the mass (or energy) of the studied system: the more massive (or energetic) the system, the more space-time is affected. Electromagnetism is agnostic to the mass but couples to electrically charged systems. Similarly, the more charged the system, the stronger the force. Likewise, the strong interaction couples to systems with a colour charge, a quantum number carried by subatomic particles. Finally, there are two types of weak interactions: the charged currents, which couple to systems with a weak isospin (another quantum number), and the neutral currents, which couple to systems with a weak isospin and/or an electric charge.

Gravitation usually dominates at macroscopic scales. It is described by general relativity and can be approximated by classical mechanics in low regime. The three other fundamental interactions usually dominate at mesoscopic and microscopic scales. They can be described by quantum mechanics and more specifically quantum field theories (QFT).

Despite the success of the Standard Model (SM) of particle physics, a QFT describing all the ordinary matter and fundamental interactions except gravitation, many questions remain open. To solve these questions, many new models beyond the Standard Model (BSM) have been proposed. This thesis mainly revolves around one class of these models: supersymmetric (SUSY) extensions of the SM [1–6]. Experimental searches conducted at the Large Hadron Collider (LHC) put stringent constraints on the masses of SUSY particles. However, these searches are pursued in a channel-by-channel approach, and most limits hold only in the context of simplified models [7–10], where only some SUSY particles with trivial mixings and branching ratios are assumed to be kinematically accessible. Reinterpretation software are then required to estimate how more realistic models are constrained by these results. This is particularly true for SUSY analyses, as even in its simplest incarnation, the Minimal Supersymmetric Standard Model (MSSM), which contains two Higgs doublets and where each SM field has a SUSY partner differing by half a spin unit, decay patterns, and thus signatures, can be quite complex. Therefore, the richness of the SUSY phenomenology, and the fact that realistic scenarios give signals in several different channels call for a global approach to elucidate regions of the parameter space yet uncovered.

The approach of this thesis is based on SModelS [11–17], a public tool designed for the fast reinterpretation of LHC searches for new physics on the basis of simplified-model results. More precisely, SModelS decomposes the signatures of full BSM scenarios into simplified-model components, which are then confronted against the experimental constraints from a large database of results. Because LHC searches for SUSY typically adopt a channel-by-channel approach, SModelS recently enabled the combination of approximately uncorrelated analyses [16], allowing not only to increase the statistical robustness of the results but to also probe different search channels at the same time and to consistently account for under- and over-fluctuations in a single, global likelihood. The main work of this thesis is to understand, through the reinterpretation tool SModelS, how the combination of analyses impacts the current LHC constraints on SUSY particles, with an emphasis on the electroweak-ino (EW-ino) sector of the MSSM, and how it can be used to search for new physics through a bottom-up approach based on dispersed signals.

This thesis is organised as follows: Chap. 1 introduces the SM and its minimal SUSY extension; Chap. 2 presents the LHC and two of its main experiments, as well as how their simplified model results constrain SUSY models; Chap. 3 describes the SModelS approach to reinterpret and combine LHC results for which reinterpretation material is available; Chap. 4 discusses the impact of the combination on the EW-ino sector of the MSSM; Chap. 5 demonstrates how the combination of analyses can be used to perform a bottom-up search for new physics; and finally, Chap. 6 summarises and concludes. As a complement, Appendix A provides an overview of the LHC constraints on long-lived SUSY particles and on R-parity SUSY models after LHC Run 2, derived using the simplified model framework; Appendix B lists all the analyses implemented in the SModelS database v3.0; and Appendix C describes the SModelS database add-ons.

The work of this thesis led to the following productions:

- **Articles in peer-reviewed journals:**
  - SModelS v2.3: Enabling global likelihood analyses, M. M. Altakach, S. Kraml, A. Lessa, S. Narasimha, T. Pascal, W. Waltenberger, SciPost Phys. 15 (5) (2023) 185. doi:10.21468/SciPostPhys.15.5.185, arXiv:2306.17676
  - Global LHC constraints on electroweak-inos with SModelS v2.3, M. M. Altakach, S. Kraml, A. Lessa, S. Narasimha, T. Pascal, T. Reymermier, W. Waltenberger, SciPost Phys. 16 (2024) 101. doi:10.21468/SciPostPhys.16.4.101, arXiv:2312.16635

- **Zenodo entries**:
  - SModelS database v2.2.0, SModelS collaboration (Apr. 2022). 10.5281/zenodo.6406087

- SModelS database v2.3.0, SModelS collaboration (May 2023). 10.5281/zenodo.7961638
- SModelS database v3.0.0, SModelS collaboration (Aug. 2024). 10.5281/zenodo.13354582
- EW-ino scan points from "SModelS v2.3: enabling global likelihood analyses" paper, M. M. Altakach, S. Kraml, A. Lessa, S. Narasimha, T. Pascal, W. Waltenberger (Aug. 2023). doi:10.5281/zenodo.8275263
- Datasets used in "Global LHC constraints on electroweak-inos with SModelS v2.3", M. M. Altakach, S. Kraml, A. Lessa, S. Narasimha, T. Pascal, T. Reymermier, W. Waltenberger, (Jan. 2024). doi:10.5281/zenodo.10471553

• **Presentations at international conferences** (as a speaker, in parallel sessions):

- "Constraining the electroweakino sector of the MSSM through the combination of orthogonal LHC searches", The 30th International Conference on Supersymmetry and Unification of Fundamental Interactions. https://indi.to/KGb5h
- "Global LHC constraints on electroweak-inos with SModelS v2.3", The 42nd International Conference on High Energy Physics. https://indi.to/Hxksp
  [*Proceedings to be published in PoS.*]

Moreover, the project described in Chap. 5 should also lead to a journal publication. In addition, the work of this thesis also led to the following productions, which will be less discussed because not the main topic of this thesis:

• SModelS v3: Going Beyond $\mathcal{Z}_2$ Topologies, M. M. Altakach, S. Kraml, A. Lessa, S. Narasimha, T. Pascal, C. Ramos, Y. Villamizar, W. Waltenberger. arXiv:2409.12942
• Data and code for "SModelS v3: Going Beyond $\mathcal{Z}_2$ Topologies", M. M. Altakach, S. Kraml, A. Lessa, S. Narasimha, T. Pascal, C. Ramos, Y. S. Villamizar, W. Waltenberger (Sep. 2024). doi:10.5281/zenodo.13784464

Grenoble, France Timothée Pascal

References

1. Haber HE, Kane GL (1985) The search for supersymmetry: probing physics beyond the standard model. Phys Rept 117:75–263. https://doi.org/10.1016/0370-1573(85)90051-1
2. Martin SP (1998) A supersymmetry primer. Adv Ser Direct High Energy Phys 1:1–98. https://doi.org/10.1142/9789812839657_0001. arXiv: hep-ph/9709356

3. Weinberg S (2000) The quantum theory of fields. Vol. 2: modern applications. Cambridge University Press, Cambridge. ISBN: 978-0-521-67055-5, 978-1-139-63263-8, 978-0-521-67055-5
4. Baer H, Tata X (2006) Weak scale supersymmetry: from superfields to scattering events. Cambridge University Press. Cambridge. ISBN: 978-0-521-29031-9, 978-0-511-19011-7, 978-0-521-29031-9, 978-0-521-85786-4
5. Dreiner HK, Haber HE, Martin SP (2023) From spinors to supersymmetry. Cambridge University Press, Cambridge. ISBN: 978-1-139-04974-0. https://doi.org/10.1017/9781139049740
6. Allanach B, Haber HE (2024) Supersymmetry, part I (theory). arXiv: 2401.03827 [hep-ph]
7. Arkani-Hamed N et al (2007) MARMOSET: the path from LHC data to the new standard model via on-shell effective theories. arXiv: hep-ph/0703088 [HEP-PH]
8. Alwall J, Schuster P, Toro N (2009) Simplified models for a first characterization of new physics at the LHC. Phys Rev D79:075020. https://doi.org/10.1103/PhysRevD.79.075020. arXiv: 0810.3921 [hep-ph]
9. Alves D (2012) Simplified models for LHC new physics searches. J Phys G39:105005. https://doi.org/10.1088/0954-3899/39/10/105005. arXiv: 1105.2838 [hep-ph]
10. Okawa H (2011) Interpretations of SUSY searches in ATLAS with simplified models. In: *Particles and fields. Proceedings, meeting of the division of the American Physical Society, DPF 2011, Providence, USA, August 9–13, 2011*. arXiv: 1110.0282 [hep-ex]
11. Kraml S et al (2013) SModelS: a tool for interpreting simplified-model results from the LHC and its application to supersymmetry. Eur Phys J C 74:2868. https://doi.org/10.1140/epjc/s10052-014-2868-5. arXiv: 1312.4175 [hep-ph]
12. Kraml S et al (2014) SModelS v1.0: a short user guide. arXiv: 1412.1745 [hep-ph]
13. Ambrogi F et al (2018) SModelS v1.1 user manual: improving simplified model constraints with efficiency maps. Comput Phys Commun 227:72–98. https://doi.org/10.1016/j.cpc.2018.02.007. arXiv: 1701.06586 [hep-ph]
14. Ambrogi F et al (2020) SModelS v1.2: long-lived particles, combination of signal regions, and other novelties. Comput Phys Commun 251:106848. https://doi.org/10.1016/j.cpc.2019.07.013. arXiv: 1811.10624 [hep-ph]
15. Alguero G et al (2002) Constraining new physics with SModelS version 2. J High Energy Phys 08:068. https://doi.org/10.1007/JHEP08(2022)068. arXiv: 2112.00769 [hep-ph]
16. Altakach MM et al (2023) SModelS v2.3: enabling global likelihood analyses. SciPost Phys 15(5):185. https://doi.org/10.21468/SciPostPhys.15.5.185. arXiv: 2306.17676 [hep-ph]
17. Altakach MM et al (2024) SModelS v3: going beyond Z_2 topologies. J High Energy Phys 11:074. https://doi.org/10.1007/JHEP11(2024)074. arXiv: 2409.12942 [hep-ph]

Acknowledgements I would like to first and foremost thank my supervisor Sabine, without whom this thesis would never have existed. I would like to thank her for her trust, for having accepted me as her student, and for her availability, even in case of close deadline. She gave me the freedom to conduct my own research and always pushed me to improve my work and to seize opportunities. I would also like to acknowledge her commitment in this work, and her countless advice, always relevant and provided when needed, as well as her rigour, whether in the production of results or in their communication.

This work has also been made possible thanks to the hard work of the whole SModelS collaboration. I would like to thank them all for their time, their pedagogy, and their kindness. I am particularly grateful to Wolfgang, who taught me the basics of statistics for high energy physics and tirelessly answered every question I asked. I would like to express my gratitude to Andre for his help on phenomenological studies and on many aspects of the SModelS software. I would also like to thank Sahana and Mohammad a lot, for their help and for their friendship.

My thanks also go to the members of my jury for their questions and their feedback on this work, and especially to Benjamin for his comments on this thesis.

Je me dois de grandement remercier Gaël, qui en plus de m'avoir initié au logiciel SModelS, m'a aidé tout au long de cette thèse et m'a hébergé durant le premier mois de celle-ci. Je le remercie pour son temps, sa patience, son écoute, sa bienveillance et son amitié.

Je remercie aussi toutes les personnes des services informatique, administratif et financier, et notamment Agnès, qui m'a guidé à travers les différents logiciels de réservation de mission, même à la dernière minute. Cette thèse n'aurait pas été la même sans leur aide précieuse.

Je souhaite également remercier ma famille, et particulièrement mes parents et grands-parents, pour leur soutien financier et émotionnel durant toutes mes études, mais surtout pour leur amour et leur sagesse. Un merci spécial à Papou, qui a su me transmettre sa passion pour les sciences.

Je remercie évidemment toustes mes ami·es, notamment les personnes incroyables avec qui j'ai eu la chance et la joie de cohabiter à Grenoble, fût-ce le temps d'un mois, d'un an, ou plus, ainsi que toustes mes proches de longue date, et toutes les chouettes personnes que j'ai rencontré au labo et qui ont su faire de ce lieu un lieu chaleureux. Un merci tout particulier à Carolina et Corentin, sans qui les week-ends de rédaction au labo sous la chaleur de l'été n'auraient pas eu la même saveur. Je ne pense pas que j'aurais pu rêver meilleurs personnes pour m'accompagner dans cette rédaction.

Pour finir, je tiens à remercier toutes les personnes derrière les logiciels gratuits qui m'ont permis de réaliser cette thèse : Python (notamment les librairies json, jsonpatch, Matplotlib, NumPy, SciPy, pandas, pyhf...), GitHub, LaTeX, TeXstudio, HEPData, Atom, Zenodo, arXiv, INSPIRE-HEP, etc. Je tiens également à remercier toutes les personnes qui se mobilisent et travaillent dur pour permettre la réinterprétation des données, ainsi que toutes celles qui participent à rendre la science plus ouverte, accessible et inclusive.

Competing Interests The authors have no competing interests to declare that are relevant to the content of this manuscript.

About the Book

The Standard Model (SM) of particle physics is undoubtedly one of the most successful models of modern physics. Nevertheless, many experimental and theoretical questions remain open and hint at the existence of beyond the Standard Model (BSM) physics. These questions include, for instance, the nature of dark matter and the hierarchy problem. Several experiments of various scales are currently searching for a manifestation of new physics. The ones carried out at the Large Hadron Collider (LHC) benefit from today's highest energy delivered by a particle accelerator. However, despite the intensive search program carried out by the LHC experiments, no compelling signal for new physics has been found.

In the absence of any discovery, the obtained results are used to set constraints on a wide variety of BSM theories. One of these theories, studied in this thesis, is the minimal supersymmetric extension of the SM (MSSM), where each SM field has a supersymmetric (SUSY) counterpart differing by half a spin unit. The constraints set by the LHC experiments on the masses of SUSY particles are quite stringent. However, these limits are derived through a channel-by-channel approach and most of the time hold only in the context of simplified models, where only a handful of SUSY particles, with trivial mixings and decay branching ratios, are assumed to be kinematically accessible. Therefore, it is necessary to globally reinterpret these results to correctly estimate how they constrain realistic BSM scenarios.

The approach adopted in this thesis is based on the reinterpretation tool SModelS, which uses simplified model results to constrain more realistic models. The work of this thesis led to the extension of the SModelS framework in several ways, on both the code and the database side. The most notable new feature is the combination of approximately uncorrelated analyses, enabling the confrontation of models against different experimental results in a consistent way through global likelihoods.

In an exemplary phenomenological application, this is used to set global LHC constraints on the electroweak-ino sector of the MSSM through the dynamical combination of 16 LHC searches for which efficiency and acceptance maps are available in the SModelS database. Particular attention is given to characterising which combinations maximise the sensitivity in different regions of the parameter space, how fluctuations in the data in individual analyses influence the global likelihood, and what is the resulting exclusion power of the combination compared to the analysis-by-analysis approach.

Another objective of this thesis is to go beyond the top-bottom approach usually adopted in BSM searches. Instead of testing each theory against the data, the data is used to guide the search for BSM physics. Signs of new physics are here searched for by constructing so-called proto-models. These are sets of simplified models not bounded by any higher-level theoretical assumptions, and by confronting them against the results in the SModelS database. This relies on the dynamical combination of LHC searches to find mutually consistent excesses from dispersed signals while being consistent with their current limits. This thesis presents an improved version of this method, with a more furnished database, a refined statistical treatment, and the possibility to probe off-shell regions of the parameter space. A first application to the LHC searches for SUSY signatures from the production of electroweak-inos with compressed mass spectra is also discussed.

Notations and Conventions

Natural units will be used throughout this thesis. That is to say

$$c = \hbar = 1\,, \tag{1}$$

with c the speed of light in the vacuum and $\hbar$ the reduced Planck constant, which leads units of space and time to be the inverse of units of energy:

$$[L] = [T] = [E]^{-1}\,. \tag{2}$$

Moreover, Einstein summation notation is implied whenever two indices repeat, e.g.:

$$F^a_{\mu\nu} F^{a\,\mu\nu} = \sum_{a,\mu,\nu} F^a_{\mu\nu} F^{a\,\mu\nu}\,. \tag{3}$$

Lorentz indices are denoted by Greek letters from the middle of the alphabet when running on space-time coordinates (e.g. $\mu = 0, 1, 2, 3$) and by Roman letters from the middle of the alphabet when running on space coordinates (e.g. $i = 1, 2, 3$). They are raised and lowered using the metric

$$g_{\mu\nu} = g^{\mu\nu} = \mathrm{diag}(+1, -1, -1, -1)^{\mu\nu}\,, \tag{4}$$

e.g.:

$$F^a_{\mu\nu} F^{a\,\mu\nu} = \sum_{a,\mu,\nu,\rho,\sigma} g^{\mu\rho} g^{\nu\sigma} F^a_{\mu\nu} F^a_{\rho\sigma}\,. \tag{5}$$

The spinor indices (denoted by Greek letters from the beginning of the alphabet) take the values 1 and 2 and are lowered and raised by the antisymmetric symbol defined by

$$\epsilon^{12} = -\epsilon^{21} = \epsilon_{21} = -\epsilon_{12} = 1\,, \quad \text{and} \quad \epsilon^{11} = \epsilon^{22} = \epsilon_{11} = \epsilon_{22} = 0\,. \tag{6}$$

Unless stated otherwise, repeated spinor indices contracted like ${}^{\alpha}{}_{\alpha}$ or ${}_{\dot\alpha}{}^{\dot\alpha}$ are suppressed. In particular, the product of two two-component spinors gives

$$\psi\xi = \psi^{\alpha}\xi_{\alpha} = \psi^{\alpha}\epsilon_{\alpha\beta}\xi^{\beta} = -\xi^{\beta}\epsilon_{\alpha\beta}\psi^{\alpha} = \xi^{\beta}\epsilon_{\beta\alpha}\psi^{\alpha} = \xi^{\beta}\psi_{\beta} = \xi\psi\,. \tag{7}$$

Likewise, $\psi^{\dagger}\xi^{\dagger} = \psi^{\dagger}_{\dot\alpha}\xi^{\dagger\dot\alpha} = \xi^{\dagger}_{\dot\alpha}\psi^{\dagger\dot\alpha} = \xi^{\dagger}\psi^{\dagger}$. The antisymmetric symbol ϵ with superscripts is equal to the Levi-Civita symbol in two dimensions. The Levi-Civita symbol is defined by (in $n > 1$ dimensions):

$$\epsilon^{a_1\cdots a_n} = \epsilon_{a_1\cdots a_n} = \begin{cases} +1 & \text{if } (a_1,\cdots,a_n) \text{ is an even permutation of } (1,\cdots,n) \\ -1 & \text{if } (a_1,\cdots,a_n) \text{ is an odd permutation of } (1,\cdots,n) \\ 0 & \text{otherwise} \end{cases}\,. \tag{8}$$

Finally, the Pauli matrices are

$$\sigma^1 = \begin{pmatrix} 0 & 1 \\ 1 & 0 \end{pmatrix}, \quad \sigma^2 = \begin{pmatrix} 0 & -i \\ i & 0 \end{pmatrix}, \quad \sigma^3 = \begin{pmatrix} 1 & 0 \\ 0 & -1 \end{pmatrix}; \tag{9}$$

and in the Weyl representation, the Dirac matrices γ^0 and γ^5 are

$$\gamma^0 = \begin{pmatrix} 0 & \mathbb{1}_2 \\ \mathbb{1}_2 & 0 \end{pmatrix}, \quad \text{and} \quad \gamma^5 = \begin{pmatrix} -\mathbb{1}_2 & 0 \\ 0 & \mathbb{1}_2 \end{pmatrix}. \tag{10}$$

Contents

Chapter 1
Introduction: Going Beyond the Standard Model of Particle Physics

Abstract The Standard Model (SM) of particle physics is a quantum field theory successfully describing all the ordinary matter in terms of fundamental fields, as well as three of the four fundamental interactions. This chapter is a short introduction to the SM, its success, its open questions, and its possible generalisations, with an emphasis on supersymmetric (SUSY) theories and the Minimal Supersymmetric Standard Model. Non-SUSY extensions of the SM, such as SU(5) and extra dimensions, are also briefly mentioned. It first introduces the SM, then discusses the SUSY formalism and, finally, provides a couple of examples of other beyond the Standard Model theories.

This chapter is a short introduction to the SM, its success, its open questions, and its possible generalisations, with an emphasis on the supersymmetric theories and the MSSM. It will first introduce the SM in Sect. 1.1, the SUSY formalism in Sect. 1.2, and examples of other BSM theories will be discussed in Sect. 1.3.

1.1 The Standard Model of Particle Physics

As briefly mentioned in the Introduction, the SM is a QFT successfully describing all the ordinary matter (as opposed to dark matter (DM) as we will see later) in terms of fundamental fields, as well as three of the four fundamental interactions.

Among all the particles that we currently consider fundamental, the first one to be observed was the electron, in 1897 [1]. Despite electromagnetism unified electricity and magnetism in 1873 [2], the fundamental particle mediating this interaction, the photon, was postulated years later, in 1905 [3], and its momentum experimentally observed in 1923 [4]. The weak interaction, on its side, took its origin in 1914, when the energy spectrum of electrons in beta decays was observed to be continuous, inconsistent with the idea that only an electron was emitted [5]. To solve this issue, it was postulated in 1930 the existence of a new particle [6], which has been experimentally confirmed in 1956 [7] and is today known as the

T. Pascal, *Beyond the Standard Model at the Large Hadron Collider: From Simplified Model Constraints to Artificial Proto-Modelling for Dispersed Signals*, Springer Theses, https://doi.org/10.1007/978-3-032-09391-2_1

electron antineutrino. It was then postulated that the observed decay was mediated by a new interaction, the weak interaction; first in terms of a four-point contact interaction in 1933 [8, 9], and later, in the 1950s, as an interaction mediated by particles called W^+, W^- (for weak) and Z. The first unambiguous signals of a $W^\pm$ and a Z came in January 1983 [10, 11] and May 1983 [12, 13], respectively. In the meantime, two particles similar to the electron, but heavier, were discovered: the muon, in 1936 [14], and the tau, in 1975 [15–18]. Furthermore, similarly to the electron (anti)neutrino, two other neutrinos were discovered: the muon neutrino, in 1962 [19], and the tau neutrino, in 2000 [20]. The electron, the muon, the tau and the neutrinos are collectively referred to as leptons. Nowadays, the electromagnetic and the weak interactions are unified, at high energy, under the framework known as the electroweak (EW) interaction.

Regarding the strong interaction, which describes the interactions between the nucleus constituents, it was proposed in 1964 that the known protons and neutrons were not fundamental but had instead a substructure made of quarks [21–25]. Later developments [26–28] extended this idea and predicted six quarks: the up, down and strange quarks, all observed in 1969 [29, 30], the charm quark, observed in 1974 [31, 32], the bottom quark, observed in 1977 [33], and the top quark, observed in 1995 [34, 35]. The particle mediating the strong interaction, the gluon, has been firstly observed in 1979 [36–41].

Finally, the particle observed in 2012 [42, 43] corresponds so far to the Higgs particle predicted by the Standard Model, hypothesised in 1964 to explain how massive particles acquire their mass [44–46]. Moreover, to all the mentioned particles must be added the antiparticles, which are the same particles as mentioned above but with an opposite electric charge (only the photon, the gluon, the Z and the Higgs particles have no antiparticle counterparts). They were predicted in 1928 [47] and the first one, the positron (the antiparticle of the electron) was first observed in 1932 [48]. Since then, the existence of all the other antiparticles has been confirmed by experimental observations.

1.1.1 Theoretical Formulation

1.1.1.1 Symmetries as a Fundamental Principle

The SM is based on symmetries, a fundamental concept in modern physics. They require the mathematical formulation of the SM to be invariant under global space-time transformations as well as under internal transformations. On the one hand, invariance under space-time symmetries means that the observables, the physical quantities that can be measured, must not depend on the space-time position, the orientation, or the (constant) velocity of the measured system. In other words, they must be invariant under the effect of the Poincaré transformations, whose algebra is (see e.g. [49])

$$[P^\mu, P^\nu] = 0\,, \tag{1.1}$$

$$[M^{\mu\nu}, P^\rho] = i(g^{\mu\rho}P^\nu - g^{\nu\rho}P^\mu)\,, \tag{1.2}$$

$$[M^{\mu\nu}, M^{\rho\sigma}] = i(g^{\mu\rho}M^{\nu\sigma} - g^{\mu\sigma}M^{\nu\rho} - g^{\nu\rho}M^{\mu\sigma} + g^{\nu\sigma}M^{\mu\rho})\,, \tag{1.3}$$

where g is the metric, P^μ are the four generators of space-time translations and $M^{\mu\nu}$ are the six generators of the Lorentz transformations.[1] The expression of P^μ and $M^{\mu\nu}$ depend on the chosen representation but they are usually taken to be 5×5 matrices. The $M^{\mu\nu}$ generators contain the rotation generators $J^i = -\frac{1}{2}\epsilon^{ijk}M^{jk}$, with ϵ the Levi-Civita symbol, and the boost generators $K^i = M^{0i}$. Each symmetry implies, through Noether's theorem, the conservation of a quantity through time. There are 10 of them:

- the symmetry under time translations implies the conservation of energy (eigenvalues of P^0);
- the symmetries under the three spatial translations imply the conservation of the momentum in the three spatial coordinates (eigenvalues of P^i);
- the symmetries under the three spatial rotations imply the conservation of the angular momentum in the three spatial coordinates (eigenvalues of J^i);
- the symmetries under the boosts in the three spatial directions imply the conservation of the "time-dependent mass moment" (which is equivalent to the law of inertia) in the three spatial coordinates (eigenvalues of K^i).

On the other hand, the internal symmetries are local gauge symmetries which are independent from space-time symmetries and under which the system is invariant. The SM is invariant under the transformations of the three gauge groups

- $SU(3)_c$, whose algebra generators are, in the fundamental representation, one half times the Gell-Mann matrices λ^a, with $a = 1, ..., 8$. The conserved quantity is the color charge.
- $SU(2)_L$, whose algebra generators are, in the fundamental representation, one half times the Pauli matrices σ^a, with $a = 1, 2, 3$. The conserved quantity is the weak isospin.
- $U(1)_Y$, whose algebra generator is a base vector of $\mathbb{R}$. The conserved quantity is the hypercharge.

The algebra generators T of the groups $SU(3)_c$ and $SU(2)_L$ satisfy the relations

$$[T^a, T^b] = if^{abc}\,T^c\,, \tag{1.4}$$

[1] There are six generators for the Lorentz transformation because $M^{\mu\nu} = -M^{\nu\mu}$.

where a sum over c is implied and f^{abc} are the totally antisymmetric structure constants.

1.1.1.2 The Field Content

The physical fundamental particles are represented by quantum fields living in the irreducible representations of positive energy of the Poincaré group, which are indexed by a non-negative finite number (the mass) and an integer or half-integer (the spin). Any system is antisymmetric under the permutations of two particles with a half-integer spin. Therefore, they obey the Pauli exclusion principle and follow the Fermi-Dirac statistics. The particles with this property are thus called fermions. The fundamental fermions are matter particles and have a spin of one half. On the contrary, any system is symmetric under the permutations of two particles with an integer spin. Such particles do not obey the Pauli exclusion principle and follow the Bose-Einstein statistics. The particles with this property are therefore called bosons. The fundamental boson with spin 0 is the Higgs field and it is a complex scalar. The fundamental bosons with spin 1[2] are particles mediating the fundamental interactions and are vectors.

The field content of the SM and their representations are shown in Table 1.1. The Higgs field is actually a Higgs doublet, composed of two complex scalar fields. The matter fermions and the Higgs doublet are charged and transform under the fundamental and anti-fundamental representations of the gauge groups. Equivalently, the vector bosons are charged and transform under the adjoint representations of the gauge groups. The fermions are divided into three families (also called generations, or flavors), corresponding to the weak isospin doublets. The three quark families

Table 1.1 Field content of the standard model. The last column corresponds to the representations of the gauge groups. Adapted from [50]

Name	Symbol	$SU(3)_C,\ SU(2)_L,\ U(1)_Y$
Quarks	$Q = (u_L\ \ d_L)^T$	$(\mathbf{3},\ \mathbf{2},\ \frac{1}{6})$
(× 3 families)	u_R	$(\bar{\mathbf{3}},\ \mathbf{1},\ \frac{2}{3})$
	d_R	$(\bar{\mathbf{3}},\ \mathbf{1},\ -\frac{1}{3})$
Leptons	$L = (\nu_{e_L}\ \ e_L)^T$	$(\mathbf{1},\ \mathbf{2},\ -\frac{1}{2})$
(× 3 families)	e_R	$(\mathbf{1},\ \mathbf{1},\ -1)$
Higgs doublet	$H = (\phi^+\ \ \phi^0)^T$	$(\mathbf{1},\ \mathbf{2},\ \frac{1}{2})$
Gluon	$G^{1,\ldots,8}$	$(\mathbf{8},\ \mathbf{1},\ 0)$
W bosons	$W^{1,2,3}$	$(\mathbf{1},\ \mathbf{3},\ 0)$
B boson	B	$(\mathbf{1},\ \mathbf{1},\ 0)$

The bold numbers are referring to the irreducible representations of Lie groups (of dimension indicated by the number)

[2] Although it would be more rigorous to talk about helicity for massless particles. In the following, the spin of such a particle will refer to its helicity.

are the (up, down), (charm, strange) and (top, bottom) doublets. The three lepton families are the (electron, electron neutrino), (muon, muon neutrino) and (tau, tau neutrino) doublets. The L, R subscripts indicate the chirality of the fields. The left- and right-handed fields can be obtained from Dirac spinors $\Psi_D = \left(\psi \ \xi^\dagger\right)^T$ through the $P_{L,R}$ projection operators:

$$\Psi_L = P_L \Psi_D = \frac{1-\gamma^5}{2}\begin{pmatrix}\psi \\ \xi^\dagger\end{pmatrix} = \begin{pmatrix}\psi \\ 0\end{pmatrix} \quad \text{and}$$
$$\Psi_R = P_R \Psi_D = \frac{1+\gamma^5}{2}\begin{pmatrix}\psi \\ \xi^\dagger\end{pmatrix} = \begin{pmatrix}0 \\ \xi^\dagger\end{pmatrix}, \tag{1.5}$$

where ψ and ξ are left- and right-handed two-component Weyl spinors, respectively, and $\gamma^5 = i\gamma^0\gamma^1\gamma^2\gamma^3$, with γ^μ the Dirac matrices in the Weyl representation.[3] Thus, two Weyl spinors are required to build a Dirac spinor. However, one can build a four-component Majorana spinor out of a single two-component Weyl spinor: $\Psi_M = \left(\psi \ \psi^\dagger\right)^T$. The Lagrangian density mass terms are then

$$\mathcal{L}_D \supset -m\overline{\Psi}_D\Psi_D = -m(\psi\xi + \psi^\dagger\xi^\dagger), \quad \text{and}$$
$$\mathcal{L}_M \supset -\tfrac{1}{2}m\overline{\Psi}_M\Psi_M = -\tfrac{1}{2}m(\psi\psi + \psi^\dagger\psi^\dagger), \tag{1.6}$$

where

$$\overline{\Psi} = \Psi^\dagger\gamma^0, \tag{1.7}$$

with γ^0 in the Weyl representation. In the SM, all the fermions are of Dirac type.

1.1.1.3 The Standard Model Lagrangian

The equations of motion can be obtained by applying the Euler-Lagrange equations to the Lagrangian density

$$\mathcal{L}_{\text{SM}} = \mathcal{L}_F + \mathcal{L}_S + \mathcal{L}_G + \mathcal{L}_Y, \tag{1.8}$$

where each term is listed below.

- $\mathcal{L}_F$ contains the kinetic terms of the fermionic fields. In terms of Dirac spinors:

$$\mathcal{L}_F = i(\overline{Q})_{ia}\not{D}(Q)^{ia} + i(\overline{u}_R)_{ia}\not{D}(u_R)^{ia} + i(\overline{d}_R)_{ia}\not{D}(d_R)^{ia}$$
$$+ i(\overline{L})_i\not{D}(L)^i + i(\overline{e}_R)_i\not{D}(e_R)^i, \tag{1.9}$$

[3] The Dirac matrices follow the algebra $\{\gamma^\mu, \gamma^\nu\} = 2g^{\mu\nu}\mathbb{1}_4$.

with $i = 1, 2, 3$ the family indices and $a = 1, 2, 3$ running on the anti-fundamental and fundamental representations of SU(3)$_c$. Here, $\not\nabla = \gamma^\mu \nabla_\mu$ and

$$\nabla_\mu = \partial_\mu + i g_a A^a_\mu T^a , \tag{1.10}$$

with a running on the adjoint representation of the gauge groups; and g_a, A and T the coupling constants, the vector boson and the algebra generators of the corresponding gauge groups in the correct representation, respectively. This covariant derivative ensures gauge invariance and contains the interaction terms between the gauge bosons and the fermions. The SU(3)$_c$ terms vanish when applied to leptons, and the SU(2)$_L$ terms vanish when applied to right-handed helicities. The U(1) generators depend on the particle. They correspond to the hypercharge Y given in Table 1.1.

- $\mathcal{L}_S$ contains the kinetic and potential terms of the scalar fields:

$$\mathcal{L}_S = |\nabla_\mu H|^2 - V(H) , \tag{1.11}$$

where ∇_μ is the same as in Eq. (1.10) but without the SU(3)$_c$ terms, and V the scalar potential is given by

$$V(H) = \mu^2 H^\dagger H + \lambda (H^\dagger H)^2 . \tag{1.12}$$

- $\mathcal{L}_G$ contains the kinetic terms of the gauge fields:

$$\mathcal{L}_G = -\frac{1}{4} F^a_{\mu\nu} F^{a\,\mu\nu} , \tag{1.13}$$

where a runs on the adjoint representation of the gauge groups and F^a are the field-strength tensors defined as

$$F^a_{\mu\nu} = \partial_\mu A^a_\nu - \partial_\nu A^a_\mu - g_a f^{abc} A^b_\mu A^c_\nu . \tag{1.14}$$

Here, $g_a f^{abc}$ does not imply a sum on a.

- $\mathcal{L}_Y$ contains the Yukawa terms that couple the Higgs field to the fermions. In terms of Dirac spinors:

$$\begin{aligned}\mathcal{L}_Y = &-(\overline{Q}^{ia})_\alpha (\mathbf{y_u})_i{}^j (H^\dagger)_\beta \epsilon^{\alpha\beta} u_{R\,ja} \\ &- (\overline{Q}^{ia})^\alpha (\mathbf{y_d})_i{}^j (H)_\alpha d_{R\,ja} - (\overline{L}^i)^\alpha (\mathbf{y_e})_i{}^j (H)_\alpha e_{R\,j} + \text{h.c.} ,\end{aligned} \tag{1.15}$$

where $i, j = 1{,}2{,}3$ are the family indices; $a = 1{,}2{,}3$ are the color indices; $\alpha, \beta = 1{,}2$ are the weak isospin indices; ϵ is the antisymmetric symbol; and $\mathbf{y}_u$, $\mathbf{y_d}$ and $\mathbf{y_e}$ are 3×3 matrices in family space called the Yukawa matrices.

1.1.1.4 Electroweak Symmetry Breaking

The electrically charged fermions and the EW gauge bosons acquire their masses through the so-called Higgs mechanism (or, historically, the Brout-Englert-Higgs (BEH) mechanism) [44–46], which spontaneously breaks the EW symmetry $\mathrm{SU}(2)_L \times \mathrm{U}(1)_Y$. To realise this, $\mu^2 < 0$ and $\lambda > 0$ are required. The resulting scalar potential has a so-called "mexican-hat" shape. Before EW symmetry breaking (EWSB), the Higgs field sits at the origin, which is symmetric in the weak isospin space, but is not the minimum. The minimum is a continuous function realising $\mu^2/2\lambda \equiv v/2$, where v is the Vacuum Expectation Value (VEV) of the Higgs field. For the photon to remain massless, we can choose to give a VEV only to the electrically neutral scalar component:

$$\langle H \rangle = \begin{pmatrix} 0 \\ v \end{pmatrix} . \tag{1.16}$$

When the Higgs field goes to the minimum of the potential, it acquires a VEV and gives to the vacuum a preferred direction in the weak isospin space, thus breaking the symmetry. Describing the system as fluctuations around the minimum allows us to write

$$H(x) = e^{-ig\theta_i \sigma^i/2} \begin{pmatrix} 0 \\ v + h(x) \end{pmatrix} , \tag{1.17}$$

with g the $\mathrm{SU}(2)_L$ coupling constant, θ three degrees of freedom associated to Nambu-Goldstone bosons, σ the Pauli matrices, and h the fourth degree of freedom. The latter is a real scalar field corresponding to the physical massive Higgs state. Inserting Eq. (1.17) in the kinetic term of (1.11) results in mass terms for the EW gauge bosons:

$$m^2_{W^\pm} = \frac{g^2 v^2}{2} , \qquad \text{and} \qquad m^2_Z = \frac{(g^2 + g'^2)v^2}{2} = \frac{m^2_{W^\pm}}{\cos^2\theta_W} , \tag{1.18}$$

where g' is the $\mathrm{U}(1)_Y$ coupling constant and θ_W is the Weinberg angle parametrising the mixing between the W^3 and B gauge eigenstates, giving rise to the γ and Z mass eigenstates:

$$W^\pm_\mu = (W^1_\mu \mp W^2_\mu)/\sqrt{2} , \tag{1.19}$$

$$\begin{pmatrix} Z_\mu \\ A_\mu \end{pmatrix} = \begin{pmatrix} \cos\theta_W & -\sin\theta_W \\ \sin\theta_W & \cos\theta_W \end{pmatrix} \begin{pmatrix} W^3_\mu \\ B_\mu \end{pmatrix} . \tag{1.20}$$

After EWSB, the electrically charged fermions acquire a mass through the Yukawa terms. After diagonalisation one obtains

$$m_{f,i} = v(y_f)_{ii} , \tag{1.21}$$

with $f = u, d, e$ running over the fermion species and $i = 1, 2, 3$ running over the generations. However, the up- and down-type quark Yukawa matrices are diagonalised by two distinct sets of unitary matrices $V_{u,d}$. This difference leads to the flavor-changing nature of the charged-current interactions, described by the Cabbibo-Kobayashi-Maskawa (CKM) matrix $V_{\rm CKM} = V_u V_d^\dagger$ [28, 51], a 3×3 unitary matrix parametrised by three mixing angles and a CP-violating phase.[4] Moreover, since there is no right-handed neutrino, the SM does not predict any mass for the neutrinos.

After EWSB, a $U(1)_{\rm EM}$ symmetry remains, implying the conservation of the electric charge defined through the Gell-Mann–Nishijima formula:

$$Q_{\rm EM} = I_3 + Y , \tag{1.22}$$

where I_3 is the weak isospin third component and Y is the hypercharge.

In total, the SM has 19 free parameters: the nine charged fermion masses, the three mixing angles and the CP-violating phase of the CKM matrix, the three gauge couplings (g_s, g and g' for $SU(3)_c$, $SU(2)_L$ and $U(1)_Y$, respectively), μ^2 and λ (which can be traded for v and m_h), and the CP-violating vacuum angle of the strong interaction $\bar{\theta}_{\rm QCD}$ (see below).[5]

1.1.2 Success of the Standard Model

The predictions of the SM are in agreement with the experimental observations realised throughout the past 40 years. Some selected examples of agreement are presented below.

First of all, not only the Higgs boson has been predicted and observed, but the measurements of its properties are in agreement with the predictions of EWSB, thus marking the success of the EW theory. Moreover, the precise measurements of the EW observables at the Large Electron–Positron (LEP) collider using the Z resonance show an extraordinary agreement with the SM [52]. The measurement of the production cross sections is still ongoing at collider experiments with an increasing precision and agreement with the SM, as shown in Fig. 1.1 for the ATLAS collaboration (a similar plot done by the CMS collaboration can be found

[4] C and P are the charge and space inversion operators, respectively.

[5] QCD stands for quantum chromodynamics, the QFT specifically describing the strong interaction.

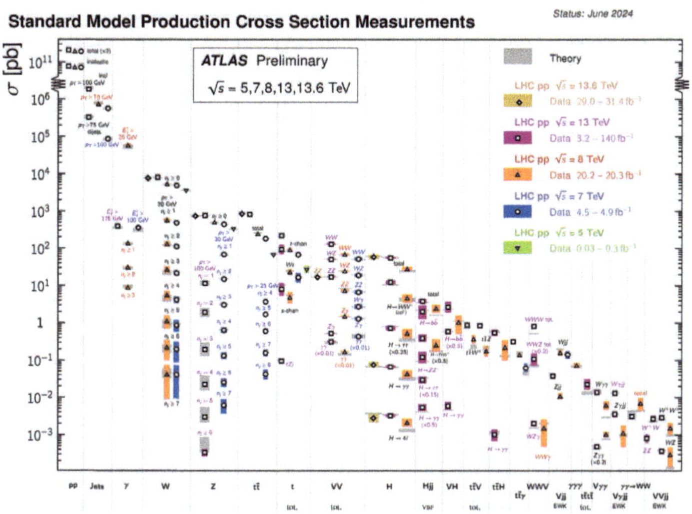

Fig. 1.1 ATLAS status on the SM production cross sections at the beginning of LHC Run 3. Taken from [54]

in [53]). The LHC as well as the ATLAS and CMS experiments will be introduced in Sect. 2.1.

Another confirmation of the SM comes from the electron g-factor, which is currently the most precisely measured SM quantity. The magnetic moment operator for a spin-1/2 electron is

$$\boldsymbol{\mu} = -g_e \mu_B \frac{\mathbf{S}}{\hbar} \tag{1.23}$$

where g_e is the electron g-factor, $\mu_B = q_e \hbar / 2m_e$ is the Bohr magneton and $\mathbf{S}$ is the spin angular momentum operator. The Dirac equation predicts $g_e = 2$. However, the measured value is a bit above 2. The success of the SM as a perturbative QFT is to increase the tree-level predicted value to the observed one by adding quantum corrections (i.e. higher-order corrections). More precisely, the corrections, quantified through the anomalous magnetic momentum $a = (g_e - 2)/2$, can be expressed as an expansion in powers of the fine structure constant $\alpha_{\text{EM}} = q_e^2/4\pi$. The measurement of α_{EM} is done by using either rubidium [55] or caesium [56], giving two values for the SM prediction a_{th}. They are both compatible with the experimentally measured value a_{exp} [57] up to 10^{-12}:

$$a_{\text{th}}(\text{Rb}) = 0.001\,159\,652\,180\,252\,(95)\,, \tag{1.24}$$

$$a_{\text{th}}(\text{Cs}) = 0.001\,159\,652\,181\,61 \quad (23)\,, \tag{1.25}$$

$$a_{\text{exp}} = 0.001\,159\,652\,180\,59 \quad (13)\,. \tag{1.26}$$

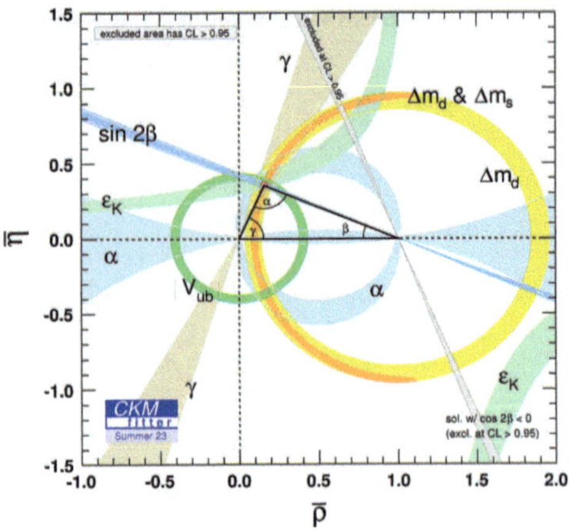

Fig. 1.2 2023 update on the CKM matrix combined constraints, in the $(\overline{\rho}, \overline{\eta})$ plane. The apex of the triangle, at $\overline{\rho} \approx 0.16$ and $\overline{\eta} \approx 0.36$, corresponds to the best-fit point. Taken from [60]. All the theoretical and experimental inputs can be found in [61]

Regarding the quarks, all the observed flavor-changing currents must be in agreement with the CKM matrix. Any deviation from it might be a sign of BSM physics. Indeed, typical BSM processes spoil the matrix unitarity. Constraints on the CKM matrix can typically be shown in the $(\overline{\rho}, \overline{\eta})$ plane (two CKM matrix parameters derived from the Wolfenstein parametrisation [58]). The results from the CKMFitter group [59] are shown in Fig. 1.2. The constraints from various experimental observations are all in agreement with the best-fit points, which preserves the matrix unitarity.

1.1.3 An Incomplete Theory

Despite its tremendous experimental success, the SM is also known to be incomplete. Notably, it fails to provide an explanation for some experimental observations and theoretical issues. Some of them are discussed below.

1.1.3.1 Dark Matter

One of the most puzzling observations is probably the experimental evidence of a weakly interacting matter not described by the SM, present in the Milky Way and in the Universe at large scale (see, e.g., [62] for a review). Historically, the first evidence for the existence of this matter came in 1933, with the observation of galaxies and galaxy clusters, whose dynamics were not understandable with only the mass of the visible matter [63]. This led to hypothesise the existence of an additional, invisible mass: the "dark" matter. Later observations of galaxy rotation curves strengthened this hypothesis [64–68]. Indeed, the star velocities in the galaxies were expected to drop as $1/\sqrt{r}$ with an increasing distance from the

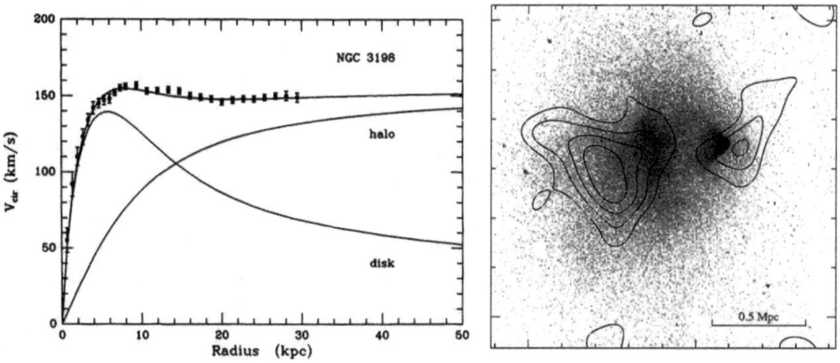

Fig. 1.3 Left: distribution of the star velocities in the galaxy NGC-3198 with respect to the distance from the centre of the same galaxy. The continuous lines are fits to obtain the correct distribution. The "halo" and "disk" lines correspond to the DM and ordinary matter components of the fit, respectively. The lines are extrapolated above 30 kpc. Taken from [68]. Right: X-ray observation of the galaxy cluster 1E0657–56, with superimposed mass contours obtained via weak lensing. Taken from [69]

galaxy centre r. What was instead observed was a constant velocity distribution at large distance from the centre, as shown in the left panel of Fig. 1.3. This behaviour can also be explained with the presence of an additional, invisible, massive matter. Today observations of galaxy rotation curves agree that DM seems to represent about 80–90% of the total mass of the galaxies.

These results are corroborated by the estimated weight of galaxy clusters through gravitational lensing. Furthermore, another evidence of DM has been observed in the galaxy cluster 1E0657–56 (dubbed the "Bullet Cluster"), where the mass distribution of the observed gas is in disagreement with the total mass distribution measured via weak lensing. The right panel of Fig. 1.3 shows the X-ray observation of 1E0657–56 by the Chandra X-Ray Observatory, superimposed with the mass contours obtained via weak lensing. The galaxy cluster 1E0657–56 is in fact two galaxy clusters having crossed each other. The majority of the mass, hypothetically made of DM, crossed the other sub-cluster without slowing down, while the observed gas, made of ordinary matter, lags behind due to its interaction with the other sub-cluster. This goes in favour that most of the mass is actually DM, and that the DM interacts only weakly with itself and with ordinary matter.

The DM effects can also be observed at large and cosmological scales. Its affects the large-scale structure formation [70], the baryon acoustic oscillations [71] and leaves an imprint in the temperature fluctuations of the cosmic microwave background. The observation of the cosmic microwave background by the Planck collaboration [72] gives one of the most precise estimations of today's DM density in the Universe (called the DM relic density, or DM relic abundance) [73]:

$$\Omega_{\mathrm{DM}}h^2 = 0.1188 \pm 0.0012\,, \tag{1.27}$$

where $\Omega_{\mathrm{DM}} = \rho_{\mathrm{DM}}/\rho_c$ is the DM relic density in units of the critical density of the Universe, h is the dimensionless Hubble constant defined as $h \equiv H_0/(100\ \mathrm{km.s^{-1}.Mpc^{-1}})$, with H_0 the current value of the Hubble parameter. The value of Eq. (1.27) has been obtained assuming $H_0 = (67.64 \pm 0.52)\ \mathrm{km.s^{-1}.Mpc^{-1}}$, as well as the typical Lambda Cold Dark Matter (ΛCDM) model, where Λ denotes the dark energy (discussed below) and cold refers to a non-relativistic DM.

Many different models of DM have been proposed and tested. However, so far, no direct or indirect observation has been able shed light on the nature of DM. It is also possible that the experimental evidence mentioned above are not evidence for a new massive object (fundamental particle or not) but are rather the manifestation of the limitations of our understanding of gravity.

1.1.3.2 Dark Energy

The first measurements of the Hubble constant H_0 were made by Lemaître [74] and Hubble [75], who observed that nearby galaxies move away from us at speeds proportional to their distances. Then, supernovae were used to constrain the H_0 expansion rate, finding that instead of slowing down due to gravity, it is actually accelerating [76, 77]. This effect can be accounted for by introducing a homogeneous field acting as a fluid with negative pressure against gravity, called dark energy (DE) and represented by Λ (the same Λ as in the ΛCDM model mentioned above). The energy density of the DE has been measured, e.g. by using the baryon acoustic oscillation measurements interpreted in a ΛCDM model with free spatial curvature [71]

$$\Omega_\Lambda = 0.651^{+0.068}_{-0.054}\,. \tag{1.28}$$

Its true nature, however, remains a mystery, neither explained by the Standard Model of cosmology, nor by the Standard Model of particle physics.

Once again, this inadequacy with our theories may be evidence for a new theory of gravitation. Many theories of modified gravity have been proposed to explain the experimental observations (DM and/or DE) by only modifying the theoretical formulation of gravity (see, e.g., [78–80] for a review), but without any significant success so far.

1.1.3.3 Hierarchy Problem

The fact that the SM cannot accommodate for gravity means that the SM is only valid below the (reduced) Planck scale $M_P \approx 2.4 \times 10^{18}$ GeV, which is the energy scale at which the gravitational effects are expected to be comparable to the quantum effects. Why such a huge difference between the fundamental Planck scale and the

EW scale ($m_{\rm EW} \approx v \approx 174$ GeV) exists is still an open question, known as the hierarchy problem.

1.1.3.4 Higgs Mass

The tree-level SM Higgs mass is a free parameter, measured to be around 125 GeV. However, quantum corrections need to be taken into account. Using momentum cutoff regularisation, one gets for each fermion f the higher-order corrections

$$\Delta m_h^2 = -\frac{|y_f|^2}{8\pi^2}\Lambda_{SM}^2 + \cdots \,, \tag{1.29}$$

where y_f is the Yukawa term of the fermion f, Λ_{SM} is the cut-off up to which the SM is valid, and the ellipses represent terms proportional to m_f^2, which grow at most logarithmically in Λ_{SM}. Each SM fermion gives such a contribution, and it should be multiplied by three for quarks to account for color charges. The Higgs mass is therefore highly dependent on the scale where new physics lies. Especially, if $\Lambda_{SM} \approx M_P$, it is clear that the Higgs mass should be way bigger. Thus, an enormous fine-tuning is required for the radiative corrections to cancel each other almost completely without invoking new physics below the Planck scale. The momentum cut-off regularisation makes the Higgs mass dependence on high energy scale explicit, but it remains true in other regularisation schemes. For instance, using the dimensional regularisation, one gets

$$\Delta m_h^2 \propto m_f^2 \,, \tag{1.30}$$

where m_f is the mass of the heaviest fermion (the top quark in the SM, but much heavier fermions possibly exist). In both cases, the Higgs mass dependency on a higher energy scale remains since it is driven by the mass of the heaviest particle coupling to the Higgs. This is true even when the coupling is indirect; for instance, if a BSM particle does not acquire a mass through the $SU(2)_L \times U(1)_Y$ breaking mechanism or the Yukawa terms but couples to the Higgs only via a gauge boson. In the SM, these large corrections are unique to the Higgs boson since it is the only scalar particle, and scalar masses are not protected by any symmetry. In order to avoid fine-tuning, all the solutions to the hierarchy problem require new physics at a rather low energy scale (not far from the EW scale). This is arguably the strongest argument in favour of new physics at the TeV scale.

1.1.3.5 Unification of Gauge Couplings

The successful unification of the weak and electromagnetic interactions (and of electricity and magnetism before that) can only suggest that the electroweak and strong interactions are unified at high energy. However, as shown in Fig. 1.4,

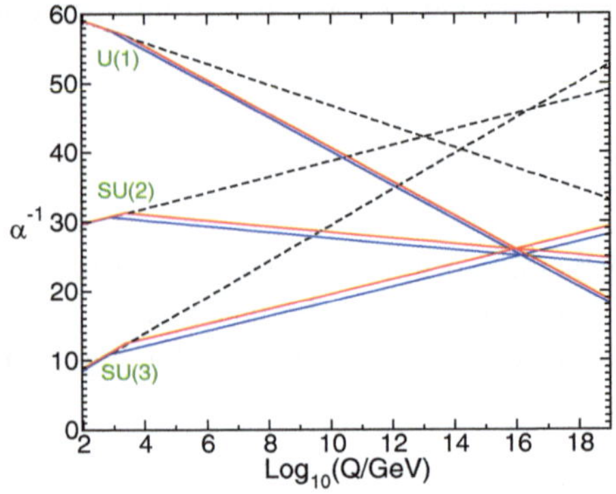

Fig. 1.4 Two-loop renormalisation group evolution of the SM inverse gauge couplings, i.e. $\alpha_i^{-1} = (g_i^2/4\pi)^{-1}$, with $g_1 = \sqrt{5/3}g'$, $g_2 = g$, and $g_3 = g_s$, as a function of the renormalisation group scale Q. The black dashed lines correspond to the SM prediction, while the full coloured lines correspond to the MSSM (defined in Sect. 1.2.2) predictions with SUSY particle masses varied between 750 GeV and 2.5 TeV. Taken from [50]

this unification cannot occur in the SM as the three gauge couplings cannot be extrapolated to a common value using the Renormalisation Group Equations (RGEs). Theories where this unification is possible are called Grand Unification Theories (GUTs) and the unification scale is therefore called the GUT scale. Moreover, GUTs typically provide a framework where all the matter fields, and the vector bosons, independently, can be described using the same representations of the gauge group. The simplest GUT (which will be briefly discussed in Sect. 1.3.1) is obtained by considering the simplest gauge symmetry group containing all the SM gauge groups at high scale, and that is broken at GUT scale.

1.1.3.6 Neutrino Masses

The absence of right-handed neutrinos in the SM prevents left-handed neutrinos from acquiring a mass. However, neutrinos have been observed to oscillate, going from one flavor to another [81–87]. It indicates that the neutrinos have a tiny mass ($\sum m_\nu < 0.072$ eV [71]) and that mass eigenstates are a mix of the flavor eigenstates. This can be explained by the Pontecorvo-Maki-Nakagawa-Sakata (PMNS) matrix [88, 89], which relates the mass eigenstates to the flavor eigenstates.

Incorporating this behaviour into the SM requires the existence of right-handed neutrinos, if they are of Dirac nature. The small left-handed neutrino masses can then be explained via the so-called see-saw mechanism and much heavier right-handed neutrinos. This mechanism could possibly relate the EW scale to the scale at which the three gauge couplings are unified, i.e. the GUT scale. However, no right-handed neutrino has been observed to date. In addition, the nature of the neutrinos (Dirac or Majorana) is still an open question, as well as their mass ordering (only their mass differences are known).

1.1.3.7 Strong CP Problem

In principle, the SM Lagrangian density allows for a CP violating term of the form

$$\mathcal{L} \supset \bar{\theta} \frac{g^2}{64\pi^2} G^{\mu\nu} \epsilon^{\mu\nu\rho\sigma} G^{\rho\sigma} , \tag{1.31}$$

with $\bar{\theta}$ a CP-violating angle and ϵ the Levi-Civita symbol. However, no experimental evidence of CP violation in the strong sector has been observed to date. This means that a very small $\bar{\theta}$ is required to match experimental constraints. The most stringent one comes from the measure of the permanent neutron electric dipole moment [90], requiring $\bar{\theta} \lesssim 10^{-9}$. One possible solution to this problem is to require an additional U(1) Peccei-Quinn symmetry [91, 92], which, when spontaneously broken, generates a Nambu-Goldstone boson, called the axion, whose VEV dynamically suppress the strong CP violating phase (for a review on axions and the strong CP problem, see [93]).

Before continuing, it is noteworthy that other experimental and theoretical open questions exist, e.g., the apparent predominance of matter over antimatter, the arbitrariness of the fermion masses and their hierarchy, why there are three fermion families, or why the quark mixings are small while the lepton mixings are large.

1.2 Supersymmetry

Although no strong experimental observation has challenged the SM predictions yet, it is now clear that the SM is an incomplete theory. Some of the issues mentioned in Sect. 1.1.3 can be resolved by introducing an additional internal symmetry relating fermionic states to bosonic states, and vice versa. Theories featuring this kind of symmetry are called SUSY theories.

In such a framework, the fermionic (bosonic) loop contributions to Δm_h^2 are systematically cancelled by an opposite contribution from bosonic (fermionic) degrees of freedom introduced by this symmetry. In that way, the Higgs boson's mass remains free from higher physics scale contributions at all orders in perturbation theory. In addition, SUSY theories can achieve electroweak and strong coupling unification at high energy, as shown in Fig. 1.4, and can provide a good DM candidate. Moreover, it is the only possible extension of the Poincaré group of symmetries in a four-dimensional relativistic quantum field theory.

This uniqueness comes from the Haag-Łopuszański-Sohnius theorem [94], circumventing the Coleman-Mandula "no-go" theorem [95], by extending the Poincaré algebra to a graded Lie algebra with additional spin-$1/2$ generators. Four-dimensional SUSY theories with more than one of these generators cannot allow for chiral fermions or parity violation as observed in the SM. Therefore, only SUSY theories with a single SUSY generator, $\mathcal{N} = 1$ SUSY theories, will be considered in the following. This introduction to $\mathcal{N} = 1$ SUSY theories is mainly based on [50]. Other good introductions to SUSY theories can be found in [96–100].

1.2.1 $\mathcal{N} = 1$ SUSY

1.2.1.1 Algebra and Superfields

In $\mathcal{N} = 1$ SUSY, each fermionic state transforms uniquely into a bosonic state, and vice versa, through a SUSY generator Q changing the spin by 1/2 unit:

$$Q|\text{Fermion}\rangle = |\text{Boson}\rangle \qquad \text{and} \qquad Q|\text{Boson}\rangle = |\text{Fermion}\rangle \, . \tag{1.32}$$

Applying Q twice on the same state leaves it unchanged. Hence, each state and its transformed counterpart form a unique pair and are called superpartners. These two bosonic and fermionic states can be gathered in the same object called supermultiplet. Furthermore, Q has to be an anti-commuting spinor with spin 1/2 and is therefore intrinsically complex. As a consequence, $Q^\dagger$ is also a SUSY generator, and each of them carries a distinct spinor index: Q_α, $Q^\dagger_{\dot{\alpha}}$, with $\alpha, \dot{\alpha} \in \{1, 2\}$. In addition to the Poincaré algebra, the SUSY algebra satisfies the relations:

$$[Q_\alpha, P^\mu] = [Q^\dagger_{\dot{\alpha}}, P^\mu] = 0 \, , \qquad [Q_\alpha, M^{\mu\nu}] = i \left(\sigma^{\mu\nu}\right)_\alpha^{\ \beta} Q_\beta \, , \tag{1.33}$$

$$\{Q_\alpha, Q_\beta\} = \{Q^\dagger_{\dot{\alpha}}, Q^\dagger_{\dot{\beta}}\} = 0 \, , \qquad \{Q_\alpha, Q^\dagger_{\dot{\alpha}}\} = 2 \left(\sigma^\mu\right)_{\alpha\dot{\alpha}} P_\mu \, , \tag{1.34}$$

with

$$\left(\sigma^{\mu\nu}\right)_\alpha^{\ \beta} = \frac{1}{4} \left(\sigma^\mu\right)_{\alpha\dot{\alpha}} \left(\overline{\sigma}^\nu\right)^{\dot{\alpha}\beta} - \frac{1}{4} \left(\sigma^\nu\right)_{\alpha\dot{\alpha}} \left(\overline{\sigma}^\mu\right)^{\dot{\alpha}\beta} \, , \tag{1.35}$$

$$\sigma^\mu = (\mathbb{1}_2, \vec{\sigma}) \qquad \text{and} \qquad \overline{\sigma}^\mu = (\mathbb{1}_2, -\vec{\sigma}) \, , \tag{1.36}$$

where $\vec{\sigma} = \left(\sigma^1, \sigma^2, \sigma^3\right)$ is the Pauli vector.

From the SUSY algebra, it is possible to show that each supermultiplet must have the same number of bosonic and fermionic degrees of freedom. Moreover, since the SUSY generators commute with P^μ and with algebra generators of the SM gauge groups, the bosonic and fermionic states of a supermultiplet must have the same mass, electric charge, weak isospin and color degrees of freedom.

Since Q and $Q^\dagger$ carry a non-zero angular momentum, SUSY transformations are space-time transformations, and SUSY can be given a geometric interpretation. The SUSY idea can be applied to the four classical bosonic dimensions t, x, y, z, leading to four additional fermionic anti-commuting coordinates θ^α, $\theta^\dagger_{\dot{\alpha}}$, with $\alpha, \dot{\alpha} \in \{1, 2\}$. The manifold built from these eight coordinates is called a superspace, and the objects living on this manifold are called superfields. A superfield joins the two superpartners of a supermultiplet in a single object[6] written in terms of superspace coordinates, and gives an adequate framework to manipulate them.

A minimal SUSY extension of the SM requires two kinds of superfields. The first one, called chiral superfield (or matter superfield, or scalar superfield), takes the form

$$\Phi(y,\theta) = \phi(y) + \sqrt{2}\theta\psi(y) + \theta\theta F(y)\,, \tag{1.37}$$

where ψ is a left-handed, two-component, spin-1/2 Weyl fermion, and ϕ is its complex scalar superpartner. Here and from now on, unless stated otherwise, spinor indices will be suppressed when contracted in a sum. The F field is also a complex scalar field but it is not propagating and is only needed to close the SUSY algebra off-shell (and to have the same number of fermionic and bosonic degrees of freedom off-shell). It is therefore called an auxiliary field. The fields are expressed in terms of the coordinate $y^\mu \equiv x^\mu + i\theta^\dagger\overline{\sigma}^\mu\theta$ for simplicity. This superfield is appropriate for describing the left-handed fermions of the SM (via ψ) and their superpartners (via ϕ). The other fermion chirality appears in the Lagrangian through[7]

$$\Phi^\dagger(y^*,\theta^\dagger) = \phi^*(y^*) + \sqrt{2}\theta^\dagger\psi^\dagger(y^*) + \theta^\dagger\theta^\dagger F^*(y^*) \tag{1.38}$$

The second kind of superfield is the vector superfield (or gauge superfield, or real superfield). In the Wess-Zumino supergauge, it takes the form

$$V(x,\theta,\theta^\dagger) = \theta\sigma^\mu\theta^\dagger A_\mu(x) + \theta^\dagger\theta^\dagger\theta\lambda(x) + \theta\theta\theta^\dagger\lambda^\dagger(x) + \frac{1}{2}\theta\theta\theta^\dagger\theta^\dagger D(x)\,. \tag{1.39}$$

Here, A_μ is a massless gauge boson, λ is its superpartner, a spin-1/2 Weyl fermion,[8] and D is a real bosonic auxiliary field. This superfield is appropriate for describing the SM bosons (via A_μ) and their superpartners (via λ).

In addition, if we include gravity in the theory, a vector superfield can be used to describe the massless spin-2 graviton and its spin-3/2 superpartner, the gravitino. Gravity can be introduced in SUSY theories by requiring the SUSY transformations to be local. The resulting theory is called supergravity.

---

[6] The irreducible representations of the algebra are not labelled by the mass and the spin anymore, but by the mass and the so-called superspin.

[7] The hermitian conjugate of a left-handed Weyl fermion is a right-handed Weyl fermion.

[8] If the superpartners of the spin-1 gauge bosons had a spin of 3/2, the theory would not be renormalisable.

1.2.1.2 The Renormalisable SUSY Lagrangian

To express the Lagrangian, it is useful to define integration over the fermionic coordinates θ and $\theta^\dagger$.[9] For a superfield $S(x, \theta, \theta^\dagger)$:

$$[S]_F + \text{h.c.} \equiv \int d^2\theta d^2\theta^\dagger \left[\delta^{(2)}(\theta^\dagger)S(x,\theta,\theta^\dagger) + \delta^{(2)}(\theta)S^\dagger(x,\theta,\theta^\dagger)\right], \tag{1.40}$$

$$[S]_D \equiv \int d^2\theta d^2\theta^\dagger S(x,\theta,\theta^\dagger). \tag{1.41}$$

They are called F- and D-terms respectively. Indeed, for a chiral superfield Φ and a vector superfield V, one has $[\Phi]_F + \text{h.c.} = F + F^*$ and $[V]_D = D/2$.

A general, renormalisable, SUSY and gauge-preserving Lagrangian density takes the form

$$\mathcal{L}_{\text{SUSY}} = \left[W(\Phi) + \frac{1}{4}\tau_a g_a^2 \mathcal{W}^{a\alpha}\mathcal{W}^a_\alpha\right]_F + \text{h.c.} + \left[\Phi^{\dagger i}\left(e^{2g_a T^a V^a}\right)_i^{\ j}\Phi_j\right]_D, \tag{1.42}$$

with a running over the adjoint representation of the gauge groups of coupling g_a and matrix entries of the generators $T_i^{a\,j}$, and i, j running over the different chiral superfields. W is the superpotential of the theory and includes the F-terms of the scalar potential as well as the interactions between scalars and fermions of the chiral superfields. It must be a holomorphic function (so it must depend only on Φ_i and not Φ_i^*), be invariant under the gauge symmetries of the theory and have dimension $[\text{mass}]^3$. In renormalisable theories, it takes the form

$$W(\Phi) = L^i\Phi_i + \frac{1}{2}M^{ij}\Phi_i\Phi_j + \frac{1}{6}y^{ijk}\Phi_i\Phi_j\Phi_k, \tag{1.43}$$

with M and y the symmetric mass matrix and the totally symmetric Yukawa coupling, respectively. The τ_a are holomorphic couplings:

$$\tau_a = \frac{1}{g_a^2} - i\frac{\Theta_a}{8\pi^2}, \tag{1.44}$$

where Θ_a is CP-violating angle giving a total derivative term in the Lagrangian density. The $\mathcal{W}^a_\alpha$ are field-strength superfields and can be expressed in component terms in the Wess-Zumino supergauge as (the dependence on the spinor index α is

[9] See, for instance, Sect. 4.1 of [50].

here made explicit for clarity)

$$\mathcal{W}^a_\alpha = \lambda^a_\alpha + \theta_\alpha D^a - \frac{i}{2}(\sigma^\mu \overline{\sigma}^\nu \theta)_\alpha F^a_{\mu\nu} + i\theta\theta(\sigma^\mu \nabla_\mu \lambda^{\dagger a})_\alpha \,. \tag{1.45}$$

Here, $F^a_{\mu\nu}$ is the usual field strength tensor defined in Eq. (1.14) and ∇_μ is the gauge-covariant derivative

$$\nabla_\mu \lambda^a = \partial_\mu \lambda^a - g_a f^{abc} A^b_\mu \lambda^c \,. \tag{1.46}$$

It is noteworthy that the equations of motion can be used to express the auxiliary fields F and D in terms of an algebraical combination of scalar fields:

$$F_i = -\, L^*_i - M^*_{ij}\phi^j - \frac{1}{2} y^*_{ijk}\phi^j\phi^k \,, \tag{1.47}$$

$$F^{*i} = -\, L^i - M^{ij}\phi_j - \frac{1}{2} y^{ijk}\phi_j\phi_k \,, \tag{1.48}$$

$$D^a = -\, g_a(\phi^{*i} T^{aj}_i \phi_j) \,. \tag{1.49}$$

It follows that the general, renormalisable, SUSY and grauge-preserving Lagrangian density can be written in terms of component fields as

$$\begin{aligned}
\mathcal{L}_{\text{SUSY}} =& \nabla_\mu \phi^{*i} \nabla_\mu \phi_i - V(\phi, \phi^*) \\
&+ i\psi^{\dagger i}\overline{\sigma}^\mu \nabla_\mu \psi_i - \frac{1}{2} M^{ij}\psi_i\psi_j - \frac{1}{2} M^*_{ij}\psi^{\dagger i}\psi^{\dagger j} \\
&- \frac{1}{2} y^{ijk}\phi_i\psi_j\psi_k - \frac{1}{2} y^*_{ijk}\phi^{*i}\psi^{\dagger j}\psi^{\dagger k} \\
&- \frac{1}{4} F^a_{\mu\nu} F^{a\,\mu\nu} + i\lambda^{\dagger a}\overline{\sigma}^\mu \nabla_\mu \lambda^a \\
&- \sqrt{2} g_a(\phi^{*i} T^{aj}_i \psi_j)\lambda^a - \sqrt{2} g_a \lambda^{\dagger a}(\psi^{\dagger i} T^{aj}_i \phi_j) \,,
\end{aligned} \tag{1.50}$$

where

$$\begin{aligned}
V(\phi, \phi^*) =& F_i F^{*i} + \frac{1}{2} D^a D^a \\
=& M^*_{ik} M^{kj}\phi^{*i}\phi_j \\
&+ \frac{1}{2} M^{il} y^*_{jkl}\phi_i\phi^{*j}\phi^{*k} + \frac{1}{2} M^*_{il} y^{jkl}\phi^{*i}\phi_j\phi_k \\
&+ \frac{1}{4} y^{ijm} y^*_{klm}\phi_i\phi_j\phi^{*k}\phi^{*l} \\
&+ \frac{1}{2} g^2_a(\phi^{*i} T^{aj}_i \phi_j)^2
\end{aligned} \tag{1.51}$$

is the scalar potential, and ∇_μ are the covariant derivatives. Their actions on ϕ and ψ are defined in Eq. (1.10), while their actions on λ^a are defined in Eq. (1.46). Their actions on ϕ^* are

$$\nabla_\mu \phi^{*i} = \partial_\mu \phi^{*i} - i g_a A_\mu^a (\phi^* T^a)^i \, . \tag{1.52}$$

1.2.1.3 Non-renormalisable SUSY Lagrangians

A non-renormalisable gauge-invariant Lagrangian density can take the form

$$\begin{aligned} \mathcal{L}_{\text{SUSY}} = & \left[W(\Phi_i) + \frac{1}{4} f_{ab}(\Phi_i) g_a g_b \mathcal{W}^{a\alpha} \mathcal{W}^b_\alpha \right]_F + \text{h.c.} \\ & + \left[K(\Phi_i, (\Phi^\dagger e^{2 g_a T^a V^a})^j) \right]_D , \end{aligned} \tag{1.53}$$

where

- W is the superpotential, defined in Eq. (1.43);
- $\mathcal{W}^a_\alpha$ are the field-strength superfields defined in Eq. (1.45);
- f_{ab} are gauge-kinetic functions, where a and b run on the adjoint representations of the gauge groups. They are symmetric under the interchange of a and b and they encode the non-renormalisable couplings of the vector supermultiplets to the chiral supermultiplets. Like the superpotential, they are holomorphic functions of chiral superfields. However, they are dimensionless and if there are less than two Abelian components in the gauge group they are proportional to $\delta_{\alpha\beta}$. In the special case of renormalisable SUSY Lagrangians at tree-level, they are independent of the chiral superfields and $f_{ab} = \delta_{ab}\tau_a$ so one gets the Lagrangian density written in Eq. (1.42);
- K is the Kähler potential. It is a function of both chiral and anti-chiral superfields, it is supergauge invariant, real and has dimension [mass]^2. For a renormalisable Lagrangian density at tree-level, $K = \Phi^{\dagger i} (e^{2 g_a T^a V^a})_i^j \Phi_j$, giving the D-term of Eq. (1.42);

1.2.1.4 Soft SUSY Breaking

A direct prediction of SUSY is that the superpartners of a superfield have the same mass. If it was true at low energy, SUSY particles would have already been detected. This means that SUSY must be spontaneously broken at some energy scale so that it is not an exact symmetry at low energy. To achieve this, the ground state must have a strictly positive energy, which means from Eq. (1.51) that there must be a (meta-)stable state where at least one auxiliary field has a non-zero VEV, i.e. $\langle F_i \rangle \neq 0$ or $\langle D^a \rangle \neq 0$. In addition, the SUSY-breaking mechanism must provide a mass for the gauginos and be nearly flavor-blind (to not conflict with experiments). Several models of spontaneous SUSY breaking have been proposed, but there is no consensus on how this should be done. However, they all require the origin of SUSY breaking to occur in a "hidden" sector (almost) decoupled from the "visible" sector of the chiral superfields containing SM matter. Nonetheless, the two sectors are linked by some flavor-blind interactions, indirectly carrying SUSY-breaking effects to the visible sector. Some SUSY-breaking models will be introduced in Sect. 1.2.2.

Even if there is no consensus on how the SUSY-breaking mechanism could occur, it is still possible to write down a renormalisable Lagrangian density containing all effective SUSY-breaking terms that could appear at low energy, regardless of the SUSY-breaking mechanism. However, in order to provide a solution for the hierarchy problem, i.e. to stay agnostic to high energy scales, SUSY must be broken "softly", that is to say the SUSY-breaking terms must only contain mass terms and coupling parameters with positive mass dimensions. The full, low-energy, renormalisable, gauge-invariant Lagrangian density is then

$$\mathcal{L} = \mathcal{L}_{\text{SUSY}} + \mathcal{L}_{\text{soft}}\,, \tag{1.54}$$

where $\mathcal{L}_{\text{SUSY}}$ is the one of Eq. (1.42) (or, equivalently Eq. (1.50)) and

$$\mathcal{L}_{\text{soft}} = -\left(\frac{1}{2}M_a\,\lambda^a\lambda^a + \frac{1}{6}a^{ijk}\phi_i\phi_j\phi_k + \frac{1}{2}b^{ij}\phi_i\phi_j + t^i\phi_i\right) + \text{h.c.} - (m^2)^i_j\phi^{j*}\phi_i \tag{1.55}$$

are the soft SUSY breaking terms. Here, M_a are the gaugino (the superpartners of the SM gauge bosons) mass parameters, a^{ijk} are trilinear couplings, b^{ij} and $(m^2)^i_j$ are scalar squared mass parameters, and t^i are "tadpole" couplings (contributions to one-point correlation functions).

1.2.1.5 The Goldstino/Gravitino

Moreover, breaking SUSY generates a Nambu-Goldstone particle, which must be a massless, neutral, Weyl fermion. It is called the goldstino and has goldstino-scalar-chiral fermion and goldstino-gaugino-gauge boson couplings. In supergravity theories, the gravitino also is massless as long as SUSY is unbroken. However, when SUSY is spontaneously broken, it acquires a mass by absorbing the goldstino, which becomes the longitudinal components (i.e. helicity $\pm 1/2$) of the gravitino. From dimensional analysis, the gravitino mass $m_{3/2}$ can be estimated as

$$m_{3/2} \sim \langle F_i \rangle / M_P \tag{1.56}$$

since the mass must vanish in the limit where SUSY is unbroken ($\langle F_i \rangle \to 0$) and where gravity is decoupled ($M_P \to \inf$) [101, 102]. Furthermore, the transverse components (i.e. helicity $\pm 3/2$) of the gravitino only have gravitational interactions. Therefore, the collider phenomenology of the gravitino relies only on its longitudinal components (the goldstino) and $\tilde{G}$ will be used to interchangeably denote the gravitino or the goldstino in the following.

1.2.2 Minimal Supersymmetric Standard Model

1.2.2.1 The Superpotential

The simplest phenomenologically viable SUSY model is the minimal SUSY extension of the SM, the MSSM. This model contains all the particles of the SM listed in Table 1.1 but also all their superpartners. Furthermore, a second Higgs chiral superfield is required to have no gauge anomaly and to generate mass terms to all the quarks and leptons. The field content of the MSSM is shown in Table 1.2. The SUSY states are denoted with a "$\sim$", the superpartners of the SM fermions get a "s-" prefix standing for "scalar-" (selectron, stop, ...) and are collectively referred to as sfermions (or sleptons and squarks to specify a type of sfermions), the superpartners of the SM bosons get a "-ino" suffix (higgsinos, gluino, winos, bino). The dagger is applied on the Weyl fermion, i.e. $\psi_R^\dagger \equiv (\psi_R)^\dagger$. The right-handed quarks and leptons and their superpartners are not displayed here but appear through Φ^*. The squarks and sleptons have no chirality but they have L and R subscripts to indicate the chirality of their superpartners.

Table 1.2 Particle content of the MSSM. The particles that are new compared to the SM are written in purple. The rows are organised in supermultiplets. The symbol in the "Names" column is the symbol for the whole chiral supermultiplet. The rows containing spin-0 and spin-1/2 particles are chiral supermultiplets, while the rows containing spin-1/2 and spin-1 particles are vector supermultiplets. Adapted from [50]

Names		Spin 0	Spin 1/2	Spin 1	$SU(3)_C,\ SU(2)_L,\ U(1)_Y$
Squarks, quarks	$\widehat{Q}$	$\widetilde{Q} = (\widetilde{u}_L\ \ \widetilde{d}_L)$	$(u_L\ \ d_L)$		$(\mathbf{3},\ \mathbf{2},\ \frac{1}{6})$
(×3 families)	$\widehat{u}$	$\widetilde{u}_R^*$	$u_R^\dagger$		$(\overline{\mathbf{3}},\ \mathbf{1},\ -\frac{2}{3})$
	$\widehat{d}$	$\widetilde{d}_R^*$	$d_R^\dagger$		$(\overline{\mathbf{3}},\ \mathbf{1},\ \frac{1}{3})$
Sleptons, leptons	$\widehat{L}$	$\widetilde{L} = (\widetilde{\nu}_{e_L}\ \ \widetilde{e}_L)$	$(\nu_{e_L}\ \ e_L)$		$(\mathbf{1},\ \mathbf{2},\ -\frac{1}{2})$
(×3 families)	$\widehat{e}$	$\widetilde{e}_R^*$	$e_R^\dagger$		$(\mathbf{1},\ \mathbf{1},\ 1)$
Higgs, higgsinos	$\widehat{H}_u$	$H_u = (H_u^+\ \ H_u^0)$	$(\widetilde{H}_u^+\ \ \widetilde{H}_u^0)$		$(\mathbf{1},\ \mathbf{2},\ \frac{1}{2})$
	$\widehat{H}_d$	$H_d = (H_d^0\ \ H_d^-)$	$(\widetilde{H}_d^0\ \ \widetilde{H}_d^-)$		$(\mathbf{1},\ \mathbf{2},\ -\frac{1}{2})$
Gluino, gluon			$\widetilde{g}$	g	$(\mathbf{8},\ \mathbf{1},\ 0)$
Winos, W bosons			$\widetilde{W}^\pm\ \ \widetilde{W}^0$	$W^\pm\ \ W^0$	$(\mathbf{1},\ \mathbf{3},\ 0)$
Bino, B boson			$\widetilde{B}$	B	$(\mathbf{1},\ \mathbf{1},\ 0)$

The MSSM superpotential is

$$
\begin{aligned}
W_{\text{MSSM}} = \ & \widehat{u}^{ia}(\mathbf{y_u})_i{}^j(\widehat{Q}_{ja})_\alpha(\widehat{H}_u)_\beta\epsilon^{\alpha\beta} - \widehat{d}^{ia}(\mathbf{y_d})_i{}^j(\widehat{Q}_{ja})_\alpha(\widehat{H}_d)_\beta\epsilon^{\alpha\beta} \\
& - \widehat{e}^{ia}(\mathbf{y_e})_i{}^j(\widehat{L}_{ja})_\alpha(\widehat{H}_d)_\beta\epsilon^{\alpha\beta} \\
& + \mu(\widehat{H}_u)_\alpha(\widehat{H}_d)_\beta\epsilon^{\alpha\beta} , \qquad (1.57)
\end{aligned}
$$

with $i, j = 1{,}2{,}3$ the family indices; $a = 1{,}2{,}3$ the color indices; $\alpha, \beta = 1{,}2$ the weak isospin indices; ϵ is the antisymmetric symbol; $\widehat{u}$, $\widehat{d}$ and $\widehat{e}$ the chiral superfields of the chiral supermultiplets defined in Table 1.2; $\widehat{Q}$, $\widehat{L}$, $\widehat{H}_u$ and $\widehat{H}_d$ the $SU(2)_L$ doublets containing the corresponding two superfields; and $\mathbf{y_u}$, $\mathbf{y_d}$ and $\mathbf{y_e}$ the 3×3 matrices in family space containing the dimensionless Yukawa couplings. After electroweak symmetry breaking, the Yukawa couplings determine the masses and CKM mixing angles of the quarks and leptons. Since the third generation of fermions is the heaviest in the SM, it is often useful to make the approximation

$$
\mathbf{y_u} \approx \begin{pmatrix} 0 & 0 & 0 \\ 0 & 0 & 0 \\ 0 & 0 & y_t \end{pmatrix}, \quad \mathbf{y_d} \approx \begin{pmatrix} 0 & 0 & 0 \\ 0 & 0 & 0 \\ 0 & 0 & y_b \end{pmatrix}, \quad \mathbf{y_e} \approx \begin{pmatrix} 0 & 0 & 0 \\ 0 & 0 & 0 \\ 0 & 0 & y_\tau \end{pmatrix}. \qquad (1.58)
$$

1.2.2.2 *R*-Parity

In addition, four terms should be added to Eq. (1.57) to get the full gauge-invariant renormalisable MSSM superpotential:

$$
W_{\Delta \text{L}=1} = \frac{1}{2}\lambda^{ijk}\widehat{L}_i\widehat{L}_j\widehat{e}_k + \lambda'^{ijk}\widehat{L}_i\widehat{Q}_j\widehat{d}_k + \mu'^i\widehat{L}_i\widehat{H}_u , \qquad (1.59)
$$

$$
W_{\Delta \text{B}=1} = \frac{1}{2}\lambda''^{ijk}\widehat{u}_i\widehat{d}_j\widehat{d}_k , \qquad (1.60)
$$

with $i, j, k = 1{,}2{,}3$ the family indices, and the $SU(2)_L$ and color indices being suppressed. However, these terms violate lepton or baryon number conservation by 1 unit. Indeed, $\widehat{L}$ and $\widehat{e}$ carry a lepton number of +1 and −1, respectively, while all the other supermultiplets have a lepton number of 0; and $\widehat{Q}$, $\widehat{d}$ and $\widehat{u}$ carry a baryon number of +1/3, −1/3, −1/3, respectively, while all the other supermultiplets have a baryon number of 0. Since no evidence of processes violating lepton or baryon number conservation by 1 unit has been experimentally observed, it is customary to add a symmetry to the MSSM preventing the existence of such terms. This new symmetry is a discrete $\mathbb{Z}_2$ (also written $\mathcal{Z}_2$) symmetry, called *R*-parity. It associates with each particle the quantity

$$
R = (-1)^{3(B-L)+2s} , \qquad (1.61)
$$

where B, L and s are the baryon number, the lepton number and the spin of the particle, respectively. Particles in the same supermultiplet have opposite R-parity numbers. All the SM particles and the Higgs bosons have $R = +1$, while all their superpartners have $R = -1$. Therefore, it is convenient to call the R-parity odd particles the SUSY particles (or sparticles for short). If R-parity is exactly conserved, no mixing between particles with different R is allowed, and only the Lagrangian terms whose product of R for all the fields in it is +1 are allowed. This directly forbids the four terms in Eqs. (1.59) and (1.60). Moreover, it has three important phenomenological implications:

- at colliders, only an even number of SUSY particles can be produced;
- each SUSY particle can only decay into an even number of other SUSY particles, plus an even or odd number of SM particles;
- the lightest SUSY particle (LSP) is stable. In particular, if it only interacts weakly with ordinary matter, it can be a good candidate to make up all or a fraction of the dark matter relic abundance observed today.

The MSSM provides several good DM candidates. The sneutrino is one of them. However, its strong coupling to the Z boson gives a DM relic abundance too low compared to the observed value. Moreover, direct DM detection experiments put strong constraints on DM made of sneutrino. Apart from the gravitino, the other good DM candidate is the lightest sparticle amongst the bino, the neutral wino, and the neutral higgsinos.

Nevertheless, even if R-parity is not conserved, the MSSM can provide a good DM candidate if a SUSY particle has a long enough lifetime (thanks to very low R-parity violating (RPV) couplings for instance).

1.2.2.3 Soft SUSY Breaking

The soft SUSY-breaking Lagrangian density of the MSSM can be written in terms of component fields as

$$
\begin{aligned}
\mathcal{L}_{\text{soft}}^{\text{MSSM}} = &-\frac{1}{2}\left(M_1\widetilde{B}\widetilde{B} + M_2\widetilde{W}^a\widetilde{W}^a + M_3\widetilde{g}^a\widetilde{g}^a + \text{h.c.}\right) \\
&-\Big(\widetilde{u}_R^{ia}(\mathbf{a_u})_i{}^j(\widetilde{Q}_{ja})_\alpha(H_u)_\beta\epsilon^{\alpha\beta} - \widetilde{d}_R^{ia}(\mathbf{a_d})_i{}^j(\widetilde{Q}_{ja})_\alpha(H_d)_\beta\epsilon^{\alpha\beta} \\
&-\widetilde{e}_R^{ia}(\mathbf{a_e})_i{}^j(\widetilde{Q}_{ja})_\alpha(H_d)_\beta\epsilon^{\alpha\beta} + \text{h.c.}\Big) \\
&-(\widetilde{Q}^{*ia})_\alpha(\mathbf{m}^2_{\widetilde{\mathbf{Q}}})_i{}^j(\widetilde{Q}_{ja})^\alpha - \widetilde{u}_R^{ia}(\mathbf{m}^2_{\widetilde{\mathbf{u}}})_i{}^j\widetilde{u}^*_{R\,ja} - \widetilde{d}_R^{ia}(\mathbf{m}^2_{\widetilde{\mathbf{d}}})_i{}^j\widetilde{d}^*_{R\,ja} \\
&-(\widetilde{L}^{*i})_\alpha(\mathbf{m}^2_{\widetilde{\mathbf{L}}})_i{}^j(\widetilde{L}_j)^\alpha - \widetilde{e}_R^i(\mathbf{m}^2_{\widetilde{\mathbf{e}}})_i{}^j\widetilde{e}^*_{R\,j} \\
&-\left(b(H_u)_\alpha(H_d)_\beta\epsilon^{\alpha\beta} + \text{h.c.}\right) - m^2_{H_u}|H_u|^2 - m^2_{H_d}|H_d|^2 ,
\end{aligned}
\tag{1.62}
$$

with a summing on the adjoint representation of the corresponding gauge group for the gaugino, and on the (anti-)fundamental representation of $SU(3)_c$ for the (anti)squarks. Here, $\widetilde{Q}$, $\widetilde{L}$, H_u and H_d are $SU(2)_L$ doublets of scalar fields, not of superfields. These soft breaking terms introduce a total of 105 free parameters, not previously in the SM, that cannot be rotated away [103]. Regardless of the SUSY-breaking mechanism, it is expected for them to be at a scale $m_{\text{soft}} \sim \frac{\langle F_i \rangle}{M}$, where M is the mass scale associated with the physics that mediates between the hidden and visible sectors (the squared mass parameters and b are expected to be at the scale m^2_{soft}). However, these 105 parameters can be reduced to only a handful before SUSY breaking occurs. These few parameters (plus the gauge and Yukawa couplings of the MSSM) are then propagated, from M to the EW scale using RGEs, to give the soft SUSY breaking parameters. Some SUSY breaking models are listed below.

- **Planck-scale-mediated SUSY breaking (PMSB)** (or gravity-mediated SUSY breaking)**:** the mediating interactions include gravity and transmit SUSY breaking effects to the visible sector around the Planck scale M_P. However, nothing enforces the interactions to be flavor-blind. A minimal model of gravity-mediated SUSY breaking, called minimal supergravity, or constrained MSSM, contains only four parameters before SUSY breaking:

  - $m_{1/2}$: a common gaugino mass: $M_1 = M_2 = M_3 = m_{1/2}$;
  - m_0^2: a common scalar mass: $\mathbf{m^2_Q} = \mathbf{m^2_{\bar{u}}} = \mathbf{m^2_{\bar{d}}} = \mathbf{m^2_L} = \mathbf{m^2_{\bar{e}}} = m_0^2\mathbb{1}$;
  - A_0: a common trilinear coupling: $\mathbf{a_u} = A_0\mathbf{y_u}$, $\mathbf{a_d} = A_0\mathbf{y_d}$, $\mathbf{a_e} = A_0\mathbf{y_e}$;
  - B_0: a term entering in the scalar squared-mass term: $b = B_0\mu$.

  By propagating these parameters with the RGEs, one finds that $M_1 : M_2 : M_3 \approx 1 : 2 : 6$, that the third generation squarks are the lightest squarks due to the large Yukawa coupling of their SM partners, and that the sleptons are usually lighter than the squarks because they do not have strong interaction loops increasing their masses.
- **Gauge-mediated SUSY breaking (GMSB):** SUSY breaking effects are transmitted through gauge interactions, making them automatically flavor-blind. It requires some chiral supermultiplets, called messengers, to couple to a non-vanishing $\langle F_i \rangle$ and to the SM gauge bosons and gauginos. These models can be parametrised with four parameters:

  - N: the number of messengers;
  - M_{mess}: the messenger mass scale;
  - Λ: an effective SUSY-breaking parameter;
  - μ: the Higgs-higgsino mass parameter.

  From them, one can obtain the SUSY-breaking parameters and use the RGEs to have their values at the EW scale. Some predictions are then that the goldstino/gravitino is the LSP, and that the squarks and gluinos need to be very heavy in order for the SM-like Higgs boson to have its observed mass of 125 GeV.

- **Anomaly-mediated SUSY breaking (AMSB):** SUSY-breaking effects are entirely transmitted by supergravity effects. The anomalous violation of a local superconformal (scale) invariance manifested in the running of the couplings causes the SUSY-breaking effects to occur in the MSSM through the renormalisation group quantities, which are flavor-blind to a good approximation because dominated by the gauge couplings. Only three parameters are necessary to build these models:
  - $m_{3/2}$: the gravitino mass, which is also the SUSY-breaking scale;
  - m_0^2: an ad-hoc scalar mass term, necessary to avoid negative mass terms for the sleptons;
  - μ: the Higgs-higgsino mass parameter.

  A unique AMSB feature is that the soft SUSY-breaking parameters, obtained from these three model parameters, hold at every renormalisation scale, to all orders in perturbation theory. It means that AMSB models are insensitive to ultra-violet physics, ensuring the low-energy MSSM soft terms to be free from flavor violation. Moreover, they predict a gravitino much larger than m_{soft}, and $M_1 : M_2 : M_3 \approx 3.3 : 1 : 10$, so the LSP might be the lightest wino or higgsino.

1.2.2.4 Electroweak Symmetry Breaking

The spontaneous breaking of the EW symmetry is different from the one occurring in the SM because of the additional Higgs doublet. The Higgs potential contains four free parameters: $|\mu|^2$, b, $m_{H_u}^2$ and $m_{H_d}^2$. Under the assumption that the potential is bounded from below and that its minimum is not obtained with $H_u^0 = H_d^0 = 0$, the spontaneous breaking of the EW symmetry can occur. These two requirements can be expressed through the inequalities

$$2b < 2|\mu|^2 + m_{H_u}^2 + m_{H_d}^2 \qquad \text{and} \qquad b^2 > (|\mu|^2 + m_{H_u}^2)(|\mu|^2 + m_{H_d}^2) \,. \tag{1.63}$$

These inequalities are not satisfied if $m_{H_u}^2 = m_{H_d}^2$, as it is the case at tree-level in minimal supergravity and GMSB models. So in these models, EWSB occurs through higher-order corrections and is called radiative EWSB.

By defining the two VEVs

$$v_u = \langle H_u^0 \rangle \qquad \text{and} \qquad v_d = \langle H_d^0 \rangle \,, \tag{1.64}$$

it is possible to express the Z boson mass as

$$v_u^2 + v_d^2 = v^2 = 2m_Z^2/(g^2 + g'^2) \approx (174 \text{ GeV})^2 \,. \tag{1.65}$$

Moreover, it is traditional to express the ratio of the two VEVs as

$$\tan\beta \equiv v_u/v_d \,. \tag{1.66}$$

Since v_u and v_d are taken real and positive by convention, then we have $0 < \beta < \pi/2$. Moreover, $\tan\beta$ should not be too large or too small in order to avoid non-perturbatively large Yukawa couplings (very roughly, $1.2 \lesssim \tan\beta \lesssim 65$).

After EWSB, three of the eight real scalar degrees of freedom of the two Higgs doublets become Nambu-Goldstone bosons (G^0, $G^\pm$) that are "eaten" by the Z and $W^\pm$ bosons to give them mass. The five remaining degrees of freedom are two CP-even neutral scalars h^0 and H^0, one CP-odd neutral scalar A^0, one positively charged scalar H^+ and its conjugate H^- ($= H^{+*}$). It is then possible to express the gauge-eigenstates in terms of the mass-eigenstates fields. If v_u, v_d minimize the tree-level potential, one gets:

$$\begin{pmatrix} H_u^0 \\ H_d^0 \end{pmatrix} = \begin{pmatrix} v_u \\ v_d \end{pmatrix} + \frac{1}{\sqrt{2}} R_\alpha \begin{pmatrix} h^0 \\ H^0 \end{pmatrix} + \frac{i}{\sqrt{2}} R_\beta \begin{pmatrix} G^0 \\ A^0 \end{pmatrix} , \tag{1.67}$$

$$\begin{pmatrix} H_u^+ \\ H_d^{-*} \end{pmatrix} = R_\beta \begin{pmatrix} G^+ \\ H^+ \end{pmatrix} , \tag{1.68}$$

where

$$R_\alpha = \begin{pmatrix} \cos\alpha & \sin\alpha \\ -\sin\alpha & \cos\alpha \end{pmatrix} \quad \text{and} \quad R_\beta = \begin{pmatrix} \sin\beta & \cos\beta \\ -\cos\beta & \sin\beta \end{pmatrix} . \tag{1.69}$$

1.2.2.5 Higgs Boson Masses

The following scalar squared-mass terms then appear in the potential:

$$V \supset \frac{1}{2} m_{h^0}^2 (h^0)^2 + \frac{1}{2} m_{H^0}^2 (H^0)^2 + \frac{1}{2} m_{A^0}^2 (A^0)^2 + \frac{1}{2} m_{H^\pm}^2 |H^+|^2 , \tag{1.70}$$

with

$$m_{A^0}^2 = 2b/\sin 2\beta = 2|\mu|^2 + m_{H_u}^2 + m_{H_d}^2 , \tag{1.71}$$

$$m_{h^0,H^0}^2 = \frac{1}{2}\left(m_{A^0}^2 + m_Z^2 \mp \sqrt{(m_{A^0}^2 - m_Z^2)^2 + 4 m_Z^2 m_{A^0}^2 \sin^2 2\beta} \right) , \tag{1.72}$$

$$m_{H^\pm}^2 = m_{A^0}^2 + m_{W^\pm}^2 , \tag{1.73}$$

and m_{h^0} the SM-like Higgs boson, by convention. The α mixing angle can be expressed at tree-level as

$$\frac{\sin 2\alpha}{\sin 2\beta} = -\left(\frac{m_{H^0}^2 + m_{h^0}^2}{m_{H^0}^2 - m_{h^0}^2} \right) \quad \text{and} \quad \frac{\tan 2\alpha}{\tan 2\beta} = \left(\frac{m_{A^0}^2 + m_Z^2}{m_{A^0}^2 - m_Z^2} \right) . \tag{1.74}$$

The mixing angle is traditionally taken negative. So, provided that $m_{A^0} > m_Z$, it follows that $-\pi/2 < \alpha < 0$.

The masses of A^0, H^0 and $H^\pm$ can be arbitrarily large. However, from Eq. (1.72), one gets

$$m_{h^0} < m_Z |\cos 2\beta| . \tag{1.75}$$

This is problematic because the observed mass of the SM-like Higgs boson is greater than the observed mass of the Z boson. This can be solved by taking quantum corrections into account. The main corrections come from the stop and top quarks. They are given by

$$\Delta m_{h^0}^2 = \frac{3}{4\pi^2} \cos^2\alpha \; y_t^2 m_t^2 \left[\ln(m_{\tilde{t}_1} m_{\tilde{t}_2}/m_t^2) + \Delta_{\text{threshold}}\right] , \tag{1.76}$$

where

$$\begin{aligned}\Delta_{\text{threshold}} &= c_{\tilde{t}}^2 s_{\tilde{t}}^2 [(m_{\tilde{t}_2}^2 - m_{\tilde{t}_1}^2)/m_t^2] \ln(m_{\tilde{t}_2}^2/m_{\tilde{t}_1}^2) \\ &+ c_{\tilde{t}}^4 s_{\tilde{t}}^4 \left[(m_{\tilde{t}_2}^2 - m_{\tilde{t}_1}^2)^2 - \frac{1}{2}(m_{\tilde{t}_2}^4 - m_{\tilde{t}_1}^4) \ln(m_{\tilde{t}_2}^2/m_{\tilde{t}_1}^2)\right] / m_t^4 ,\end{aligned} \tag{1.77}$$

with $m_{\tilde{t}_1}$ and $m_{\tilde{t}_2}$ the masses of the two stop mass eigenstates (defined below in Eq. (1.82)), and $c_{\tilde{t}}$ and $s_{\tilde{t}}$ the cosine and sine of their mixing angle, respectively. One can therefore see that the correct m_{h^0} can be obtained if the stop quarks have a large mixing angle and/or if they are heavy.

Finally, if $m_{A^0} \gg m_Z$, h^0 has the same couplings as the observed SM-like Higgs boson, $\alpha \approx \beta - \pi/2$, and A^0, H^0 and $H^\pm$ are much heavier than h^0.

1.2.2.6 Gauge Boson Masses

The photon and the gluon are still massless, and the Z and $W^\pm$ masses still follow Eq. (1.18) but with $v^2 = v_u^2 + v_d^2$.

1.2.2.7 Fermion Masses

The fermion masses are also computed with $v^2 = v_u^2 + v_d^2$, but now also depend on $\tan\beta$. At tree-level, one gets

$$m_u = y_u v \sin\beta , \qquad m_d = y_d v \cos\beta , \qquad m_e = y_e v \cos\beta . \tag{1.78}$$

1.2.2.8 Gluino Mass

The $SU(3)_c$ remains unbroken and no other MSSM particle shares the same quantum numbers with the gluinos, so they do not mix and at tree-level:

$$m_{\tilde{g}} = |M_3| \,. \tag{1.79}$$

1.2.2.9 Sfermion Masses

We will here take the example of the stops, but the masses of the other sfermions can be obtained in the same way. After EWSB, four sources contribute to the squared sfermion mass matrices. For the stops, these contributions come from (in the $(\tilde{t}_L, \tilde{t}_R)$ basis):

- the superpotential terms: the stop mass matrix gets a diagonal contribution identical to the Yukawa mass for the top quark, and off-diagonal contributions $m_t \mu \cot\beta$ from the Yukawa couplings;
- the SUSY-breaking trilinear term: an off-diagonal contribution arises from EW symmetry-breaking. It is common to parametrize is quantity in terms of fermion mass, i.e. to use $A_t \equiv a_t / y_t$ instead of a_t;
- the SUSY-breaking scalar masses: $m^2_{\tilde{t}_L}$ and $m^2_{\tilde{t}_R}$ enter diagonally. Actually, since these terms arise before $SU(2)_L$ is broken, the two left-handed sfermion of the same generation share the same squared-mass parameter;
- the D-terms: they come from squark-Higgs cross terms after EWSB and give diagonal contributions depending on the sfermion charges.

The resulting squared stop mass matrix can then be written as

$$\mathbf{m}^2_{\tilde{\mathbf{t}}} = \begin{pmatrix} m^2_{\tilde{t}_L} + m^2_t + D(\tilde{t}_L) & m_t(-A_t + \mu\cot\beta) \\ m_t(-A_t + \mu\cot\beta) & m^2_{\tilde{t}_R} + m^2_t + D(\tilde{t}_R) \end{pmatrix} , \tag{1.80}$$

with

$$D = m_Z^2 \cos 2\beta \left(I_3 - Q_{\mathrm{EM}} \sin^2(\theta_W) \right) . \tag{1.81}$$

The off-diagonal terms depend on the SM fermion mass. Since the top quark has a large mass, a large mixing between $\tilde{t}_L$ and $\tilde{t}_R$ is allowed. It can be parametrised by defining the stop mixing angle $\theta_{\tilde{t}}$ as

$$\begin{pmatrix} \tilde{t}_1 \\ \tilde{t}_2 \end{pmatrix} = \begin{pmatrix} \cos\theta_{\tilde{t}} & \sin\theta_{\tilde{t}} \\ -\sin\theta_{\tilde{t}} & \cos\theta_{\tilde{t}} \end{pmatrix} \begin{pmatrix} \tilde{t}_L \\ \tilde{t}_R \end{pmatrix} , \tag{1.82}$$

such that $m_{\tilde{t}_1} < m_{\tilde{t}_2}$. Since the other SM fermions have a much lower mass, the mixing of the other sfermions are often neglected.

1.2.2.10 Neutralino and Chargino Masses[10]

The higgsinos and the EW gauginos also mix after EWSB. They give rise to new mass eigenstates called neutralinos and charginos, represented by $\tilde{\chi}^0_{1\ldots4}$ and $\tilde{\chi}^{\pm}_{1,2}$, respectively. They are collectively referred to as EW-inos. Concretely, neutralinos are linear combinations of the bino $\widetilde{B}$, neutral wino $\widetilde{W}^0$ and neutral higgsino $\widetilde{H}^0_u$ and $\widetilde{H}^0_d$ gauge eigenstates, while charginos are linear combinations of the electrically charged wino ($\widetilde{W}^+$ and $\widetilde{W}^-$) and electrically charged higgsino ($\widetilde{H}^+_u$ and $\widetilde{H}^-_d$) gauge eigenstates.

In the basis $\psi^0 = (\widetilde{B}, \widetilde{W}^0, \widetilde{H}^0_d, \widetilde{H}^0_u)^T$, $\psi^{\pm} = (\widetilde{W}^+, \widetilde{H}^+_u, \widetilde{W}^-, \widetilde{H}^-_d)^T$, the relevant part of the Lagrangian is

$$\mathcal{L}_{m_{\tilde{\chi}}} = -\frac{1}{2}(\psi^0)^T \mathcal{M}_N \psi^0 - \frac{1}{2}(\psi^{\pm})^T \mathcal{M}_C \psi^{\pm} + \text{h.c.}, \tag{1.83}$$

where

$$\mathcal{M}_N = \begin{pmatrix} M_1 & 0 & -c_\beta\, s_W\, m_Z & s_\beta\, s_W\, m_Z \\ 0 & M_2 & c_\beta\, c_W\, m_Z & -s_\beta\, c_W\, m_Z \\ -c_\beta\, s_W\, m_Z & c_\beta\, c_W\, m_Z & 0 & -\mu \\ s_\beta\, s_W\, m_Z & -s_\beta\, c_W\, m_Z & -\mu & 0 \end{pmatrix}, \tag{1.84}$$

and

$$\mathcal{M}_C = \begin{pmatrix} \mathbf{0} & \mathbf{X}^T \\ \mathbf{X} & \mathbf{0} \end{pmatrix}, \qquad \mathbf{X} = \begin{pmatrix} M_2 & \sqrt{2}\, s_\beta\, m_{W^{\pm}} \\ \sqrt{2}\, c_\beta\, m_{W^{\pm}} & \mu \end{pmatrix}, \tag{1.85}$$

are the neutralino and chargino mass matrices. Here, M_1, M_2 and μ are the bino, wino and higgsino mass parameters, respectively, and $s_\beta = \sin\beta$, $c_\beta = \cos\beta$, $s_W = \sin\theta_W$, and $c_W = \cos\theta_W$. The chargino mass matrix is written in 2×2 block form for convenience.

The mass eigenstates are related to the gauge eigenstates by the unitary matrices N, U and V, which diagonalise the mass matrices:

$$N^* \mathcal{M}_N N^{-1} = \text{diag}\left(m_{\tilde{\chi}^0_1}, \ldots, m_{\tilde{\chi}^0_4}\right), \tag{1.86}$$

$$U^* \mathbf{X} V^{-1} = \text{diag}\left(m_{\tilde{\chi}^{\pm}_1}, m_{\tilde{\chi}^{\pm}_2}\right), \tag{1.87}$$

so that

$$\tilde{\chi}^0_i = N_{ij}\psi^0_j, \qquad \begin{pmatrix} \tilde{\chi}^+_1 \\ \tilde{\chi}^+_2 \end{pmatrix} = V \begin{pmatrix} \widetilde{W}^+ \\ \widetilde{H}^+_u \end{pmatrix}, \quad \begin{pmatrix} \tilde{\chi}^-_1 \\ \tilde{\chi}^-_2 \end{pmatrix} = U \begin{pmatrix} \widetilde{W}^- \\ \widetilde{H}^-_d \end{pmatrix}. \tag{1.88}$$

---

[10] Taken from [104].

By convention, the physical states are mass ordered: $|m_{\tilde{\chi}_i^0}| < |m_{\tilde{\chi}_j^0}|$ for $i < j$ and $m_{\tilde{\chi}_1^\pm} < m_{\tilde{\chi}_2^\pm}$. As mentioned above, the lightest neutralino, $\tilde{\chi}_1^0$, is typically also the LSP and the dark matter candidate.

If the mass parameters $|M_1|$, $|M_2|$ and $|\mu|$ in Eqs. (1.84) and (1.85) are sufficiently different from each other, the mixing is small and one ends up with a bino-like neutralino with a mass of about $|M_1|$, an almost mass-degenerate pair of wino-like chargino/neutralino with a mass of about $|M_2|$, and a triplet of higgsino-like neutralinos and chargino with a mass of about $|\mu|$.[11] The bino, wino and higgsino contents of neutralino $\tilde{\chi}_i^0$ are given by $|N_{i1}|^2$, $|N_{i2}|^2$ and $|N_{i3}|^2 + |N_{i4}|^2$, respectively. Likewise, the wino and higgsino admixtures of chargino $\tilde{\chi}_i^+$ are given by $|V_{i1}|^2$ and $|V_{i2}|^2$.

In practice, loop corrections are important, and the masses and mixing angles of the neutralinos and charginos are best computed numerically (see, e.g., SOFT-SUSY [105–111] or SPHENO [112, 113]).

1.2.2.11 The Phenomenological MSSM

Despite our ignorance of the SUSY-breaking mechanism, the large number of free parameters of the MSSM can be drastically reduced by making some assumptions based on phenomenological observations. The resulting model is called the phenomenological MSSM (pMSSM) and is based on the following assumptions described in [114]:

- no new source of CP-violation (all phases in the soft SUSY-breaking terms are suppressed);
- no flavor-changing neutral current (off-diagonal terms in the sfermion mass matrices and trilinear couplings are suppressed);
- first and second generation universality (assume that the soft SUSY-breaking scalar masses are identical for the first and second generations), and neglect the trilinear couplings of the first and second generations because the third generation has higher Yukawa couplings.

Moreover, some parameters of the Higgs sector can be related to SM parameters, so the pMSSM contains the following 19 parameters:

- $\tan\beta$: the ratio of the two Higgs VEVs;
- m_{A^0}: the mass of the pseudoscalar Higgs boson;
- μ: the Higgs-higgsino mass parameter;
- M_1, M_2, M_3: the wino, bino and gluino mass parameters, respectively;

[11] Actually, the winos and the higgsinos form a triplet and a quadruplet, respectively, but due to the mass degeneracy $m_{\tilde{\chi}_{1,2}^+} = m_{\tilde{\chi}_{1,2}^-}$ the $\tilde{\chi}_1^\pm$ and $\tilde{\chi}_2^\pm$ are each considered a single state.

- $m_{\tilde{q}}$, $m_{\tilde{u}_R}$, $m_{\tilde{d}_R}$, $m_{\tilde{\ell}}$, $m_{\tilde{e}_R}$: the mass matrices of the first and second sfermion generations;
- $m_{\tilde{Q}}$, $m_{\tilde{t}_R}$, $m_{\tilde{b}_R}$, $m_{\tilde{L}}$, $m_{\tilde{\tau}_R}$: the mass matrices of the third sfermion generation;
- a_t, a_b, a_τ: the top, bottom and tau trilinear couplings, respectively.

1.2.2.12 The μ-Problem

Minimizing the Higgs potential and including radiative corrections, one gets

$$m_Z^2/2 = \frac{m_{H_d}^2 + \Sigma_d^d - (m_{H_u}^2 + \Sigma_u^u)\tan^2\beta}{\tan^2\beta - 1} - \mu^2\,, \tag{1.89}$$

where Σ_d^d and Σ_u^u contain loop corrections mainly sensitive to the gaugino and third generation sfermion masses (see [115] and Appendix of [116]). Therefore, the Z boson mass relies on parameters from different origins: μ^2 comes from the superpotential while all the other terms are, or depend on, soft SUSY-breaking parameters. Nothing in the theory connects these parameters. However, they must cancel exactly to give the observed (and rather low) Z boson mass. This fine-tuning issue is called the "μ-problem". To avoid too much fine-tuning, each term on the right-hand side of Eq. (1.89) must be low. Models where this is true are called "natural" SUSY models. Actually, it as been argued that these terms should be naturally low (less than $30 \times m_Z^2/2$, see [117] for a review) for anthropic reasons (life cannot emerge in universes with a v more than 4 times bigger than the observed value) [118].

1.2.3 Non-minimal SUSY Extensions of the SM

1.2.3.1 The NMSSM

A way to circumvent the μ-problem is to connect μ to the SUSY-breaking parameters. This is the case in the simplest extension of the MSSM, the Next-to-Minimal Supersymmetric Standard Model (NMSSM), where a chiral superfield $\widehat{S}$ containing a complex scalar field S and its fermionic partner, the singlino $\tilde{S}$, is added. In this model, μ is induced by the SUSY-breaking terms, similar to the Yukawa mass terms of fermions.

The NMSSM superpotential is

$$W_{\text{NMSSM}} = W_{\text{MSSM}} + \lambda\widehat{S}(\widehat{H}_u)_\alpha(\widehat{H}_d)_\beta\epsilon^{\alpha\beta} + \frac{1}{3}\kappa\widehat{S}^3 + \frac{1}{2}\mu_{\widehat{S}}\widehat{S}^2 + \xi\widehat{S}; \tag{1.90}$$

and the soft SUSY-breaking Lagrangian density is

$$\mathcal{L}_{\text{soft}}^{\text{NMSSM}} = \mathcal{L}_{\text{soft}}^{\text{MSSM}} - \left(a_\lambda S (H_u)_\alpha (H_d)_\beta \epsilon^{\alpha\beta} - \frac{1}{3} a_\kappa S^3 + \frac{1}{2} b_S S^2 + tS + \text{h.c.} \right)$$
$$- m_{\tilde{S}}^2 |S|^2 \,. \tag{1.91}$$

If all the dimensionful parameters are removed from the superpotential, i.e. $\mu = \mu_{\widehat{S}} = \xi = 0$, and if $b = b_S = t = 0$ (by, e.g., requiring the superpotential and Lagrangian to be invariant under a $\mathbb{Z}_3$ symmetry), it is possible to have phenomenologically viable VEVs for S, H_u^0 and H_d^0. Moreover, if the SUSY-breaking mechanism does not introduce new CP violating phase, λ, κ, a_λ and a_κ are real and the model is free from large CP violation.

From EWSB, μ arises when S acquires a VEV, naturally of order m_{soft}:

$$\mu = \lambda \langle S \rangle \sim m_{\text{soft}} \,, \tag{1.92}$$

thus being on the same scale as the other soft SUSY-breaking terms and resolving the μ-problem. Furthermore, after EWSB, the S will mix with the H_u and H_d to give one CP-even and one CP-odd neutral scalar in addition to the ones of the MSSM; and the $\tilde{S}$ will mix with the neutral higgsinos and EW gauginos to give an additional neutralino.

1.2.3.2 The Kim-Nilles Mechanism

Another solution to the μ-problem can be obtained through the Kim-Nilles mechanism, which introduces non-renormalisable terms in the superpotential [119]. In these models, the MSSM is extended with two $SU(3)_c \times SU(2)_L \times U(1)_Y$ singlet chiral superfields $\widehat{S}$ and $\widehat{S'}$ (the second one is needed to ensure the required spontaneous SUSY breaking with a stable vacuum). A possible superpotential then does not contain the μ and b terms but contains the terms

$$W \supset \frac{\lambda_\mu}{2M_P} \widehat{S}^2 (\widehat{H}_u)_\alpha (\widehat{H}_d)_\beta \epsilon^{\alpha\beta} + \frac{\lambda_{\widehat{S}}}{4M_P} \widehat{S}^2 \widehat{S'}^2 \,, \tag{1.93}$$

and the soft SUSY-breaking Lagrangian density includes the terms

$$\mathcal{L}_{\text{soft}} \supset - \left(\frac{a_b}{M_P} S^2 (H_u)_\alpha (H_d)_\beta \epsilon^{\alpha\beta} - \frac{a_S}{4M_P} S^2 S'^2 + \text{h.c.} \right) - m_{\tilde{S}}^2 |S|^2 - m_{\tilde{S}'}^2 |S'|^2 \,, \tag{1.94}$$

Table 1.3 Peccei-Quinn charges of the chiral superfields

	$\widehat{Q}$	$\widehat{L}$	$\widehat{H}_u$	$\widehat{H}_d$	$\widehat{S}$	$\widehat{S}'$	$\widehat{u}$	$\widehat{d}$	$\widehat{e}$
Peccei-Quinn charge	-1	-1	$+1$	$+1$	-1	$+1$	0	0	0

where $a_b, a_S \sim m_{\text{soft}}$ and $m_S^2, m_{S'}^2 \sim m_{\text{soft}}^2$. If a_S is sufficiently large or if m_S^2 and $m_{S'}^2$ are negative, then S acquires a VEV after EWSB of order

$$\langle S \rangle \sim \sqrt{m_{\text{soft}} M_P}\,, \tag{1.95}$$

and the low-energy effective theory will contain the usual μ and b terms:

$$\mu = \frac{\lambda_\mu}{2M_P}\langle S \rangle^2 \sim m_{\text{soft}} \qquad \text{and} \qquad b = \frac{a_b}{M_P}\langle S \rangle^2 \sim m_{\text{soft}}^2\,. \tag{1.96}$$

Furthermore, when the MSSM superpotential does not include the μ term, it has a U(1) Peccei-Quinn symmetry, with the chiral superfields having the charges listed in Table 1.3. This symmetry is broken by an $SU(3)_c$ anomaly and the soft SUSY-breaking mass terms of S and S'. The spontaneous breaking of the symmetry dynamically provides a solution to the strong CP problem by giving rise to a pseudo-scalar degree of freedom, the axion,[12] that is a mixture of S and S', as well as a scalar degree of freedom that is called the saxion. The fermionic, R-parity odd, superpartner is the axino and can be a good DM candidate if it is the LSP.

1.2.3.3 The Giudice-Masiero Mechanism

Alternatively, non-renormalisable terms can be added through the D-term of the Lagrangian density. The Giudice-Masiero mechanism [122] consists of introducing non-renormalisable contributions to the Kähler potential, for instance:

$$K \supset \frac{\lambda_\mu}{M_P}\widehat{X}^\dagger(\widehat{H}_u)_\alpha(\widehat{H}_d)_\beta\epsilon^{\alpha\beta} + \text{h.c.}\,, \tag{1.97}$$

with λ_μ a dimensionless coupling and $\widehat{X}$ a chiral superfield with a Peccei-Quinn charge of $+2$. If its auxiliary field F_X obtains a VEV (which always happens in supergravity [122]), it induces spontaneous SUSY breaking and the low-energy effective superpotential obtains an effective μ:

$$\mu_{\text{eff}} = \frac{\lambda_\mu}{M_P}\langle F_X^* \rangle\,. \tag{1.98}$$

[12] More specifically, a DFSZ axion [120, 121].

As requested, $\mu \sim m_{3/2} \sim m_{\text{soft}}$ if $\langle F_X^* \rangle \sim M_P m_{\text{soft}}$ as it is typically the case in PMSB models.[13] However, in order for this to work, the MSSM μ needs to be absent from the superpotential in the first place. This is usually achieved via an additional symmetry (see [123] for an example). Unfortunately, the added terms do not have any direct effect on the phenomenology and these models are therefore difficult to challenge experimentally.

1.2.3.4 Vectorlike Quarks and Leptons

Other ways to extend the MSSM, but not directly related to the μ-problem, were proposed. For instance, one can add chiral superfields that are self-conjugate (vectorlike) representations of the SM gauge groups. New fermions evading current detection can then be introduced without requiring extremely large Yukawa couplings to the Higgs VEVs. Such large couplings would indeed lead to very large corrections to the SM-like Higgs boson production cross section and to the EW precision observable, contradicting current observations. If these new vectorlike supermultiplets live in the fundamental representation of $SU(3)_c$ or $SU(2)_L$, or if they are charged under $U(1)_Y$, then supermultiplets of opposite gauge quantum numbers must also be present. If we symbolise these pairs by Φ_i and $\overline{\Phi}_i$, then the superpotential is allowed to have mass terms free from Higgs boson interactions:

$$W \supset M_i \Phi_i \overline{\Phi}_i \, . \tag{1.99}$$

If any of the three mechanisms discussed above actually occurs and dynamically produces a μ term, it is reasonable to suppose that this mechanism is also the one responsible for the generation of M_i. The lightest of these new particles could cause problems as stable relics from thermal production in the early universe. To avoid this, a solution is to assume Φ_i or $\overline{\Phi}_i$ to have the same quantum numbers as one of the MSSM quark and lepton chiral superfields. This would allow small Yukawa couplings to the Higgs bosons and the new vectorlike fermions to decay into SM fermions.

Vectorlike chiral superfields also have the particularity to enhance the tree-level h^0 mass up to the observed value of 125 GeV. This is convenient for models that would otherwise feature a too low m_{h^0}, such as GMSB models which, unless all the superpartners are extremely heavy, predict small stop mixing angle (Δm_{h^0} depends on the stop masses and mixing angle, see Eq. (1.76)). Reciprocally, adding vectorlike quarks with large Yukawa couplings can allow GMSB models to have a h^0 of 125 GeV while allowing the MSSM sparticles to be low in mass [124–127].

---

[13] Similarly, a b term can also be generated through non-renormalisable terms in the Kähler potential.

Table 1.4 Additional chiral superfields to add to the MSSM to obtain the MDGSSM

Names		Spin 0	Spin 1/2	$SU(3)_C,\ SU(2)_L,\ U(1)_Y$
DG-octet	$\widehat{O}_g$	O_g	$\widetilde{g}'$	$(\mathbf{8},\ \mathbf{1}\ ,\ 0)$
DG-triplet	$\widehat{T}$	$T^0\ \ T^{\pm}$	$\widetilde{W}'^{\pm}\ \ \widetilde{W}'^0$	$(\mathbf{1},\ \mathbf{3},\ 0)$
DG-singlet	$\widehat{S}$	S	$\widetilde{B}'$	$(\mathbf{1},\ \mathbf{1},\ 0)$

The bold numbers are referring to the irreducible representations of Lie groups (of dimension indicated by the number)

1.2.3.5 The MDGSSM

Another possible extension of the MSSM is the Dirac Gaugino Supersymmetric Standard Model (DGSSM). In the MSSM, the gauginos are of Majorana type: they are of only one type of chirality (conventionally the left-handed component of a Dirac gaugino). They acquire a mass through a Majorana mass term (see Eqs. (1.6) and (1.62)). However, it is possible to make them of Dirac type by adding extra superfields, in the adjoint representation of the gauge groups, whose spin-1/2 component can be identified as a right-handed gaugino.

The minimal extension of this kind, the minimal DGSSM (MDGSSM) enriches the MSSM with chiral superfields in the adjoint representation of the gauge groups as described in Table 1.4. Since they share the same quantum numbers, the MSSM gauginos and the newly added Weyl fermions can mix together. For instance:

$$\mathcal{L} \supset -\begin{pmatrix}\widetilde{g} & \widetilde{g}'\end{pmatrix}\begin{pmatrix}M_3 & M_D \\ M_D & M\end{pmatrix}\begin{pmatrix}\widetilde{g} \\ \widetilde{g}'\end{pmatrix} + \text{h.c.}\,, \tag{1.100}$$

where the gauginos are pure Dirac if $M_3 = M = 0$, are pure Majorana if $M_D = 0$, or are otherwise a mix. Models featuring Dirac gauginos enhance the tree-level h^0 mass, possibly preserve an R-symmetry (i.e. a continuous symmetry whose generator does not commute with the SUSY generators), can be associated with $\mathcal{N} = 2$ SUSY theories, and increase the production cross section of the gluinos and the EW-inos, while reducing the one of the squarks (possibly explaining why they still evade detection) (see e.g. [128–131] and references therein). In the EW-ino sector, two neutralinos and one chargino are added to the EW-ino content of the MSSM. Moreover, the mass splitting between the bino-like and wino-like EW-inos is small, inducing potentially long-lived sparticles.

1.2.3.6 Others

Many other extensions of the MSSM are of course possible. One of the most well-known is for instance the MSSM+RN [132, 133], where the MSSM is extended with a chiral superfield containing a right-handed neutrino and a right-handed sneutrino, yielding Dirac neutrino masses. The two sneutrino chiralities can mix to allow a mostly right-handed sneutrino to be the LSP and a good DM candidate.

Another well-known extension is the UMSSM, where a U(1)′ gauge group is added to the MSSM gauge groups. It predicts $m_{h^0} = 125$ GeV without relying on large contributions from the stops, h^0 tree-level couplings similar to a SM Higgs boson, a solution to the μ-problem by generating μ through the VEV of the singlet field responsible of the breaking of the U(1)′ symmetry, and, as in the MSSM+RN model, a neutralino or a sneutrino as a good DM candidate.

1.3 Non-SUSY Extensions of the Standard Model

Of course, multiple alternatives to SUSY extensions of the SM exist. They are usually proposed to address a specific theoretical or observational issue. Some of them are listed here, without any intention of completeness. More details and models can be found in [134–136].

1.3.1 SU(5)

In addition to compacting and generalising the SM, a GUT unifies the description of the matter and boson fields, and unifies the strong, weak, and electromagnetic interactions under a single gauge coupling. However, below the GUT scale, the gauge symmetry group of a GUT should break up into some combination of the groups that comprise the SM gauge groups. To achieve this, the SM gauge groups need to be subgroups of a bigger Lie group. The simplest Lie group that can contain $SU(3)_c \times SU(2)_L \times U(1)_Y$ and that has a complex representation (necessary in order to have chiral representations, as the anti-fundamental representations in the SM) is SU(5).

In SU(5), the left-handed fermions are the irreducible representations **10** and $\overline{\mathbf{5}}$. In terms of $SU(3)_c \times SU(2)_L \times U(1)_Y$ irreducible representations, this gives:

$$\mathbf{10} = \underbrace{(\mathbf{3}, \mathbf{2}, \tfrac{1}{6})}_{Q} + \underbrace{(\overline{\mathbf{3}}, \mathbf{1}, -\tfrac{2}{3})}_{u_R^\dagger} + \underbrace{(\mathbf{1}, \mathbf{1}, 1)}_{e_R^\dagger}\,, \tag{1.101}$$

$$\overline{\mathbf{5}} = \underbrace{(\overline{\mathbf{3}}, \mathbf{1}, \tfrac{1}{3})}_{d_R^\dagger} + \underbrace{(\mathbf{1}, \mathbf{2}, -\tfrac{1}{3})}_{L}\,. \tag{1.102}$$

Moreover, a $\nu_R^\dagger$ can be added via $\mathbf{1} = (\mathbf{1}, \mathbf{1}, 0)$. Regarding the gauge bosons, the smallest irreducible representation that can generate the SM gauge bosons is the adjoint representation **24**:

$$\mathbf{24} = \underbrace{(\mathbf{8}, \mathbf{1}, 0)}_{g} + \underbrace{(\mathbf{1}, \mathbf{3}, 0)}_{W^{1,2,3}} + \underbrace{(\mathbf{1}, \mathbf{1}, 0)}_{B} + \underbrace{(\mathbf{3}, \mathbf{2}, -\tfrac{5}{6}) + (\bar{\mathbf{3}}, \mathbf{2}, \tfrac{5}{6})}_{X^{1,2,3}\ \bar{X}^{1,2,3}\ Y^{1,2,3}\ \bar{Y}^{1,2,3}} \quad , \tag{1.103}$$

where g, $W^{1,2,3}$ and B are massless vector fields that can be identified with the SM gauge bosons, whereas X, $\bar{X}$, Y and $\bar{Y}$ are 12 new massless bosons called leptoquarks. They are called this way because they couple to the strong and EW forces, and can therefore connect quarks and leptons together. They can also turn a quark directly into an antiquark. They carry an electric charge of $\pm\frac{1}{3}, \pm\frac{4}{3}$, as well as baryon and lepton numbers. The baryon and lepton numbers are not conserved individually, but their difference is, which leads to proton decay through leptoquark exchange.

The breaking of SU(5) into $SU(3)_c \times SU(2)_L \times U(1)_Y$ can be achieved by introducing a 24-plet Σ that acquires a VEV at the GUT scale. It can be done in such a way that only the leptoquarks become massive, by "eating" the massless Nambu-Goldstone bosons from Σ, leaving the SM vector bosons massless. The $W^\pm$ and Z become massive in a similar fashion as in the SM, when $SU(3)_c \times SU(2)_L \times U(1)_Y \rightarrow SU(3)_c \times U(1)_{EM}$. The Higgs boson can be generated by either the **5** or the $\bar{\mathbf{5}}$ irreducible representation:

$$\mathbf{5} = \underbrace{(\mathbf{3}, \mathbf{1}, \tfrac{1}{3})}_{\phi_3} + \underbrace{(\mathbf{1}, \mathbf{2}, -\tfrac{1}{2})}_{\phi_2} , \tag{1.104}$$

with ϕ_3 an $SU(3)_c$ triplet and ϕ_2 the SM $SU(2)_L$ Higgs doublet. However, the color triplet mediates proton decay. Its mass needs to be heavier than 10^{13} GeV to satisfy current experimental constraints. A fine-tuning is thus required to have ϕ_3 at the GUT scale and ϕ_2 at the EW scale. This is known as the doublet-triplet splitting problem. Moreover, the Yukawa sector predicts fermion masses in contradiction to experimental observations. To palliate this, non-renormalisable Yukawa couplings need to be introduced, or the Higgs sector expanded by adding the **45** irreducible representation.

1.3.2 Extra Dimensions

Models with additional space-time dimensions were found to have interesting properties. For instance, they could solve some SM issues and/or provide a framework required to unify quantum mechanics and gravity (e.g. through string theory). However, we experience a four-dimension universe, meaning that these extra dimensions need to be "compactified" over a circle.

In the Randall-Sundrum model of warped extra dimensions [137], the conserved interval with one extra dimension is

$$ds^2 = e^{-2kr_c\phi} g_{\mu\nu} dx^\mu dx^\nu + r_c^2 d\phi^2\,, \tag{1.105}$$

with k a scale of the order of the Planck scale, r_c is the radius of the compactified dimension, $0 \leq \phi \leq \pi$ is the coordinate of the extra dimension, and x^μ are the usual space-time coordinates. The four-dimensional fields are located at the fixed points $\phi = 0, \pi$. The "warp" term $e^{-2kr_c\phi}$ can be used to naturally explain the mass difference between the EW and the Planck scales, and thus provide a solution to the hierarchy problem. Indeed, in this model, the Planck scale, the symmetry-breaking scale and the physical mass can be written

$$M_P^2 = \frac{M^3}{k}(1 - e^{-2kr_c\pi})\,, \qquad v \equiv e^{-kr_c\pi} v_0\,, \qquad \text{and} \qquad m \equiv e^{-kr_c\pi} m_0\,, \tag{1.106}$$

where M, k, r_c, v_0 and m_0 are fundamental parameters. Unlike v and m, M_P depends only weakly on r_c in the large kr_c limit. Thanks to an exponential geometric factor $e^{-2kr_c\phi}$, no extremely large r_c is needed to explain the huge difference between the scales. In other words, it is possible to generate a large observed hierarchy without requiring a hierarchy between the fundamental scales. In fact, if all the fundamental parameters are around the Planck scale (r_c is the inverse of the fundamental parameter μ_c, which is the compactification scale, so $r_c \sim M_P^{-1}$), the Planck scale can have its estimated value while the physical mass can be at the TeV scale only by requiring $kr_c \approx 11$. Moreover, the mass hierarchy between the SM fermions can be explained by considering different localisations of the corresponding fields in the five-dimensional bulk, see e.g. [138]. These models are called bulk-RS models.

The hierarchy problem can be addressed differently in the ADD model [139], where

$$M_P^2 \sim M^{2+n} r_c^n\,, \tag{1.107}$$

with n is the number of extra compact spatial dimensions. By requiring $M \sim m_{\rm EW}$ and M_P to be at its estimated value,

$$r_c \sim 10^{\frac{30}{n}-17}\,\text{cm} \times \left(\frac{1\,\text{TeV}}{m_{\rm EW}}\right)^{1+\frac{2}{n}}\,. \tag{1.108}$$

Stringent constraints on ADD models are set by astrophysical observations and require $n \geq 4$ to solve the hierarchy problem [140].

A third class of models are the Universal Extra Dimension models (UED) [141, 142]. In these models, all SM fields can propagate in the bulk of the extra dimensions and they generate the so-called Kaluza-Klein resonances (or KK excited states), which are infinite sums of massive complex fields. Each KK state has a mass given by $m_j = p_j/r_c$, with $p_j^2 = j_1^2 + \cdots + j_n^2$, where n is the number of extra dimensions and j_i are the Kaluza-Klein numbers representing the conserved quantised momentum in the compactified dimension i. The states are usually degenerate in mass, but higher-order corrections can break this degeneracy. Moreover, KK states can only be pair-produced and decay to a lighter excited state. The lightest KK state is therefore a good DM candidate. Furthermore, not only KK states are present, but also the so-called 0-modes. The 0-modes fermions are chiral, while the KK states are vectorlike. UED models can be constrained by collider experiments, see e.g. [143, 144].

References

1. Thomson JJ (1897) XL. Cathode rays. London Edinburgh Dublin Philos Mag J Sci 44(269):293–316. https://doi.org/10.1080/14786449708621070
2. Maxwell JC (1873) A treatise on electricity and magnetism. Nature 7:478–480. https://doi.org/10.1038/007478a0
3. Einstein A (1905) Über einen die Erzeugung und Verwandlung des Lichtes betreffenden heuris- tischen Gesichtspunkt. Ann Phys 322(6):132–148 (1905). https://doi.org/10.1002/andp.19053220607
4. Compton AH (1923) A quantum theory of the scattering of x-rays by light elements. Phys. Rev. 21:483–502. https://doi.org/10.1103/PhysRev.21.483
5. Chadwick J (1914) The intensity distribution in the magnetic spectrum of βparticles from radium (B + C). Verh Phys Gesell 16:383–391
6. Pauli W (1930) Dear radioactive ladies and gentlemen. Phys Today 31N9:27
7. Cowan CL et al (1956) Detection of the free neutrino: a confirmation. Science 124:103–104. https://doi.org/10.1126/science.124.3212.103
8. Fermi E (1933) Tentativo di una teoria dell'emissione dei raggi beta. Ric Sci 4:491–495
9. Fermi E (1934) An attempt of a theory of beta radiation. 1. Z Phys 88:161–177. https://doi.org/10.1007/BF01351864
10. Arnison G et al (1983) Experimental observation of isolated large transverse energy electrons with associated missing energy at $s = 540$ GeV. Phys Lett B 122:103–116. https://doi.org/10.1016/0370-2693(83)91177-2
11. Banner M et al (1983) Observation of single isolated electrons of high transverse momentum in events with missing transverse energy at the CERN anti-p p collider. Phys Lett B 122:476–485. https://doi.org/10.1016/0370-2693(83)91605-2
12. Arnison G et al (1983) Experimental observation of lepton pairs of invariant mass around 95-GeV/c**2 at the CERN SPS collider. Phys Lett B **126**:398–410. https://doi.org/10.1016/0370-2693(83)90188-0
13. Bagnaia P et al (1983) Evidence for $Z^0 \to e^+e^-$ at the CERN $\bar{p}p$ collider. Phys Lett B 129:130–140. https://doi.org/10.1016/0370-2693(83)90744-X
14. Neddermeyer SH, Anderson CD (1937) Note on the nature of cosmic ray particles. Phys Rev 51:884–886. https://doi.org/10.1103/PhysRev.51.884
15. Perl ML et al (1975) Evidence for anomalous lepton production in e+ - e- annihilation. Phys Rev Lett 35:1489–1492. https://doi.org/10.1103/PhysRevLett.35.1489

16. Feldman GJ et al (1977) Inclusive anomalous muon production in e+ e- annihilation. Phys Rev Lett 38:117. [Erratum: Phys.Rev.Lett. 38, 576 (1977)]. https://doi.org/10.1103/PhysRevLett.38.117
17. Badtke DH et al (1978) Inclusive muon production in e^+e^-annihilation with $\sqrt{s} = 7.3$-GeV. Phys Rev Lett 40:827. https://doi.org/10.1103/PhysRevLett.40.827
18. Burmester J et al (1977) Anomalous muon production in e+ e- annihilation as evidence for heavy leptons. Phys Lett B 68:297. https://doi.org/10.1016/0370-2693(77)90292-1
19. Danby G et al (1962) Observation of high-energy neutrino reactions and the existence of two kinds of neutrinos. Phys Rev Lett 9:36–44. https://doi.org/10.1103/PhysRevLett.9.36
20. Burmester J et al (2001) Observation of tau neutrino interactions. Phys Lett B 504:218–224. https://doi.org/10.1016/S0370-2693(01)00307-0. arXiv:hep-ex/0012035
21. Ne'eman Y (1961) Derivation of strong interactions from a gauge invariance. Nucl Phys 26:222–229. Ed. by R. Ruffini and Y. Verbin. https://doi.org/10.1016/0029-5582(61)90134-1
22. Gell-Mann M (1961) The eightfold way: a theory of strong interaction symmetry. https://doi.org/10.2172/4008239
23. Gell-Mann M (1964) A schematic model of baryons and mesons. Phys Lett 8:214–215. https://doi.org/10.1016/S0031-9163(64)92001-3
24. Zweig G (1964) An SU(3) model for strong interaction symmetry and its breaking. Version 1. https://doi.org/10.17181/CERN-TH-401
25. Zweig G (1964) An SU(3) model for strong interaction symmetry and its breaking. Version 2. In: Lichtenberg DB, Rosen SP (eds) Developments in the quark theory of hadrons, vol 1, pp 22–101. https://doi.org/10.17181/CERN-TH-412
26. Bjorken JD, Glashow SL (1964) Elementary particles and SU(4). Phys Lett 11:255–257. https://doi.org/10.1016/0031-9163(64)90433-0
27. Glashow SL, Iliopoulos J, Maiani L (1970) Weak interactions with Lepton-Hadron symmetry. Phys Rev D 2:1285–1292. https://doi.org/10.1103/PhysRevD.2.1285
28. Kobayashi M, Maskawa T (1973) CP violation in the renormalizable theory of weak interaction. Prog Theor Phys 49:652–657. https://doi.org/10.1143/PTP.49.652
29. Bloom ED et al (1969) High-energy inelastic e p scattering at 6-degrees and 10-degrees. Phys Rev Lett 23:930–934. https://doi.org/10.1103/PhysRevLett.23.930
30. Breidenbach M et al (1969) Observed behavior of highly inelastic electron-proton scattering. Phys Rev Lett 23:935–939. https://doi.org/10.1103/PhysRevLett.23.935
31. Aubert JJ et al (1974) Experimental observation of a heavy particle J. Phys Rev Lett 33:1404–1406. https://doi.org/10.1103/PhysRevLett.33.1404
32. Augustin JE et al (1974) Discovery of a narrow resonance in e^+e^- annihilation. Phys Rev Lett 33:1406–1408. https://doi.org/10.1103/PhysRevLett.33.1406
33. Herb SW et al (1977) Observation of a dimuon resonance at 9.5-GeV in 400-GeV proton-nucleus collisions. Phys Rev Lett 39:252–255. https://doi.org/10.1103/PhysRevLett.39.252
34. Abe F et al (1995) Observation of top quark production in $\bar{p}p$ collisions. Phys Rev Lett 74:2626–2631. https://doi.org/10.1103/PhysRevLett.74.2626. arXiv:hep-ex/9503002
35. Abachi S et al (1995) Observation of the top quark. Phys Rev Lett 74:2632–2637. https://doi.org/10.1103/PhysRevLett.74.2632. arXiv:hep-ex/9503003
36. Berger C et al (1979) Jet analysis of the Υ (9.46) decay into charged hadrons. Phys Lett B 82:449–455. https://doi.org/10.1016/0370-2693(79)90265-X
37. Berger C et al (1981) Topology of the Υ decay. Z Phys C 8:101. https://doi.org/10.1007/BF01547873
38. Brandelik R et al (1979) Evidence for planar events in e+ e- annihilation at high-energies. Phys Lett B 86:243–249. https://doi.org/10.1016/0370-2693(79)90830-X
39. Barber DP et al (1979) Discovery of three jet events and a test of quantum chromodynamics at PETRA energies. Phys Rev Lett 43:830. https://doi.org/10.1103/PhysRevLett.43.830
40. Berger C et al (1979) Evidence for gluon bremsstrahlung in e+ e- annihilations at high-energies. Phys Lett B 86418–425. https://doi.org/10.1016/0370-2693(79)90869-4

41. Bartel W et al (1980) Observation of planar three jet events in e+ e- annihilation and evidence for gluon bremsstrahlung. Phys Lett B 91:142–147. https://doi.org/10.1016/0370-2693(80)90680-2
42. Aad G et al (2012) Observation of a new particle in the search for the standard model higgs boson with the ATLAS detector at the LHC. Phys Lett B 716:1–29. https://doi.org/10.1016/j.physletb.2012.08.020. arXiv:1207.7214 [hep-ex]
43. Chatrchyan S et al (2012) Observation of a new boson at a mass of 125 GeV with the CMS experiment at the LHC. Phys Lett B 716:30–61. https://doi.org/10.1016/j.physletb.2012.08.021. arXiv:1207.7235 [hep-ex]
44. Englert F, Brout R (1964) Broken symmetry and the mass of gauge vector mesons. Phys Rev Lett 13:321–323. https://doi.org/10.1103/PhysRevLett.13.321
45. Higgs PW (1964) Broken symmetries and the masses of gauge bosons. Phys Rev Lett 13:508–509. https://doi.org/10.1103/PhysRevLett.13.508
46. Guralnik GS, Hagen CR, Kibble TWB (1964) Global conservation laws and massless particles. Phys Rev Lett 13:585–587. https://doi.org/10.1103/PhysRevLett.13.585
47. Dirac PAM (1928) The quantum theory of the electron. Proc Roy Soc Lond A 117:610–624. https://doi.org/10.1098/rspa.1928.0023
48. Anderson CD (1932) The apparent existence of easily deflectable positives. Science 76:238–239. https://doi.org/10.1126/science.76.1967.238
49. Ohlsson T (2012) Relativistic quantum physics: from advanced quantum mechanics to introductory quantum field theory. Cambridge University Press, Cambridge. ISBN: 978-1-139-21072-0, 978-0-521-76726-2
50. Martin SP (1998) A supersymmetry primer. Adv Ser Direct High Energy Phys 18:1–98. https://doi.org/10.1142/9789812839657.0001. arXiv:hep-ph/9709356
51. Cabibbo N (1963) Unitary symmetry and leptonic decays. Phys Rev Lett 10:531–533. https://doi.org/10.1103/PhysRevLett.10.531
52. Schael S et al (2006) Precision electroweak measurements on the Z resonance. Phys Rept 427:257–454. https://doi.org/10.1016/j.physrep.2005.12.006. arXiv:hep-ex/0509008
53. CMS Collaboration (2023) CMS standard model summary plots. https://twiki.cern.ch/twiki/bin/view/CMSPublic/PhysicsResultsSUS. Accessed 15 Aug 2024
54. ATLAS Collaboration (2024) Standard model summary plots June 2024. https://atlas.web.cern.ch/Atlas/GROUPS/PHYSICS/PUBNOTES/ATL-PHYS-PUB-2024-011/. Accessed 15 Aug 2024
55. Morel L et al (2020) Determination of the fine-structure constant with an accuracy of 81 parts per trillion. Nature 588(7836):61–65. https://doi.org/10.1038/s41586-020-2964-7
56. Parker RH et al (2018) Measurement of the fine-structure constant as a test of the standard model. Science 360:191. https://doi.org/10.1126/science.aap7706. arXiv:1812.04130 [physics.atom-ph]
57. Fan X et al (2023) Measurement of the electron magnetic moment. Phys Rev Lett 130(7):071801. https://doi.org/1103/PhysRevLett.130.071801. arXiv:2209.13084 [physics.atom-ph]
58. Wolfenstein L (1983) Parametrization of the Kobayashi-Maskawa matrix. Phys Rev Lett 51:1945. https://doi.org/10.1103/PhysRevLett.51.1945
59. Charles J et al (2005) CP violation and the CKM matrix: assessing the impact of the asymmetric Bfactories. Eur Phys C 41(1):1–131. https://doi.org/10.1140/epjc/s2005-02169-1. arXiv:hep-ph/0406184
60. Silva LV (2024) 2023 update of the extraction of the CKM matrix elements. In: 12th international workshop on the CKM unitarity triangle. arXiv:2405.08046 [hep-ph]
61. The CKMFitter Collaboration. http://ckmfitter.in2p3.fr/. Accessed 15 August 2024
62. Arbey A, Mahmoudi F (2021) Dark matter and the early universe: a review. Prog Part Nucl Phys 119:103865. https://doi.org/10.1016/j.ppnp.2021.103865. arXiv:2104.11488 [hep-ph]
63. Zwicky F (1933) Die Rotverschiebung von extragalaktischen Nebeln. Helv Phys Acta 6:110–127. https://doi.org/10.1007/s10714-008-0707-4

64. Rubin VC, Kent Ford W Jr (1970) Rotation of the Andromeda Nebula from a spectroscopic survey of emission regions. Astrophys J 159:379–403. https://doi.org/10.1086/150317
65. Freeman KC () On the disks of spiral and SO Galaxies. Astrophys J 160:811. https://doi.org/10.1086/150474
66. Bosma A (1978) The distribution and kinematics of neutral hydrogen in spiral galaxies of various morphological types. Ph.D. Thesis. University of Groningen, Netherlands
67. Rubin VC, Thonnard N, Ford WK Jr (1980) Rotational properties of 21 SC galaxies with a large range of luminosities and radii, from NGC 4605 (R = 4kpc) to UGC 2885 (R = 122 kpc). Astrophys J 238:471. https://doi.org/10.1086/158003
68. van Albada TS et al (1985) The distribution of dark matter in the spiral galaxy NGC-3198. Astrophys J 295:305–313. ©AAS. Reproduced with permission. https://doi.org/10.1086/163375
69. Markevitch M et al (2003) Direct constraints on the dark matter self-interaction cross-section from the merging galaxy cluster 1E0657-56. Astrophys J 606:819–824. © AAS. Reproduced with permission. https://doi.org/10.1086/383178. arXiv:astro-ph/0309303
70. Springel V, Frenk CS, White SDM (2006) The large-scale structure of the universe. Nature 440:1137. https://doi.org/10.1038/nature04805. arXiv:astro-ph/0604561
71. Adame AG et al (2024) DESI 2024 VI: cosmological constraints from the measurements of baryon acoustic oscillations. arXiv:2404.03002 [astro-ph.CO]
72. Aghanim N et al. (2020) Planck 2018 results. VI. Cosmological parameters. Astron Astrophys 641:A6. [Erratum: Astron. Astrophys. 652, C4 (2021)]. https://doi.org/10.1051/0004-6361/201833910. arXiv:1807.06209 [astro-ph.CO]
73. Tristram M et al (2024) Cosmological parameters derived from the final Planck data release (PR4). Astron Astrophys 682:A37. https://doi.org/10.1051/0004-6361/202348015. arXiv:2309.10034 [astro-ph.CO]
74. Lemaitre G (1927) A homogeneous universe of constant mass and growing radius accounting for the radial velocity of extragalactic nebulae. Ann Soc Sci Bruxelles A 47:49–59. https://doi.org/10.1007/s10714-013-1548-3
75. Hubble E (1929) A relation between distance and radial velocity among extra-galactic nebulae. Proc Nat Acad Sci 15:168–173. https://doi.org/10.1073/pnas.15.3.168
76. Riess AG et al (1998) Observational evidence from supernovae for an accelerating universe and a cosmological constant. Astron J 116:1009–1038. https://doi.org/10.1086/300499. arXiv:astro-ph/9805201
77. Perlmutter S et al (1999) Measurements of Ω and Λ from 42 high redshift supernovae. Astrophys J 517:565–586. https://doi.org/10.1086/307221. arXiv:astro-ph/9812133
78. Clifton T et al (2012) Modified gravity and cosmology. Phys Rept 513:1–189. https://doi.org/10.1016/j.physrep.2012.01.001. arXiv:1106.2476 [astro-ph.CO]
79. Shankaranarayanan S, Johnson JP (2022) Modified theories of gravity: why, how and what? Gen Rel Grav 54(5):44. https://doi.org/10.1007/s10714-022-02927-2. arXiv:2204.06533 [gr-qc]
80. Pascoli G (2024) A comparative study of MOND and MOG theories versus the κ-model: an application to galaxy clusters. Can J Phys 102(2):69–84. https://doi.org/10.1139/cjp-2023-0159. arXiv:2307.01555 [astro-ph.CO]
81. Davis R Jr, Harmer DS, Hoffman KC (1968) Search for neutrinos from the sun. Phys Rev Lett 20:1205–1209. https://doi.org/10.1103/PhysRevLett.20.1205
82. Fukuda Y et al (1998) Evidence for oscillation of atmospheric neutrinos. Phys Rev Lett 81:1562–1567. https://doi.org/10.1103/PhysRevLett.81.1562. arXiv:hep-ex/9807003
83. Ahmad QR et al (2001) Measurement of the rate of $\nu_e + d \to p + e^-$ interactions produced by ^{8}B solar neutrinos at the sudbury neutrino observatory. Phys Rev Lett 87:071301. https://doi.org/10.1103/PhysRevLett.87.071301. arXiv:nucl-ex/0106015
84. Abe Y et al (2012) Indication of reactor $\bar{\nu}_e$ disappearance in the double chooz experiment. Phys Rev Lett 108:131801. https://doi.org/10.1103/PhysRevLett.108.131801. arXiv:1112.6353 [hep-ex]

85. An FP et al (2012) Observation of electron-antineutrino disappearance at Daya Bay. Phys Rev Lett 108:171803. https://doi.org/10.1103/PhysRevLett.108.171803. arXiv:1203.1669 [hep-ex]
86. Ahn JK et al (2012) Observation of reactor electron antineutrino disappearance in the RENO experiment. In: Phys Rev Lett 108:191802. https://doi.org/10.1103/PhysRevLett.108.191802. arXiv:1204.0626 [hep-ex]
87. Abe K et al (2013) Evidence of electron neutrino appearance in a muon neutrino beam. Phys Rev D 88(3):032002. https://doi.org/10.1103/PhysRevD.88.032002. arXiv:1304.0841 [hep-ex]
88. Pontecorvo B (1957) Inverse beta processes and nonconservation of lepton charge. Zh. Eksp. Teor. Fiz. 34:247.
89. Maki Z, Nakagawa M, Sakata S (1962) Remarks on the unified model of elementary particles. Prog Theor Phys 28:870–880. https://doi.org/10.1143/PTP.28.870
90. Abel C et al (2020) Measurement of the permanent electric dipole moment of the neutron. Phys Rev Lett 124(8):081803. https://doi.org/10.1103/PhysRevLett.124.081803. arXiv:2001.11966 [hep-ex]
91. Peccei RD, Quinn HR (1977) CP conservation in the presence of instantons. Phys Rev Lett 38:1440–1443. https://doi.org/10.1103/PhysRevLett.38.1440
92. Peccei RD, Quinn HR (1977) Constraints imposed by CP conservation in the presence of instantons. Phys Rev D 16:1791–1797. https://doi.org/10.1103/PhysRevD.16.1791
93. Peccei RD (2008) The strong CP problem and axions. Lect Notes Phys 741:3–17. https://doi.org/10.1007/978-3-540-73518-2_1. arXiv:hep-ph/0607268
94. Haag R, Lopuszanski JT, Sohnius M (1975) All possible generators of super-symmetries of the S matrix. Nucl Phys B 88:257. https://doi.org/10.1016/0550-3213(75)90279-5
95. Coleman SR, Mandula J (1967) All possible symmetries of the S matrix. Phys Rev 159:1251–1256. https://doi.org/10.1103/PhysRev.159.1251
96. Haber HE, Kane GL (1985) The search for supersymmetry: probing physics beyond the standard model. Phys Rept 117:75–263. https://doi.org/10.1016/0370-1573(85)90051-1
97. Weinberg S (2000) The quantum theory of fields. Vol. 2: modern applications. Cambridge University Press, Cambridge. ISBN: 978-0-521-67055-5, 978-1-139-63263-8, 978-0-521-67055-5
98. Baer H, Tata X (2006) Weak scale supersymmetry: from superfields to scattering events. Cambridge University Press, Cambridge. ISBN: 978-0-521-29031-9, 978-0-511-19011-7, 978-0-521-29031-9, 978-0-521-85786-4
99. Dreiner HK, Haber HE, Martin SP (2023) From spinors to supersymmetry. Cambridge University Press, Cambridge. ISBN: 978-1-139-04974-0. https://doi.org/10.1017/9781139049740
100. Allanach B, Haber HE (2024) Supersymmetry, part I (theory). arXiv:2401.03827 [hep-ph]
101. Deser S, Zumino B (1977) Broken supersymmetry and supergravity. Phys Rev Lett 38:1433–1436. https://doi.org/10.1103/PhysRevLett.38.1433
102. Cremmer E et al (1978) Super-higgs effect in supergravity with general scalar interactions. Phys Lett B 79:231–234. https://doi.org/10.1016/0370-2693(78)90230-7
103. Dimopoulos S, Sutter DW (1995) The supersymmetric flavor problem. Nucl Phys B 452:496–512. https://doi.org/10.1016/0550-3213(95)00421-N. arXiv:hep-ph/9504415
104. Altakach MM et al (2024) Global LHC constraints on electroweak-inos with SModelS v2.3. SciPost Phys 16:101. https://doi.org/10.21468/SciPostPhys.16.4.101. arXiv:2312.16635 [hep-ph]
105. Allanach BC (2002) SOFTSUSY: a program for calculating supersymmetric spectra. Comput Phys Commun 143:305–331. https://doi.org/10.1016/S0010-4655(01)00460-X. arXiv:hep-ph/0104145 [hep-ph]
106. Allanach BC, Bernhardt MA (2010) Including R-parity violation in the numerical computation of the spectrum of the minimal supersymmetric standard model: SOFTSUSY. Comput Phys Commun181:232–245. https://doi.org/10.1016/j.cpc.2009.09.015. arXiv:0903.1805 [hep-ph]

107. Allanach BC, Kom CH, Hanussek M (2012) Computation of neutrino masses in R- parity violating supersymmetry: SOFTSUSY3.2. Comput Phys Commun 183:785–793. https://doi.org/10.1016/j.cpc.2011.11.024. arXiv:1109.3735 [hep-ph]
108. Allanach BC et al (2014) Next-to-minimal SOFTSUSY. Comput Phys Commun 185:2322–2339. [Erratum: Comput. Phys. Commun. 250, 107044 (2020)]. https://doi.org/10.1016/j.cpc.2014.04.015. arXiv:1311.7659 [hep-ph]
109. Allanach BC, Bednyakov A, de Austri RR (2015) Higher order corrections and unification in the minimal supersymmetric standard model: SOFTSUSY3.5. Comput Phys Commun 189:192–206. https://doi.org/10.1016/j.cpc.2014.12.006. arXiv:1407.6130 [hep-ph]
110. Allanach BC et al (2017) The inclusion of two-loop SUSYQCD corrections to gluino and squark pole masses in the minimal and next-to-minimal supersymmetric standard model: SOFTSUSY3.7. Comput Phys Commun 219:339–345. https://doi.org/10.1016/j.cpc.2017.05.006. arXiv:1601.06657 [hep-ph]
111. Allanach BC, Cridge T (2017) The calculation of sparticle and higgs decays in the minimal and next-to-minimal supersymmetric standard models: SOFTSUSY4.0. arXiv:1703.09717 [hep-ph]
112. Porod W (2003) SPheno, a program for calculating supersymmetric spectra, SUSY particle decays and SUSY particle production at e+ e-colliders. Comput Phys Commun 153:275–315. https://doi.org/10.1016/S0010-4655(03)00222-4. arXiv:hep-ph/0301101
113. Porod W, Staub F (2012) SPheno 3.1: extensions including flavour, CP-phases and models beyond the MSSM. Comput Phys Commun 183:2458–2469. https://doi.org/10.1016/j.cpc.2012.05.021. arXiv:1104.1573 [hep-ph]
114. Djouadi A et al (1998) The minimal supersymmetric standard model: group summary report. GDR (Groupement De Recherche) Supersymetrie. arXiv:hep-ph/9901246
115. Dedes A, Pietro S (2003) Two loop corrections to radiative electroweak symmetry breaking in the MSSM. Nucl Phys B 657:333–354. https://doi.org/10.1016/S0550-3213(03)00173-1. arXiv:hep-ph/0212132
116. Baer H, Barger V, Martinez D (2022) Comparison of SUSY spectra generators for natural SUSY and string landscape predictions. Eur Phys J C 82(2):172. https://doi.org/10.1140/epjc/s10052-022-10141-2. arXiv:2111.03096 [hep-ph]
117. Baer H et al (2020) Status of weak scale supersymmetry after LHC run 2 and ton-scale noble liquid WIMP searches. Eur Phys J ST 229(21):3085–3141. https://doi.org/10.1140/epjst/e2020-000020-x. arXiv:2002.03013 [hep-ph]
118. Agrawal V et al (1998) Anthropic considerations in multiple domain theories and the scale of electroweak symmetry breaking. Phys Rev Lett 80:1822–1825. https://doi.org/10.1103/PhysRevLett.80.1822. arXiv:hep-ph/9801253
119. Kim JE, Nilles HP (1984) The mu problem and the strong CP problem. Phys Lett B 138:150–154. https://doi.org/10.1016/0370-2693(84)91890-2
120. Zhitnitsky AR (1980) On possible suppression of the axion hadron interactions (in Russian). Sov J Nucl Phys 31:260
121. Dine M, Fischler W, Srednicki M (1981) A simple solution to the strong CP problem with a harmless axion. Phys Lett B 104:199–202. https://doi.org/10.1016/0370-2693(81)90590-6
122. Giudice GF, Masiero A (1988) A natural solution to the mu problem in supergravity theories. Phys Lett B 206:480–484. https://doi.org/10.1016/0370-2693(88)91613-9
123. Chen M-C et al (2013) The mu term and neutrino masses. Nucl Phys B 866:157–176. https://doi.org/10.1016/j.nuclphysb.2012.08.018. arXiv:1206.5375 [hep-ph]
124. Endo M et al (2011) Higgs mass and muon anomalous magnetic moment in supersymmetric models with vector-like matters. Phys Rev D 84:075017. https://doi.org/10.1103/PhysRevD.84.075017. arXiv:1108.3071 [hep-ph]
125. Endo M et al (2012) Higgs mass, muon g-2, and LHC prospects in gauge mediation models with vector-like matters. Phys Rev D 85:095012. https://doi.org/10.1103/PhysRevD.85.095012. arXiv:1112.5653 [hep-ph]
126. Endo M et al (2012) Vacuum stability bound on extended GMSB models. J High Energy Phys 06:060. https://doi.org/10.1007/JHEP06(2012)060. arXiv:1202.2751 [hep-ph]

127. Martin SP, Wells JD (2012) Implications of gauge-mediated supersymmetry breaking with vector-like quarks and a ~ 125 GeV Higgs Boson. Phys Rev D 86:035017. https://doi.org/10.1103/PhysRevD.86.035017. arXiv:1206.2956 [hep-ph]
128. Benakli K, Goodsell MD, Maier A-K (2011) Generating mu and Bmu in models with Dirac Gauginos. Nucl Phys B 851:445–461. https://doi.org/10.1016/j.nuclphysb.2011.06.001. arXiv:1104.2695 [hep-ph]
129. Benakli K (2011) Dirac gauginos: a user manual. Fortsch Phys 59:1079–1082. Ed. by Konstantinos Anagnostopoulos, Nikos Irges, and George Zoupanos. https://doi.org/10.1002/prop.201100071. arXiv:1106.1649 [hep-ph]
130. Chalons G et al (2019) LHC limits on gluinos and squarks in the minimal Dirac gaugino model. J High Energy Phys 04:113. https://doi.org/10.1007/JHEP04(2019)113. arXiv:1812.09293 [hep-ph]
131. Goodsell MD et al (2020) Constraining electroweakinos in the minimal dirac gaugino model. SciPost Phys 9(4):047. https://doi.org/10.21468/SciPostPhys.9.4.047. arXiv:2007.08498 [hep-ph]
132. Arkani-Hamed N et al (2001) Small neutrino masses from supersymmetry breaking. Phys Rev D 64:115011. https://doi.org/10.1103/PhysRevD.64.115011. arXiv:hep-ph/0006312
133. Borzumati F, Nomura Y (2001) Low scale seesaw mechanisms for light neutrinos. Phys Rev D 64 (2001), p. 053005. https://doi.org/10.1103/PhysRevD.64.053005. arXiv:hep-ph/0007018
134. Ross GG (1995) Beyond the standard model. CERN, Meyrin. https://doi.org/10.5170/CERN-1995-004.119
135. Lykken JD (2010) Beyond the standard model. arXiv:1005.1676 [hep-ph]
136. Dobrescu, BA (2014) Beyond the standard model. In: 2011 European school of high-energy physics. CERN, Geneva, pp 119–149. https://doi.org/10.5170/CERN-2014-003.119
137. Randall L, Sundrum R (1999) A large mass hierarchy from a small extra dimension. Phys Rev Lett 83:3370–3373. https://doi.org/10.1103/PhysRevLett.83.3370. arXiv:hep-ph/9905221
138. Gherghetta T, Pomarol A (2000) Bulk fields and supersymmetry in a slice of AdS. Nucl Phys B 586:141–162. https://doi.org/10.1016/S0550-3213(00)00392-8. arXiv:hep-ph/0003129
139. Arkani-Hamed N, Dimopoulos S, Dvali GR (1998) The Hierarchy problem and new dimensions at a millimeter. Phys Lett B 429:263–272. https://doi.org/10.1016/S0370-2693(98)00466-3. arXiv:hep-ph/9803315
140. Uehara Y (2002) A mini review of constraints on extra dimensions. Mod Phys Lett A 17:1551–1558. https://doi.org/10.1142/S0217732302007971. arXiv:hep-ph/0203244
141. Appelquist T, Cheng H-C, Dobrescu BA (2001) Bounds on universal extra dimensions. Phys Rev D 64:035002. https://doi.org/10.1103/PhysRevD.64.035002 arXiv:hep-ph/0012100
142. Kong K, Matchev KT (2007) Phenomenology of universal extra dimensions. AIP Conf Proc 903(1):451–454. https://doi.org/10.1063/1.2735221. arXiv:hep-ph/0610057
143. Deutschmann N, Flacke T, Kim JS (2017) Current LHC constraints on minimal universal extra dimensions. Phys Lett B 771:515–520. https://doi.org/10.1016/j.physletb.2017.06.004. arXiv:1702.00410 [hep-ph]
144. Beuria J et al (2018) LHC collider phenomenology of minimal universal extra dimensions. Comput Phys Commun 226:187–205. https://doi.org/10.1016/j.cpc.2017.12.021. arXiv:1702.00413 [hep-ph]

Chapter 2
Searching for SUSY at the Large Hadron Collider

Abstract Many experiments are running worldwide to probe the fundamental nature of our Universe and to confront theories against experimental data. The work of this thesis mostly relies on the results obtained by two of them: ATLAS and CMS, both conducted at the Large Hadron Collider (LHC). This chapter first introduces the LHC, ATLAS and CMS. It then discusses how the Minimal Supersymmetric Standard Model could manifest itself in them, and how LHC direct searches for Supersymmetry use the simplified model framework to constrain it.

Many experiments are running worldwide to probe the fundamental nature of our Universe and to confront theories against experimental data. The present work mostly relies on the results obtained by two of them, conducted at the LHC. This chapter will thus introduce the LHC and two of its major experiments in Sect. 2.1, how the MSSM could manifest itself in them in Sect. 2.2, and how well it is constrained by direct searches for SUSY at the LHC in Sect. 2.3.

2.1 The Large Hadron Collider and its Experiments

2.1.1 The LHC

The LHC is a 100 m underground circular hadron collider of 27 km in circumference, operating at 1.9 K, and based at the European Council for Nuclear Research (abbreviated CERN, for Conseil Européen pour la Recherche Nucléaire) [1], located at the border between France and Switzerland. A schematic representation of the main CERN accelerators and detectors is shown in Fig. 2.1. The protons (p) or heavy ion (Pb) are first injected in a linear accelerator, then transmitted to three circular accelerators: the Proton Synchrotron Booster (PSB), the Proton Synchrotron (PS) and the Super Proton Synchrotron (SPS), before reaching the LHC, where they are accelerated to their maximal velocity. For protons, this corresponded to

T. Pascal, *Beyond the Standard Model at the Large Hadron Collider: From Simplified Model Constraints to Artificial Proto-Modelling for Dispersed Signals*, Springer Theses, https://doi.org/10.1007/978-3-032-09391-2_2

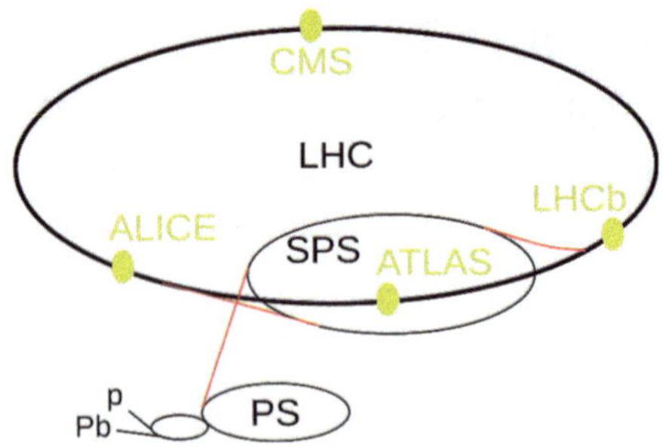

Fig. 2.1 Schematic representation of the main CERN accelerators and the locations of the four major LHC experiments. Taken from [7]

around 4 TeV (approximately 99.999997% of the speed of light) during the Run 1, and around 6.5 TeV (approximately 99.999999% of the speed of light) during the Run 2. The Run 3, which started in 2022 and is planned to stop at the end of 2025, reaches an unprecedented energy of around 6.8 TeV per proton. Two beams (each beam containing many proton bunches) are simultaneously accelerated this way, but in opposite directions. Collisions happen when these two beams cross each other, resulting in an energy in the center-of-mass frame of twice the energy of one particle. During Run 2, each proton beam contained 2808 bunches, each containing around 1.2×10^{11} protons (before colliding), resulting in around 10^9 collisions per second (it corresponds to around 1000 charged particles emerging from the interaction region every 25 ns) and in an integrated luminosity of around 160 fb^{-1} (only a fraction is actually recorded and can be used for physics studies).

Since the collision energy is way above the energy scale of quantum chromodynamics (QCD) (the scale above which the strong interaction can be expressed as a perturbative QFT, $\Lambda_{\text{QCD}} \sim 200$ MeV), the partons (i.e. quarks and gluons) constituting the hadrons, can be considered as free particles. It is therefore possible to describe hadronic interactions as partonic interactions. This parton model framework relies on the QCD factorisation theorem [2, 3] which factorises the universal, non-perturbative, "long-distance" physics with the computable and process-dependent "short-distance" physics. The cross section of two hadrons H_1 and H_2 carrying momenta P_1 and P_2 respectively is then given by

$$\sigma^{H_1,H_2}(P_1, P_2) = \sum_{i,j} \int_0^1 dx_1 \int_0^1 dx_2 \ \ f_i^{(H_1)}(x_1, \mu_F^2) \ f_j^{(H_2)}(x_2, \mu_F^2) \ \hat{\sigma}_{ij} \times (x_1 P_1, x_2 P_2, \alpha_s(\mu_R^2), \mu_F^2).$$

Here, x_1 and x_2 are the fraction of energy carried by the parton from H_1 and H_2, respectively; $f_i^{(H)}$ is the parton density function for parton i in hadron H. More precisely, given the hadron H with momentum P, the probability to find in H the parton i with momentum between xP and $(x + dx)P$ is $dx\ f_i^{(H)}(x, \mu_F^2)$, with μ_F the factorisation scale. Finally, $\hat{\sigma}_{ij}(x_1 P_1, x_2 P_2, \alpha_s(\mu_R^2), \mu_F^2)$ describes the hadronic process considered at the partonic level, with partons i and j carrying respective momenta $x_1 P_1$ and $x_2 P_2$. It is evaluated at the factorisation scale $\mu_{\rm F}$ and the strong interaction coupling α_s is evaluated at the renormalisation scale $\mu_{\rm R}$. The advantage is that the parton density functions are universal, that is to say, they do not depend upon the specific process considered and are fitted from data. Their evolution with respect to $\mu_{\rm F}$ is given by the Dokshitzer-Gribov-Lipatov-Altarelli-Parisi (DGLAP) equation [4–6].

These interactions happen at four collision points, each being at a different location and surrounded by a specific detector to record the collision products. These four detectors are A Large Ion Collider Experiment (ALICE), A Toroidal LHC Apparatus (ATLAS), Compact Muon Solenoid (CMS), and LHC-beauty (LHCb). Unlike the other three, ALICE is specifically designed to the study of QCD matter created in collisions between lead nuclei. The LHCb detector on the other hand is designed to search for indirect evidence of new physics in CP violation and rare decays of beauty and charm hadrons. It is looking for new particle effects in processes that are precisely predicted in the SM. The other two detectors, ATLAS and CMS, are general-purpose detectors designed to precisely measure the properties of all electrons, muons, photons, and (charged and neutral) hadrons emerging from proton–proton collisions. The work presented in this manuscript relies on the results published by the ATLAS and CMS collaborations using proton-proton collision data. The following will then focus on these two detectors and their results.

2.1.2 ATLAS and CMS

The ATLAS and CMS detectors are both designed to maximise their detection and reconstruction sensitivities to all kind of particles. To this end, they adopted a "traditional" high-energy experiment structure, made of several cylindrical layers, each covering the previous one and aiming to detect a specific type of particle, with a near-complete coverage in solid angle. A schematic representation of these different layers and of the particles they intend to detect is shown in Fig. 2.2. The four depicted layers are:

- **The tracking system:** it is designed to measure particle trajectories and momenta as close as possible to the interaction region.
- **The electromagnetic calorimeter:** its purpose is to identify photons and electrons, and to fully reconstruct their positions and energies by inducing and stopping their electromagnetic showers.
- **The hadronic calorimeter:** it has the same objective but for (charged and neutral) hadrons, and therefore plays a key role in identifying, reconstructing and measuring the energy of hadronic jets.
- **The muon system:** it specifically targets muons (which are too massive to leave significant energy in the calorimeters) and is optimised in the reconstruction of their tracks and momenta.

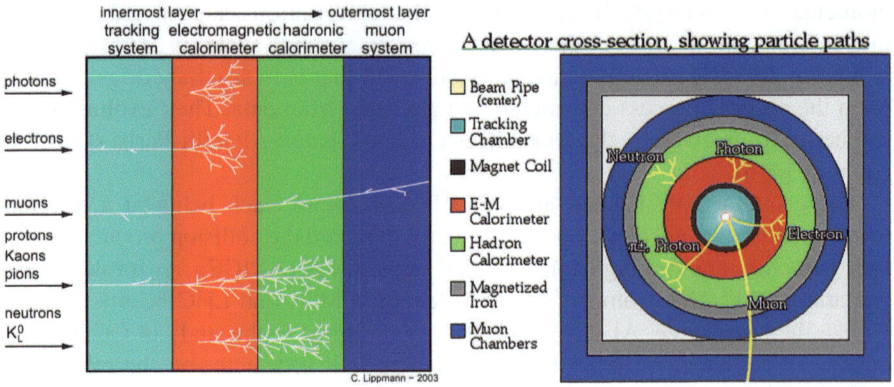

Fig. 2.2 Schematic representation, in the transversal plane, of the different detector layers of ATLAS and CMS, with the tracks left by different types of particles. The picture in the right panel is based on the CMS design. Taken from [8] (left) and [9] (right)

These layers are immersed in a high-intensity magnetic field bending the track of electrically charged particles. Knowing the bending of the tracks allows one to reconstruct the particle's electric charge and momentum (the higher the momentum, the less curved the trajectory). The particles that interact only weakly with matter (such as neutrinos) fully escape the detector without leaving any track. They are thus referred to as invisible particles.

Despite being similar, the ATLAS and CMS detectors remain complementary. The main difference in their designs comes from the way they produce and maintain a stable magnetic field. ATLAS comprises a thin superconducting solenoid magnet surrounding the inner tracking detectors and producing a 2 T axial magnetic field. In addition, two large endcap toroid magnets and eight barrel toroid magnets arranged with an azimuthal symmetry around the calorimeters are installed. The total detector is 44 m long and 25 m height for a weight of 7×10^3 tonnes. On the other hand, CMS uses a superconducting solenoid that surrounds the calorimeters and produces a 3.8 T axial magnetic field. The latter is returned by a steel yoke made of five wheels and two endcaps, composed of three disks each. With an overall length of 28.7 m and an overall diameter of 15 m for a total weight of 14×10^3 tonnes, CMS is more compact than ATLAS but is also heavier.

In addition, both detectors are equipped with an online trigger system and event filter fast enough to process events in real-time and to select only the ones of interest. The selected events are then saved in the CERN computer centre for further analysis.

Events are reconstructed and classified based on several quantities. They typically are the number of reconstructed photons, electrons, muons, taus, t-jets, b-jets and light jets (induced by u, d, s or c quarks), as well as their tracks, momenta and masses. Due to the beams travelling in the symmetry axis of the cylinder-shaped detectors, only the particle momenta in the transverse plane, labelled $\vec{p}_T$, can be fully reconstructed.

Two other observables are also of interest: the stransverse mass (m_{T2}) [10, 11] and the total missing transverse energy ($\not{E}_T$, or interchangeably E_T^{miss} or p_T^{miss}, for missing transverse momentum). The first one is useful to compute the masses of pair-produced particles, where each of them decays into one invisible and one directly observable particle. The second one is used to quantify the energy carried out by the particles not interacting with the detectors. Since the momentum is conserved by the collision and the initial state is free from any transverse momentum, the total missing transverse energy (carried by invisible particles) can be computed using the recorded transverse momenta

$$\not{E}_T = \left\| \vec{p}_T^{\,\text{miss}} \right\| = \left\| \sum_i^{\text{invisible}} \vec{p}_T^{\,i} \right\| = - \left\| \sum_j^{\text{visible}} \vec{p}_T^{\,j} \right\| ,$$

where the first sum is on the number of invisible particles and the second is on the number of reconstructed tracks.

2.2 MSSM Phenomenology at the LHC

The search for SUSY has been one of the main priorities of the LHC for several decades. Because the LHC collides SM particles, and the conservation of R-parity is more appealing regarding the current experimental observations (existence of a DM particle, conservation of the baryon and lepton numbers), the dominant processes are assumed to produce exactly and only two SUSY particles, i.e. $pp \to XY$, with X and Y two SUSY particles, identical or not. If R-parity is conserved, each SUSY particle will then decay, at leading order, into at least one SM particle and exactly one SUSY particle, producing a decay chain that stops with the LSP. This decay chain is most of the time assumed to be prompt, i.e. assumed to end before any SUSY particle that is not the LSP could enter the detector. The different ATLAS and CMS analyses are therefore mostly looking for signatures of SUSY particles through the tracks left in the detector by SM particles.

At the LHC, the search for new physics is made through analyses usually adopting a channel-by-channel approach. They are targeting specific final states (e.g. three leptons and missing transverse momentum), hoping to find a sizeable excess with respect to the predicted SM background in at least one of them. If this happened, understanding well enough particle physics phenomenology at the LHC would be primordial to explain the origin of the observed excess. However, if no excess shows up, exclusion limits are set on certain models that can produce the targeted final states. It is then once again very important to understand well enough particle physics phenomenology at the LHC to correctly constrain the appropriate models. Constraints are usually derived assuming simplified models [12–15].[1] This is particularly the case in SUSY searches, for its phenomenology is very rich.

In a simplified model, one assumes that only a handful of particles are relevant to the phenomenology. It is fully characterised by the minimal set of parameters that are the masses of the particles, their production cross sections, their branching ratios and their widths. The branching ratios are most of the time taken to be 1, i.e. that only one decay channel is accessible to the decaying mother particle. This goes along the channel-by-channel approach. Even if the simplified model framework is not a model-independent approach, it has the advantage of quantifying how well particle masses and rates are constrained by the data, not how well they are constrained based on theoretical parameters. The most important drawback is the difficulty of quantifying how a full, UV-complete, theory is constrained by the data. Multiple reinterpretation tools emerged to overcome this issue, aiming at constraining more complex models using experimental results (see, e.g., [16, 17] for a review). One of these tools has been extensively used in the work presented in this manuscript and is introduced in Chap. 3.

Since simplified models assume that only a few SUSY particles at accessible at the LHC energy, their production cross sections are typically computed without any other SUSY particle that could generate additional amplitudes. Figure 2.3 shows the most relevant production cross sections of SUSY particles with respect to their

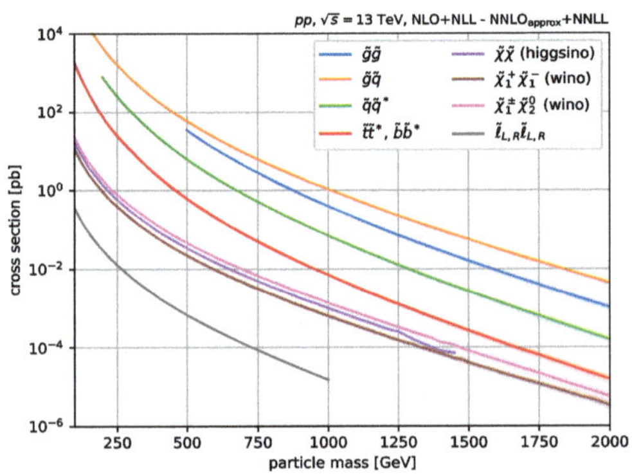

Fig. 2.3 Most relevant production cross sections of SUSY particles as a function of their masses, for pp collisions with a centre-of-mass energy of 13 TeV. The cross sections are computed assuming all the other SUSY particles are decoupled. The $\tilde{g}\tilde{q}$ cross section assumes the mass degeneracy of the $\tilde{g}$ and of the $\tilde{u}, \tilde{d}, \tilde{s}, \tilde{c}$ squarks. Moreover, it comprises the production of $(pp \to \tilde{g}\tilde{q} \to) \tilde{q}\tilde{q}q$, which is therefore not considered in the computation of the $\tilde{q}\tilde{q}$ next-to-leading-order (NLO) cross section. The $\tilde{q}\tilde{q}^*$ cross section assumes the mass degeneracy of the $\tilde{u}, \tilde{d}, \tilde{s}, \tilde{c}, \tilde{b}$ squarks. "higgsino" and "wino" refer to the electroweak-ino scenarios discussed in the text. The values for the $\tilde{t}\tilde{t}^*$, $\tilde{b}\tilde{b}^*$ cross section corresponds to one of the two processes, not both, with all the other SUSY particles decoupled. Taken from [18]

[1] Simplified models are also very useful if an excess is recorded to help rule out candidate models.

masses, assuming that the other SUSY particles are decoupled and are thus not taken into account in the computation.

2.2.1 Gluinos and Squarks

The production cross sections of colored particles are driven by the coupling of the strong interaction α_s (and no SUSY parameter except the masses), resulting in higher values compared to the production cross sections of particles coupling only through the electroweak force. The gluino and squark production processes at play in a hadron collider are

$$gg \to \tilde{g}\tilde{g},\ \tilde{q}\tilde{q}^* ; \tag{2.1}$$

$$gq \to \tilde{g}\tilde{q} ; \tag{2.2}$$

$$g\overline{q} \to \tilde{g}\tilde{q}^* ; \tag{2.3}$$

$$q\overline{q} \to \tilde{g}\tilde{g},\ \tilde{q}\tilde{q}^* ; \tag{2.4}$$

$$qq \to \tilde{q}\tilde{q} . \tag{2.5}$$

The corresponding leading-order (LO) Feynman diagrams are shown in Fig. 2.4. If the squarks are decoupled, gluinos can be produced in the s-channel from the fusion of two gluons or of one quark and one antiquark, but also in the t- and u-channels from the collision of two gluons via a virtual gluino. The resulting cross section is the blue line in Fig. 2.3.

On the contrary, if squarks are kinematically allowed but not gluinos, the corresponding diagrams can be obtained from the previous ones by replacing gluinos with squarks, and by including the four-point process gluino-gluino-squark-squark. The resulting cross sections are the green and red lines in Fig. 2.3.

If both gluinos and squarks are kinematically allowed, all the previous processes remain accessible and are complemented with the ones involving squark or gluino mediators that were previously forbidden. In addition, the simultaneous production of a squark and a gluino is now possible through three processes, giving the orange line in Fig. 2.3. The impact of the squark mass on the $\tilde{g}\tilde{g}$ and $\tilde{g}\tilde{q}$ production cross sections is exemplified in Fig. 2.5. On the one hand, the $\tilde{g}\tilde{g}$ cross section is reduced when introducing amplitudes involving virtual $\tilde{q}$, due to their destructive interference. On the other hand, when the $\tilde{q}$ mass is decreased below the $\tilde{g}$ mass, the $\tilde{g}\tilde{q}$ cross section increases because the produced $\tilde{q}$ (off- and on-shell) remains "light" even with increasing $m_{\tilde{g}}$. If instead the $\tilde{q}$ mass had been increased, the cross section would have decreased.

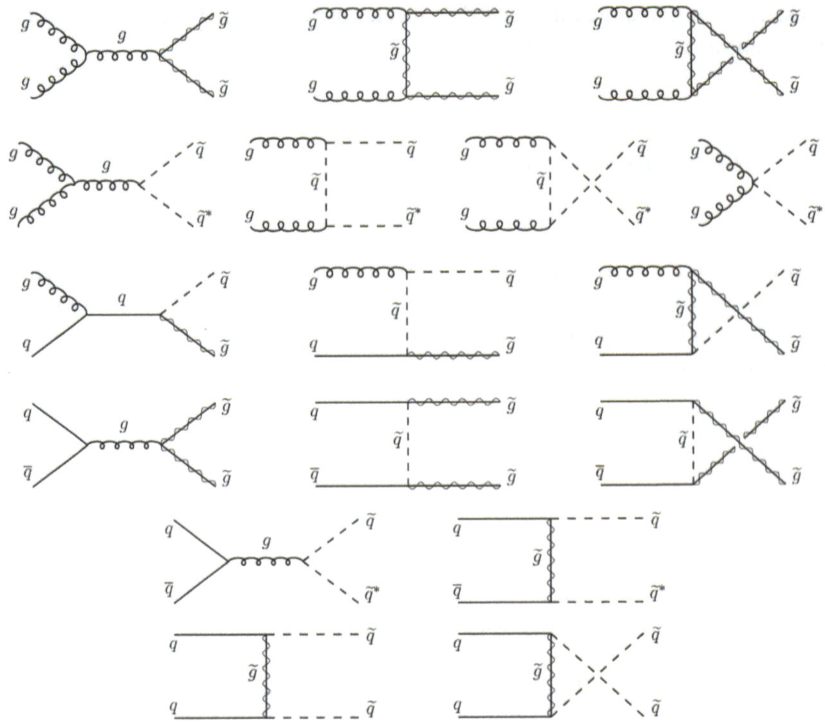

Fig. 2.4 Leading-order Feynman diagrams for the production of gluinos and squarks at hadron colliders. Taken from [19]

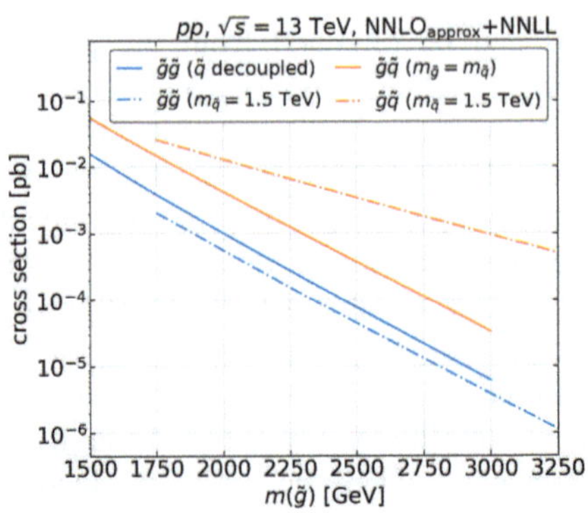

Fig. 2.5 Evolution of the $\tilde{g}\tilde{g}$ and $\tilde{g}\tilde{q}$ production cross sections with respect to the gluino mass for *pp* collisions with a centre-of-mass energy of 13 TeV. The cross sections are computed for different squark masses, as detailed in the legend. An eight-fold squark mass degeneracy is assumed ($\tilde{q} = \{\tilde{u}_{\mathrm{L,R}},\ \tilde{d}_{\mathrm{L,R}},\ \tilde{s}_{\mathrm{L,R}},\ \tilde{c}_{\mathrm{L,R}}\}$). Values taken from [18]

Regarding decays, the gluino decays always involve a (virtual or on-shell) squark. If

- $\mathbf{m_{\tilde{g}}} > \mathbf{m_{\tilde{q}}}$: the dominant decay is the two-body decay

$$\tilde{g} \rightarrow q\tilde{q} \,. \tag{2.6}$$

- $\mathbf{m_{\tilde{g}}} < \mathbf{m_{\tilde{q}}}$: the gluino can only undergo a three-body decay through off-shell $\tilde{q}$ as

$$\tilde{g} \rightarrow qq'\tilde{\chi}_i^{\pm},\; qq\tilde{\chi}_j^0 \,, \tag{2.7}$$

  with $i \in \{1, 2\}$ and $j \in \{1, 2, 3, 4\}$.
- $\mathbf{m_{\tilde{g}}} - \mathbf{m_{\tilde{\chi}_1^0}}$ is close to 0: the $\tilde{g}$ can undergo a two-body decay as

$$\tilde{g} \rightarrow g\tilde{\chi}_1^0 \tag{2.8}$$

  through a $q\tilde{q}$ loop. In that case, the $\tilde{g}$ is assumed to hadronise with a SM quark to form a color-singlet state known as R-hadron [20]. The latter does not decay promptly in the interaction region but farther in the detector or beyond it.

If not the LSP, each produced SUSY particle will further decay into other particles, producing a decay chain stopping with the LSP if R-parity is conserved, thus creating potentially long decay chains.

Regarding squarks, if

- $\mathbf{m_{\tilde{q}}} > \mathbf{m_{\tilde{g}}}$: the preferred decay for squarks is the two-body decay

$$\tilde{q} \rightarrow q\tilde{g} \,. \tag{2.9}$$

- $\mathbf{m_{\tilde{\chi}}} < \mathbf{m_{\tilde{q}}} < \mathbf{m_{\tilde{g}}}$: the $\tilde{q}_{\mathrm{R}}$ preferably decays into the bino-like neutralino and a same-flavor quark:

$$\tilde{q}_{\mathrm{R}} \rightarrow q\tilde{\chi}_{\tilde{B}}^0 \,, \tag{2.10}$$

  where $\tilde{\chi}_{\tilde{B}}^0$ is the bino-like neutralino. The $\tilde{q}_{\mathrm{L}}$ preferably decays into the wino-like chargino and a different-flavor quark or the neutralino and a same-flavor quark:

$$\tilde{q}_{\mathrm{L}} \rightarrow q'\tilde{\chi}_{\tilde{W}}^{\pm},\; q\tilde{\chi}_{\tilde{W}}^0 \,, \tag{2.11}$$

  where $\tilde{\chi}_{\tilde{W}}^{\pm}$ and $\tilde{\chi}_{\tilde{W}}^0$ are the wino-like chargino and neutralino, respectively. This is because the winos only couple to $\tilde{q}_{\mathrm{L}}$ (and $\tilde{\ell}_{\mathrm{L}}$), and the $\mathrm{SU(2)_L}$ gauge coupling is greater than the $\mathrm{U(1)}_Y$ gauge coupling. The $\tilde{q}$ decays to higgsinos-like neutralinos and chargino are less important due to the weak couplings, except for the $\tilde{b}$ and $\tilde{t}$ whose Yukawa couplings are significant.

It is worth mentioning that even when the $\tilde{\chi}_1^0$ is the only other SUSY particle (as often assumed in simplified models), the $\tilde{t}_1$ can already produce various signatures due to the large top quark mass. If

- $\mathbf{\Delta m(\tilde{t}_1, \tilde{\chi}_1^0) = m_{\tilde{t}_1} - m_{\tilde{\chi}_1^0} > m_t}$: the $\tilde{t}_1$ will undergo a two-body decay:

$$\tilde{t}_1 \to t\tilde{\chi}_1^0 , \tag{2.12}$$

followed by an on-shell decay of the t quark into an on-shell W boson and a b quark.

- $\mathbf{m_t > \Delta m(\tilde{t}_1, \tilde{\chi}_1^0) > m_{W^\pm} + m_b}$: the $\tilde{t}_1$ will undergo a three-body decay:

$$\tilde{t}_1 \to Wb\tilde{\chi}_1^0 \tag{2.13}$$

through an off-shell t quark.

- $\mathbf{m_{W^\pm} + m_b > \Delta m(\tilde{t}_1, \tilde{\chi}_1^0) > m_b}$: the $\tilde{t}_1$ will undergo a four-body decay:

$$\tilde{t}_1 \to ff'b\tilde{\chi}_1^0 \tag{2.14}$$

through an off-shell t quark, itself decaying into a b quark and an off-shell W, itself decaying into two fermions: f and f'.

On top of missing transverse momentum ($\tilde{\chi}_1^0$ evades the detector without leaving any track) and jets, the final state can therefore contain 0, 1 or 2 leptons.

2.2.2 *Sleptons*

As far as sleptons are concerned, they have the lowest production cross section, as shown by the grey line in Fig. 2.3. They can only be produced at a hadron collider through the production of electroweak vector bosons in the s-channel via $q\bar{q}$ annihilation, where the two quarks are either of the same flavor or one is an up-type quark and the other is a down-type quark, i.e.

$$q\overline{q} \to \tilde{\ell}^+\tilde{\ell}^-,\ \tilde{\nu}_\ell\tilde{\nu}_\ell^* ; \tag{2.15}$$

$$u\overline{d} \to \tilde{\ell}_L^+\tilde{\nu}_l ; \tag{2.16}$$

$$d\overline{u} \to \tilde{\ell}_L^-\tilde{\nu}_l^* . \tag{2.17}$$

The corresponding production diagrams are shown in Fig. 2.6, except for the process (2.17) whose diagram can be obtained by charge conjugation of the rightmost one.

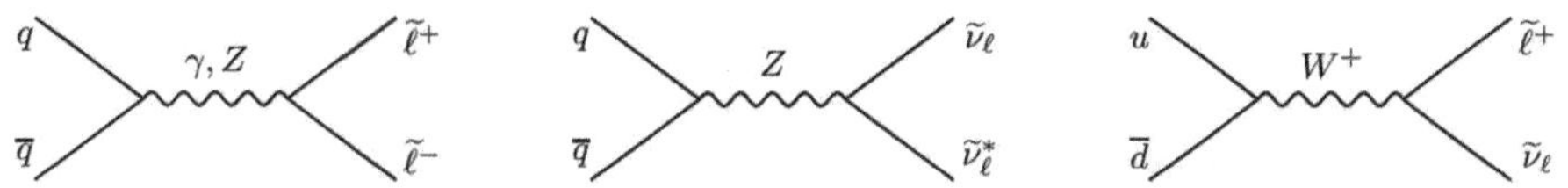

Fig. 2.6 Leading-order Feynman diagrams for the production of sleptons at hadron colliders. One additional diagram can be obtained by charge-conjugating the rightmost one. Taken from [19]

If kinematically allowed, the sleptons decays are the following:

$$\tilde{\ell}^{\pm} \to \nu\tilde{\chi}_i^{\pm},\ \ell^{\pm}\tilde{\chi}_j^0\,; \tag{2.18}$$

$$\tilde{\nu} \to \ell^{\mp}\tilde{\chi}_i^{\pm},\ \nu\tilde{\chi}_j^0\,, \tag{2.19}$$

with $i \in \{1, 2\}$ and $j \in \{1, 2, 3, 4\}$. Similarly to the squarks, the $\tilde{\ell}_{\mathrm{R}}$ prefer to decay into the bino-like neutralino, and the $\tilde{\ell}_{\mathrm{L}}$ prefer to decay into the wino-like neutralino and chargino.

2.2.3 *Electroweak-inos*

Let us now turn to the collider signatures expected from EW-ino production at the LHC. The EW-ino sector of the MSSM is of peculiar interest because EW-inos play a crucial role in gauge coupling unification in the MSSM [21–23], they are intimately tied to an understanding of the EW scale through the Higgs mixing parameter μ [24, 25] (see also [19]), and they are of primordial interest in the context of SUSY dark matter [26]. Even if all the rest of the SUSY spectrum is decoupled, light EW-inos are a well-motivated and hugely interesting possibility for BSM physics. Furthermore, with the gain in energy and in luminosity accompanying the Run 2 of the LHC, the EW-ino sector could be more easily probed and was the main focus of ATLAS and CMS SUSY search programs. Indeed, with the stringent constraints put on the masses of the colored SUSY particles (briefly discussed in Sects. 2.3.1 and 2.3.2), if SUSY is realised in nature and is accessible at LHC energies, it is more likely to manifest itself in the EW-ino sector. The main work of this thesis, discussed in Chap. 4, therefore focuses on this sector. Particular attention will thus be given in describing its LHC signatures. The following discussion is mostly coming from [27].

If the squarks are decoupled, EW-ino production is identical to that of sleptons. They can only be produced at a hadron collider through the production of electroweak vector bosons in the s-channel via $q\bar{q}$ annihilation, where the two quarks are either of the same flavor or one is an up-type quark and the other is a down-type quark. These diagrams are shown in the left column of Fig. 2.7. If the squarks are not decoupled, additional t-channel diagrams are allowed with a mediator squark, as shown in the middle and right columns of Fig. 2.7. In the end, all the possible

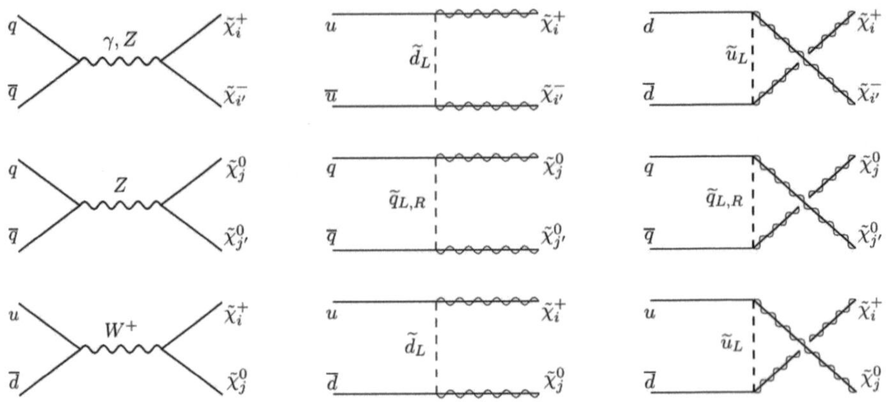

Fig. 2.7 Leading-order Feynman diagrams for the production of EW-inos at hadron colliders, with $i, i' \in \{1, 2\}$ and $j, j' \in \{1, 2, 3, 4\}$. Three more valid diagrams not displayed here can be obtained by charge-conjugating the lower line. Only the gaugino component of the $\tilde{\chi}^{\pm}$ and $\tilde{\chi}^0$ couple to squarks, this is why the EW-inos on the t-channel diagrams are drawn as wavy lines superimposed on solid. Taken from [19]

processes are

$$q\overline{q} \to \tilde{\chi}_i^{\pm}\tilde{\chi}_{i'}^{\mp},\ \tilde{\chi}_j^0\tilde{\chi}_{j'}^0\,; \tag{2.20}$$

$$u\overline{d} \to \tilde{\chi}_i^{+}\tilde{\chi}_j^0\,; \tag{2.21}$$

$$d\overline{u} \to \tilde{\chi}_i^{-}\tilde{\chi}_j^0\,, \tag{2.22}$$

with $i, i' \in \{1, 2\}$ and $j, j' \in \{1, 2, 3, 4\}$. The process (2.22) is not shown in Fig. 2.7 but can be obtained by charge conjugation of the three lower diagrams.

Since no theoretical principle guide the hierarchy between the bino, the wino and higgsino mass scales, the production cross section for EW-inos depends on their hierarchy and mixing. Thus, the phenomenology of the EW-ino sector is model-dependent and very rich in itself.

A common assumption is then to consider at least one decoupled scale and the allowed neutralinos and charginos to be pure states of the remaining one(s). The production cross sections at the 13 TeV LHC are shown in Fig. 2.8, assuming pure wino or pure higgsino states with degenerate masses (the denotation as $\tilde{\chi}_1^{\pm}$, $\tilde{\chi}_1^0$, etc. is completely arbitrary) and decoupled squarks. As can be seen, winos are produced much more copiously than higgsinos of the same mass. The production of pure binos, on the other hand, is not shown in the plot, as it is negligible compared to the other modes. Similarly, the pair production of $\tilde{\chi}_1^0$ vanishes if the $\tilde{\chi}_1^0$ is a pure higgsino.

For the discussion of signatures, the focus will hence be made on decays of wino-like or higgsino-like EW-inos into lighter ones. Figure 2.9 shows examples of four scenarios with different LHC signatures which are discussed below.

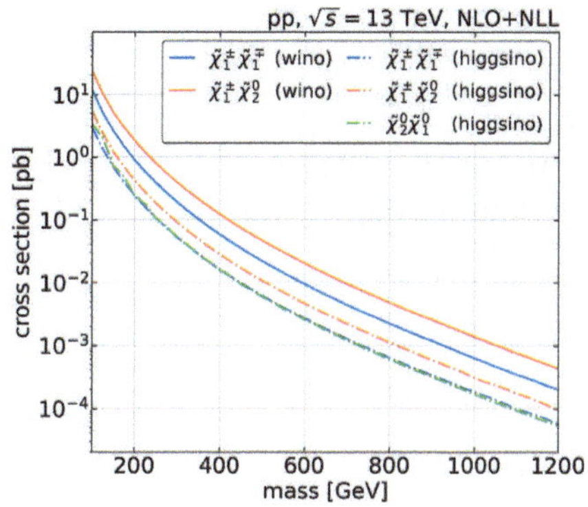

Fig. 2.8 Production cross sections of pure wino or higgsino states with degenerate masses and decoupled squarks. The production cross sections for higgsino-like $\tilde{\chi}_1^\pm \tilde{\chi}_1^\mp$ and $\tilde{\chi}_2^0 \tilde{\chi}_1^0$ are almost identical. Values taken from [18]

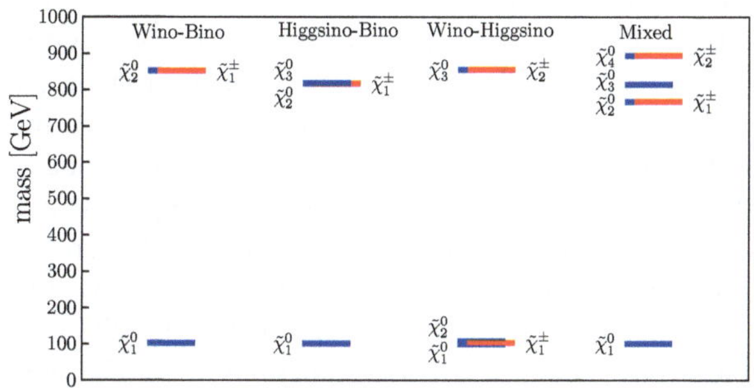

Fig. 2.9 Illustrative spectra for the different physics scenarios discussed in Sect. 2.2.3. The mass parameters that set the (heavy, light) scales are, for wino-bino: (M_2, M_1), for higgsino-bino: (μ, M_1), for wino-higgsino: (M_2, μ); for the mixed scenario, $(M_2 = \mu, M_1)$ was used. The 'heavy' scale is always set to 800 GeV, and the 'light' scale to 100 GeV. In the first three scenarios, the third mass parameter is decoupled. To guide the eye, neutralinos are indicated in blue and charginos in red

2.2.3.1 Wino-Bino Scenario

The case most commonly considered in LHC studies is the production of wino-like $\tilde{\chi}_1^\pm$ and $\tilde{\chi}_2^0$ (with $m_{\tilde{\chi}_1^\pm} = m_{\tilde{\chi}_2^0}$) which decay to a bino-like $\tilde{\chi}_1^0$. This is realised for $\mu \gg M_2 > M_1$; this will be referred to as the wino-bino scenario in the following. An example is shown in the left-most spectrum in Fig. 2.9. In the absence of intermediate sfermions, the available 2-body decay modes of the $\tilde{\chi}_1^\pm$ and $\tilde{\chi}_2^0$ are

$$\tilde{\chi}_1^\pm \to W^\pm \tilde{\chi}_1^0; \tag{2.23}$$

$$\tilde{\chi}_2^0 \to Z\tilde{\chi}_1^0,\ h\tilde{\chi}_1^0, \tag{2.24}$$

where h is the SM-like Higgs boson with a mass of 125 GeV (the other Higgs bosons are assumed to be heavy). If the 2-body decays are kinematically forbidden, $\tilde{\chi}_1^\pm \to f\bar{f}' + \tilde{\chi}_1^0$ and $\tilde{\chi}_2^0 \to f\bar{f} + \tilde{\chi}_1^0$ via off-shell W and Z respectively. The

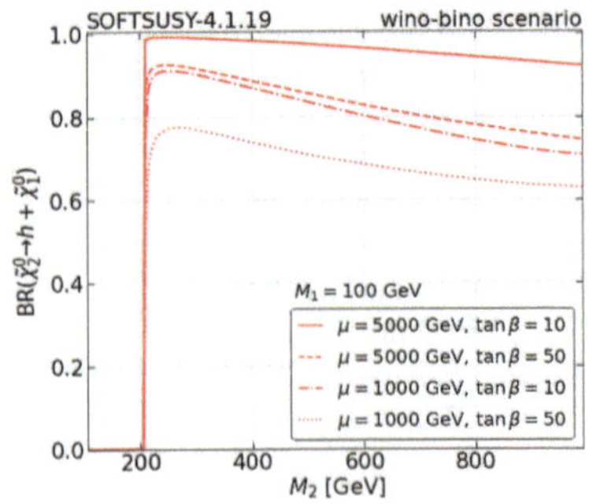

Fig. 2.10 Decay branching ratios of a wino-like $\tilde{\chi}_2^0$ into a bino-like $\tilde{\chi}_1^0$ plus SM-like Higgs boson h as a function of M_2, for $M_1 = 100$ GeV and different values of $\tan\beta$ and μ. Here, $\mathrm{BR}(\tilde{\chi}_2^0 \to Z + \tilde{\chi}_1^0) = 1 - \mathrm{BR}(\tilde{\chi}_2^0 \to h + \tilde{\chi}_1^0)$

signals looked for are thus primarily $W^\pm Z + \not{E}_T$ and $W^\pm h + \not{E}_T$ from $pp \to \tilde{\chi}_1^\pm \tilde{\chi}_2^0$ production, as well as $W^+W^- + \not{E}_T$ from $pp \to \tilde{\chi}_1^+ \tilde{\chi}_1^-$ production. Other modes, like $pp \to \tilde{\chi}_2^0 \tilde{\chi}_2^0$, $\tilde{\chi}_1^\pm \tilde{\chi}_1^0$, $\tilde{\chi}_2^0 \tilde{\chi}_1^0$ production, are less important, partly because of smaller production cross sections and partly because the resulting signals suffer from larger SM backgrounds. The W and Z bosons are typically looked for in leptonic final states ($W \to \ell\nu$, $Z \to \ell^+\ell^-$), while Higgs bosons are identified through $b\bar{b}$ or $\gamma\gamma$ final states. Only very recently, for the full Run 2 dataset, analyses were carried out in fully hadronic final states using boosted boson tagging [28, 29].

The $\tilde{\chi}_1^\pm$ decay into $W^\pm \tilde{\chi}_1^0$ (or $f\bar{f}'\tilde{\chi}_1^0$) proceeds through the left current and, as the only available decay mode, has 100% branching ratio. The branching ratios of the $\tilde{\chi}_2^0$ decays, however, depend on the details of the scenario. While $\mathrm{BR}(\tilde{\chi}_2^0 \to Z\tilde{\chi}_1^0) \simeq 1$ for $m_{\tilde{\chi}_2^0} - m_{\tilde{\chi}_1^0} < m_h$, the decay into Higgs bosons quickly becomes dominant once kinematically allowed. The exact branching ratios depend also on the values of μ and $\tan\beta$, since neutralinos couple to Z bosons only through their higgsino components.[2] For illustration, Fig. 2.10 shows the decay branching ratios of a wino-like $\tilde{\chi}_2^0$ into a SM-like Higgs boson and a bino-like $\tilde{\chi}_1^0$ as a function of $M_2 \simeq m_{\tilde{\chi}_2^0}$, for $m_{\tilde{\chi}_1^0} \simeq M_1 = 100$ GeV and varying values of $\tan\beta$ and μ. We see from this figure that limits under the assumption of 100% signal in either $\tilde{\chi}_1^\pm \tilde{\chi}_2^0 \to W^\pm Z + \not{E}_T$ or $\tilde{\chi}_1^\pm \tilde{\chi}_2^0 \to W^\pm h + \not{E}_T$ are unrealistic for most of the parameter space. Even for the $W^\pm h + \not{E}_T$ channel, for which the difference in signal compared to the simplified model result can be small, the combination of results with the $W^\pm Z + \not{E}_T$ channel, included in the phenomenological study in Chap. 4, should be interesting. Moreover, $\tilde{\chi}_1^+ \tilde{\chi}_1^- \to W^+W^- + \not{E}_T$ is always present in addition and needs to be included for realistic constraints.

2.2.3.2 Higgsino-Bino Scenario

For $M_2 \gg \mu > M_1$, the next-to-lightest mass scale is set by the higgsinos, with the lightest state being bino-like, as illustrated by the 2nd spectrum from the left

[2] Concretely, the $\tilde{\chi}_i^0 \tilde{\chi}_j^0 Z$ coupling is given by $\frac{ig}{2c_W}(N_{i4}N_{j4}^\star - N_{i3}N_{j3}^\star)$, see [30, 31].

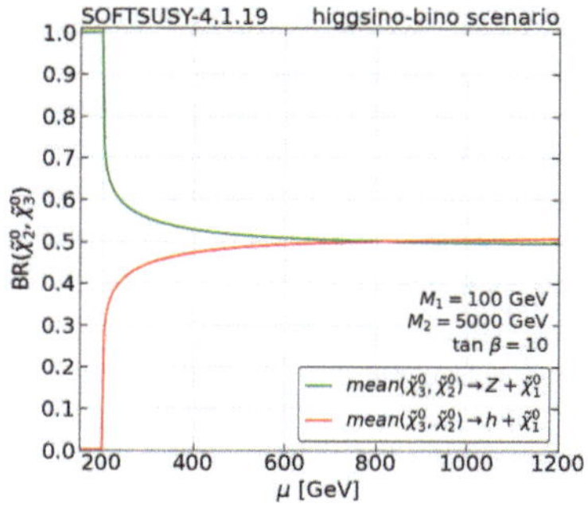

Fig. 2.11 Decay branching ratios of higgsino-like $\tilde{\chi}_{2,3}^0$ into a bino-like $\tilde{\chi}_1^0$ plus a Higgs or Z boson as a function of μ, for $M_1 = 100$ GeV, $M_2 = 5000$ GeV and $\tan\beta = 10$. Since the $\tilde{\chi}_{2,3}^0$ are indistinguishable, only the total rates of decays into $h + \tilde{\chi}_1^0$ and $Z + \tilde{\chi}_1^0$ are shown

in Fig. 2.9. In this higgsino-bino scenario, the EW-ino signatures arise from $pp \to \tilde{\chi}_1^+\tilde{\chi}_1^-$, $\tilde{\chi}_1^\pm\tilde{\chi}_{2,3}^0$ and $\tilde{\chi}_2^0\tilde{\chi}_3^0$ production followed by

$$\tilde{\chi}_1^\pm \to W^\pm\tilde{\chi}_1^0\,; \tag{2.25}$$

$$\tilde{\chi}_{2,3}^0 \to Z\tilde{\chi}_1^0,\; h\tilde{\chi}_1^0\,. \tag{2.26}$$

The $\tilde{\chi}_{2,3}^0$ are practically degenerate in mass. For large enough μ, one of them decays dominantly (roughly to 70–80%) into $Z\tilde{\chi}_1^0$, while the other decays dominantly into $h\tilde{\chi}_1^0$ (also roughly to 70–80%), giving almost equal rates of Z and h. The total EW-ino signal is therefore again a mix of WW, WZ, Wh and $Zh + \not{E}_T$ signatures, with a small addition of ZZ and $hh + \not{E}_T$. The relative rates of Z or h bosons from $\tilde{\chi}_{2,3}^0$ decays are shown in Fig. 2.11; they are almost insensitive to variations of M_2 and $\tan\beta$.

2.2.3.3 Wino-Higgsino Scenario

If the LSP is higgsino-like instead of bino-like, things become much more complicated. The wino-higgsino scenario (or the higgsino-LSP scenario) is realised for $M_1 \gg M_2 > \mu$. In this case, the $\tilde{\chi}_{1,2}^0$ and $\tilde{\chi}_1^\pm$ are a triplet of higgsino-like states with $m_{\tilde{\chi}_1^0} < m_{\tilde{\chi}_1^\pm} < m_{\tilde{\chi}_2^0}$. The mass differences between them are generally small, $m_{\tilde{\chi}_1^\pm} - m_{\tilde{\chi}_1^0} \lesssim$ few GeV, and $m_{\tilde{\chi}_2^0} - m_{\tilde{\chi}_1^0}$ is in the range of about 3–30 GeV, see the third spectrum from the left in Fig. 2.9. The wino-like states are now the $\tilde{\chi}_2^\pm$ and $\tilde{\chi}_3^0$, and they have a large variety of possible decay modes:

$$\tilde{\chi}_2^\pm \to W^\pm\tilde{\chi}_2^0,\; Z\tilde{\chi}_1^\pm,\; h\tilde{\chi}_1^\pm,\; W^\pm\tilde{\chi}_1^0\,; \tag{2.27}$$

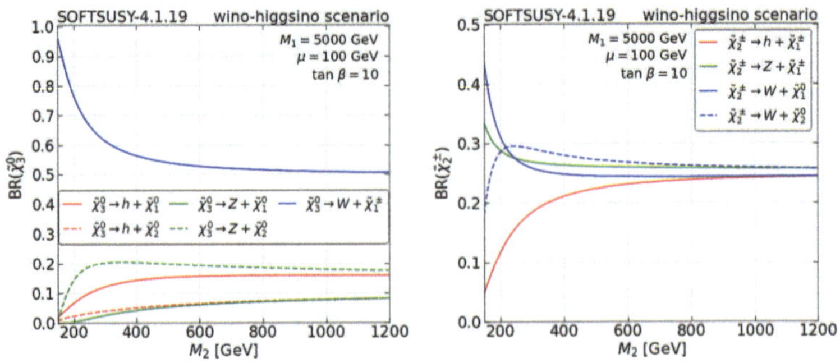

Fig. 2.12 Decay branching ratios of $\tilde{\chi}_3^0$ (left) and $\tilde{\chi}_2^\pm$ (right) in the wino-higgsino scenario as a function of M_2, for $\mu = 100$ GeV, $M_1 = 5000$ GeV, and $\tan\beta = 10$

$$\tilde{\chi}_3^0 \to Z\tilde{\chi}_2^0,\ h\tilde{\chi}_2^0,\ W^\mp\tilde{\chi}_1^\pm,\ Z\tilde{\chi}_1^0,\ h\tilde{\chi}_1^0; \tag{2.28}$$

followed by $\tilde{\chi}_2^0 \to ff'\tilde{\chi}_1^\pm,\ f\bar{f}\tilde{\chi}_1^0$ and/or $\tilde{\chi}_1^\pm \to f\bar{f}'\tilde{\chi}_1^0$ transitions if the $\tilde{\chi}_2^\pm$ or $\tilde{\chi}_3^0$ decay is not directly into the LSP. Since $pp \to \tilde{\chi}_3^0\tilde{\chi}_3^0$ production is suppressed, only $pp \to \tilde{\chi}_2^\pm\tilde{\chi}_3^0$ and $pp \to \tilde{\chi}_2^+\tilde{\chi}_2^-$ are relevant and lead to a complicated mix of W^+W^-, $W^\pm Z$, $W^\pm h$, ZZ, Zh, hh plus $\not{E}_T$ final states, often accompanied by additional soft jets or leptons.

The relevant branching ratios are depicted in Fig. 2.12 as a function of the wino mass parameter, for $\mu = 100$ GeV, $M_1 = 5000$ GeV and $\tan\beta = 10$. Variations with M_1 and $\tan\beta$ are small and do not change the overall picture of a multitude of relevant final states. Indeed, the EW-ino signal is split almost democratically into the different di-boson+$\not{E}_T$ and di-boson + X+$\not{E}_T$ channels, X meaning additional soft jets and/or leptons. It is therefore expected that individual simplified model results will be too weak to significantly constrain this scenario; the statistical combination of results from different analyses, performed in Chap. 4, should be of great advantage for obtaining sensitive limits.[3]

2.2.3.4 Wino-LSP Scenarios

When M_2 is the smallest mass parameter, the doublet of wino-like $\tilde{\chi}_1^0$ and $\tilde{\chi}_1^\pm$ are the lightest states. The $\tilde{\chi}_1^0$ is still the LSP, but the mass difference to the $\tilde{\chi}_1^\pm$ is so small (only about 150–160 MeV) that the $\tilde{\chi}_1^\pm$ becomes long-lived on collider scales. The signatures to look for are then 1–2 disappearing tracks from $\tilde{\chi}_1^\pm\tilde{\chi}_1^0$ and $\tilde{\chi}_1^+\tilde{\chi}_1^-$ production, respectively.

[3] The exceptions are perhaps constraints from the fully hadronic searches [28, 29], which are sensitive to a variety of boosted boson final states.

2.2.3.5 Mixed Scenarios

If two (or all three) EW-ino mass parameters are close to each other, we obtain mixed scenarios, in which the properties of the relevant charginos and neutralinos are not well approximated as being bino, wino or higgsino. One consequence of such mixing is that more states are relevant at the same time and they share their couplings. Moreover, the mass splittings among similar states increase.

For example, for $M_1 = 100$ GeV and $M_2 = \mu = 800$ GeV (with $\tan\beta = 10$), we have a bino-like $\tilde{\chi}_1^0$ of mass $m_{\tilde{\chi}_1^0} = 105$ GeV; the heavier states have masses of $m_{\tilde{\chi}_{2,3,4}^0} = 787, 824, 922$ GeV and $m_{\tilde{\chi}_{1,2}^\pm} = 787, 922$ GeV with the $\tilde{\chi}_2^0$ ($\tilde{\chi}_4^0$) being 42% (58%) wino and 58% (42%) higgsino, see the right-most spectrum in Fig. 2.9. The charginos are also highly mixed, having both wino and higgsino components. Only the $\tilde{\chi}_3^0$ is almost pure higgsino. Consequently, the production of all $\tilde{\chi}_{1,2}^\pm$ and $\tilde{\chi}_{2,3,4}^0$ combinations is important and adds to the complexity of the EW-ino signal. In particular, while the $\tilde{\chi}_1^\pm$ and $\tilde{\chi}_{2,3}^0$ decay directly into the $\tilde{\chi}_1^0$ (plus W, h or Z bosons), the $\tilde{\chi}_2^\pm$ decays to 44% into $W^\pm\tilde{\chi}_2^0$, 31% into $Z\tilde{\chi}_1^\pm$ and 5% into $h\tilde{\chi}_1^\pm$, only 12% going into $W^\pm\tilde{\chi}_1^0$. Likewise, the $\tilde{\chi}_4^0$ decays to almost 80% into $W^\pm\tilde{\chi}_1^\mp$; only about 10% of the $\tilde{\chi}_4^0$ decays go into $h\tilde{\chi}_1^0$ and 4% into $Z\tilde{\chi}_1^0$.

In a similar fashion, when the mass scale of the LSP is set by two parameters with very similar values, e.g. $M_2 \simeq M_1 \ll \mu$ or $\mu \simeq M_1 \ll M_2$, this leads to quite complicated scenarios. The signatures can be similar to those of the wino-higgsino case discussed above, but with more variety of nearly-degenerate states. These scenarios will not be further discussed here, but only note that $M_1 \simeq M_2$ can give a mostly wino-like LSP with a large enough mass splitting such that the $\tilde{\chi}_1^\pm$ decays promptly.

2.2.4 *Decays to the Goldstino/Gravitino*

Usually, the $\tilde{\chi}_1^0$ is assumed to be the LSP, but in GMSB models the LSP is instead the goldstino/gravitino $\tilde{G}$. In such models, all SUSY particles can decay into the $\tilde{G}$, but other decay channels remain predominant. The only SUSY particle with a sizeable branching ratio into the $\tilde{G}$ is the next-to-lightest SUSY particle (NLSP) if R-parity is conserved. Most GMSB models have an NLSP originating from a right-handed electrically charged slepton or from the bino. If R-parity is conserved and the NLSP is a

- **bino-like $\tilde{\chi}_1^0$**: the possible decays are

$$\tilde{\chi}_1^0 \to \gamma\tilde{G},\ Z\tilde{G},\ h\tilde{G},\ A\tilde{G},\ H\tilde{G}\,. \tag{2.29}$$

  However, only the first one is guaranteed to be kinematically allowed, while the last two are unlikely to be.

- $\tilde{\tau}_1$: the possible decay is

$$\tilde{\tau}_1 \to \tau\tilde{G}\,. \tag{2.30}$$

However, if the mass splitting between the $\tilde{e}_R$, $\tilde{\mu}_R$ and the $\tilde{\tau}_1$ is small,[4] the $\tilde{e}_R \to e\tau\tilde{\tau}_1$ and $\tilde{\mu}_R \to \mu\tau\tilde{\tau}_1$ decays are suppressed and the three sleptons are effective NLSP whose decay are

$$\tilde{\ell} \to \ell\tilde{G}\,. \tag{2.31}$$

This is the "co-NLSP" scenario. The $\tilde{\chi}_1^0$ can also act as an effective NLSP if its mass is slightly above $\tilde{\tau}_1$. (The opposite is also true, if the slepton masses are slightly above $\tilde{\chi}_1^0$.)

If the $\tilde{G}$ is heavier than $\mathcal{O}(1)$ keV, the NLSP decay is so reduced that it is likely to escape the detector without decaying, appearing as an effective LSP for the detector.

2.3 LHC Searches for Supersymmetry: Status at the End of Run 2

Since the LHC has been operating, numerous analyses in search of SUSY signatures have been carried out, notably by the ATLAS and CMS collaborations. To give an idea of the current constraints on the MSSM, an overview of the experimental results obtained by ATLAS and CMS will be discussed in the following.[5] However, before looking at the experimental results, it is worth describing the analysis structure that is typically used to obtain them. Most of the time, analyses are designed in such a way that the events they select can be grouped in different regions.

The signal regions (SRs) are defined to maximise the sensitivity to the searched signal with respect to SM backgrounds (predicted by Monte Carlo simulations). To correctly assess the number of SM events populating the SRs, orthogonal regions are designed to select only controlled SM events. They are the control regions (CRs). Additionally, validation regions (VRs) can be designed to verify that the SM predictions agree with the data without any influence from the SRs. Their topology is as close as possible to the SRs but with only a small signal contamination.

---

[4] In GMSB models, the $\tilde{e}_R$, $\tilde{\mu}_R$ and $\tilde{\tau}_R$ have the same mass. The $\tilde{e}_R$ and $\tilde{\mu}_R$ do not mix with their left-handed counterparts due to the small electron and muon Yukawa couplings and can be considered as degenerate, unmixed mass eigenstates. However, the tau has a Yukawa coupling high enough for the $\tilde{\tau}_R$ and $\tilde{\tau}_L$ to mix, giving a $\tilde{\tau}_1$ with a lower mass than of the $\tilde{e}_R$ and $\tilde{\mu}_R$. Nonetheless, this mass difference can be lower than m_τ.

[5] The MSSM is also constrained by other experiments than ATLAS and CMS, but these constraints will not be discussed in this thesis.

To test if new physics has been recorded, the probability that the observed data in the SRs does not come from the SM is computed. This test is realised through the CL_s prescription [32] and the profile likelihood ratio test statistic [33]. Similarly, limits can also be derived by evaluating how unlikely it is that the observed data can be explained by a given model. Usually, upper limits on the cross sections are set to a 95% confidence level (CL) (corresponding to a CL_s value of 0.05).

Despite the large number of searches for SUSY signatures at the LHC, no evidence for SUSY has shown up in the data. Therefore, only constraints on the masses, production cross sections, branching ratios and widths of various SUSY particles could be set. It is important to keep in mind, however, that these limits are derived using simplified models and hold only in this context. The following overview will hence present ATLAS and CMS constraints interpreted in terms of simplified models. Due to the large number of analyses, only a selection of them will be presented and the following list of results is therefore not exhaustive. The focus will be made on searches targeting prompt signatures of R-parity conserving (RPC) SUSY models. A similar overview for models assuming long-lived particles (LLPs) and/or non-zero RPV couplings is presented in Appendix A.

This section takes its inspiration from the ATLAS [34] and CMS [35] summaries of SUSY searches after Run 2 (for Run 1, see [36] for ATLAS and [37] for CMS). Another review of experimental constraints can also be found in Chapter 89 ("Supersymmetry, Part II (Experiment)") of [38].

Unless stated otherwise, the following analyses use the full luminosity recorded by their experiments during Run 2 and the $\tilde{\chi}_1^0$ is assumed to be the LSP. Since it interacts only weakly with ordinary matter, it is assumed to escape the detector without leaving any track. Therefore, all the following analyses required events with $\not{E}_T$ in their final states.

2.3.1 Gluino Production

Gluinos are the SUSY particles with the most stringent constraints. It comes from their high production cross section and their relatively easy-to-detect decay products.

Figures 2.13 and 2.14 summarise ATLAS and CMS constraints on the pair production of $\tilde{g}$ in the $m_{\tilde{\chi}_1^0}$ vs $m_{\tilde{g}}$ plane. The ATLAS plot contains limits coming from different analyses investigating different models, where the $\tilde{g}$ can either have no choice but to decay into the LSP, or to first decay into other SUSY particles that will further decay into the LSP ($\tilde{\chi}_1^0$ or $\tilde{G}$). On the contrary, the CMS plots show limits set by analyses constraining the model but through different final states. In this way, the CMS plots also exemplify the model dependence on the limit setting.

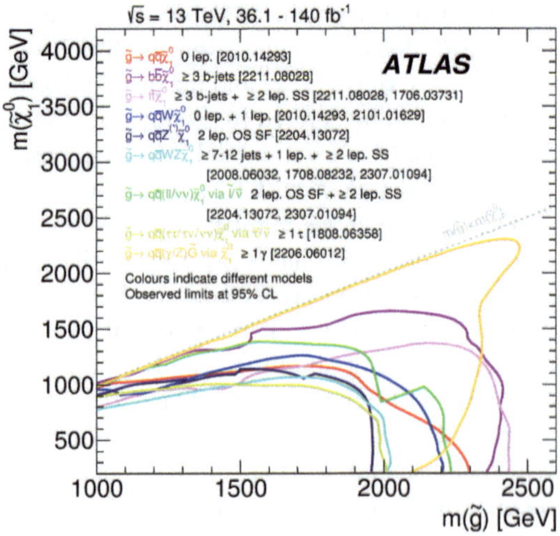

Fig. 2.13 Summary of ATLAS limits, after LHC Run 2, on the $\tilde{g}$ and the $\tilde{\chi}_1^0$ masses with $\tilde{g}$ decaying directly or indirectly, through a decay chain involving other SUSY particles, into the LSP ($\tilde{\chi}_1^0$ or $\tilde{G}$). Only the direct production of $\tilde{g}\tilde{g}$ is considered. Taken from [34]

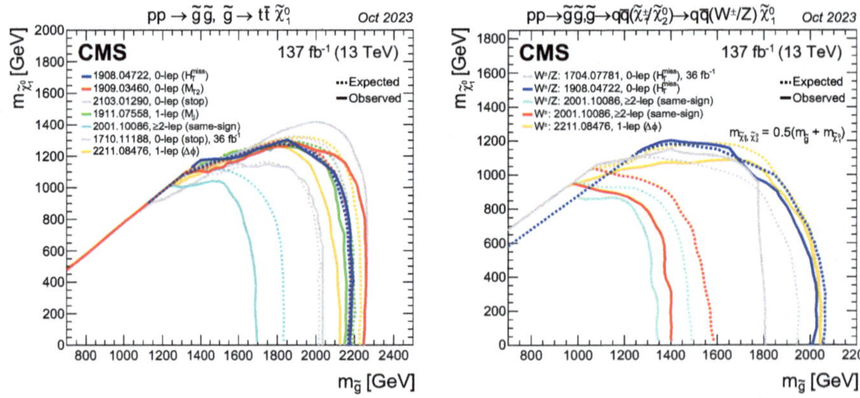

Fig. 2.14 Summary of CMS limits, after LHC Run 2, on the $\tilde{g}$ and the $\tilde{\chi}_1^0$ masses, with $\tilde{g}$ decaying directly (left) or indirectly, through the intermediate production of $\tilde{\chi}_1^{\pm}$ or $\tilde{\chi}_2^0$ (right), into the $\tilde{\chi}_1^0$. Only the direct production of $\tilde{g}\tilde{g}$ is considered. Taken from [35], where exclusion limits assuming other gluino decays are also available

Indeed, the analysis searching for 0 leptons with an emphasis on H_T^{miss}[6] [39] (dark blue lines), the one searching for 2 or more same-sign leptons [40] (light blue lines), and the one searching for 1 lepton with an emphasis on $\Delta\phi$[7] [41] (gold lines) are

[6] H_T^{miss} is the magnitude of the negative of the sum of the transverse momentum vectors of the reconstructed jets: $H_T^{\text{miss}} = \left\| -\sum_i^{jets} \vec{p}_T^{\,i} \right\|$.

[7] $\Delta\phi$ is the absolute value of the azimuthal angle between the transverse momentum vector of a hypothetically W boson decaying leptonically and the transverse momentum of that lepton: $\Delta\phi = \sphericalangle(\vec{p}_T^{\,l}, \vec{p}_T^{\,W} = \vec{p}_T^{\,l} + \vec{p}_T^{\,\text{miss}})$.

all constraining both models, but the model producing intermediate EW-inos is less constrained, up to ~ 350 GeV, than the other one.

The limits also depend on the mass difference between the mother and the daughter particles. On the one hand, for $m_{\tilde{\chi}_1^0} \approx 0$ GeV, gluinos are excluded up to ~ 2.45 TeV and ~ 2.25 TeV by ATLAS and CMS, respectively. On the other hand, more compressed mass spectra (i.e. $m_{\tilde{g}} \approx m_{\tilde{\chi}_1^0}$) will result in the production of softer SM particles, which are harder to detect and to separate from SM events, thus decreasing the search sensitivity and lowering the constraint on the models. It is the case for most of the SUSY searches and is not specific to gluinos. Apart from models where the $\tilde{\chi}_1^0$ is not the LSP, no model is excluded by either ATLAS or CMS if $m_{\tilde{\chi}_1^0} \gtrsim 1.7$ TeV ($m_{\tilde{\chi}_1^0} \gtrsim 1.4$ TeV when considering indirect gluino decays only).

The limit set by the ATLAS search for 1 or more γ (orange line in Fig. 2.13) does not decrease when $m_{\tilde{g}} - m_{\tilde{\chi}_1^0}$ diminishes because the $\tilde{\chi}_1^0$ is here an intermediate particle (decaying into a nearly massless $\tilde{G}$). The decay products of the $\tilde{\chi}_1^0$ can be highly energetic, even when $m_{\tilde{g}} \approx m_{\tilde{\chi}_1^0}$.

2.3.2 *Squark Production*

At the end of Run 2, the second most important constraints are set on squark production. As shown in Fig. 2.15, squark masses up to ~ 1.8 TeV are excluded when assuming an eightfold squark mass degeneracy (i.e. $\tilde{q} = \{\tilde{u}, \tilde{d}, \tilde{s}, \tilde{c}\}$) as well

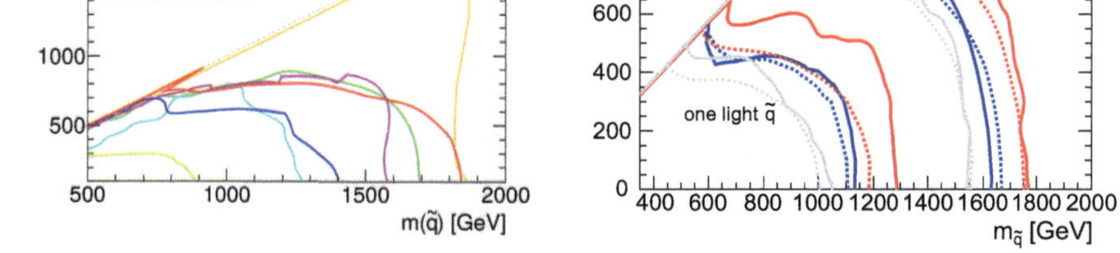

Fig. 2.15 Summary of ATLAS (left) [34] and CMS (right) [35] limits on direct $\tilde{q}\tilde{q}^*$ production, in the $m_{\tilde{\chi}_1^0}$ vs. $m_{\tilde{q}}$ plane, after LHC Run 2. An eightfold squark mass degeneracy ($\tilde{q} = \{\tilde{u}, \tilde{d}, \tilde{s}, \tilde{c}\}$) is assumed in both plots. Additionally, the CMS plot shows the results should only one of these eight states (one flavor and one chirality) be accessible at the LHC

as a direct decay to a nearly massless $\tilde{\chi}_1^0$. This limit is $\sim$ 150 GeV lower when considering an indirect squark decay into the $\tilde{\chi}_1^0$.

Once again, when $m_{\tilde{q}} - m_{\tilde{\chi}_1^0}$ is low, the sensitivity and exclusion limits are much lower. For a mass splitting of a few GeV, the $\tilde{q}$ and the $\tilde{\chi}_1^0$ are both excluded up to $\sim$ 900 GeV by ATLAS and up to $\sim$ 1 TeV by CMS. However, every model assuming an eightfold squark mass degeneracy, as well as a $\tilde{\chi}_1^0$ LSP, is still allowed if the $\tilde{\chi}_1^0$ is above that value, regardless of the $\tilde{q}$ mass. This limit is reduced when considering only one $\tilde{q}$ chirality. In that case, the limits are approximately 450 GeV to 650 GeV weaker.

Regarding third-generation squarks ($\tilde{b}$ and $\tilde{t}$), dedicated analyses were performed. Both ATLAS and CMS exclude a $\tilde{b}_1$ mass up to around 1.27 GeV for models in which $\tilde{b}_1 \rightarrow b\tilde{\chi}_1^0$ and $m_{\tilde{\chi}_1^0} \approx 0$ GeV. For compressed spectra, the exclusion limits are reduced to $\sim$ 700 GeV for both $m_{\tilde{b}_1}$ and $m_{\tilde{\chi}_1^0}$. All the models featuring higher masses are then still allowed. The limits are reduced when assuming $\tilde{b}_1$ to indirectly decay into the LSP.

The $\tilde{t}_1$ pair production has been extensively targeted by SUSY searches during the first two LHC runs. As discussed in Sect. 2.2.1, even when assuming the $\tilde{t}_1$ and the $\tilde{\chi}_1^0$ to be the only SUSY particles, the final states can contain $\not{E}_T$, jets, and 0, 1 or 2 leptons.

ATLAS limits for $\tilde{t}_1$ pair production are shown in Fig. 2.16. The kinematic regions mentioned in Sect. 2.2.1 are also indicated with dashed grey lines. The inset shows a particularly challenging phase-space region where $\Delta m(\tilde{t}_1, \tilde{\chi}_1^0) =$

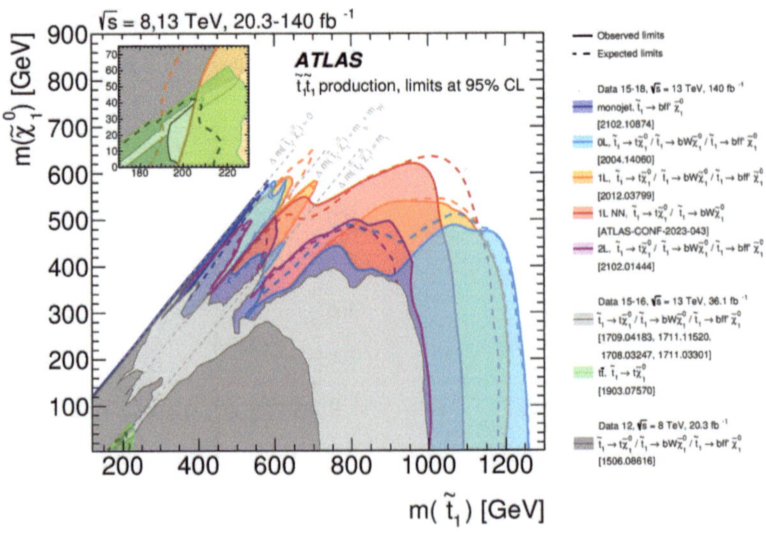

Fig. 2.16 Summary of ATLAS limits on direct $\tilde{t}_1\tilde{t}_1^*$ production, in the $m_{\tilde{\chi}_1^0}$ vs. $m_{\tilde{t}_1}$ plane, after LHC Run 2. Only direct decays to the $\tilde{\chi}_1^0$ are considered. Taken from [34]

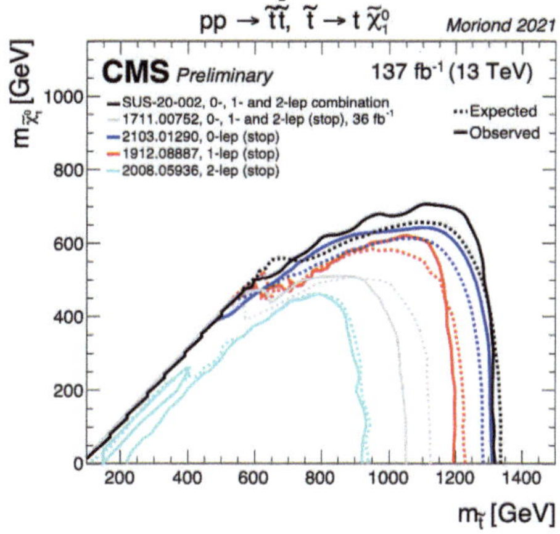

Fig. 2.17 Summary of CMS limits on direct $\tilde{t}_1\tilde{t}_1^*$ production, in the $m_{\tilde{\chi}_1^0}$ vs. $m_{\tilde{t}_1}$ plane, after LHC Run 2. Only $\tilde{t}_1 \to t\tilde{\chi}_1^0$ is considered. The black lines correspond to the statistical combination of the three analyses using the full luminosity of LHC Run 2 (blue, red and teal lines). Taken from [35]

$m_{\tilde{t}_1} - m_{\tilde{\chi}_1^0} \approx m_t$ and $\tilde{t}_1\tilde{t}_1^* \to t\bar{t}\tilde{\chi}_1^0\tilde{\chi}_1^0$, leading to a final state and kinematics very difficult to disentangle from the $t\bar{t}$ production of the SM, especially if $m_{\tilde{\chi}_1^0} \approx 0$ GeV. For $m_{\tilde{\chi}_1^0} \approx 0$ GeV, $\tilde{t}_1$ are excluded up to approximately 1.25 TeV, while for compressed mass spectra, masses up to $(m_{\tilde{t}_1}, m_{\tilde{\chi}_1^0}) \approx$ (700 GeV, 600 GeV) are excluded. No model featuring a $\tilde{\chi}_1^0$ above ~ 620 GeV is excluded.

The CMS summary plot of constraints on models where $\tilde{t}_1 \to t\tilde{\chi}_1^0$ is shown in Fig. 2.17. Similarly to ATLAS, the 0 lepton channel is the most constraining one for low $m_{\tilde{\chi}_1^0}$, excluding $m_{\tilde{t}_1}$ up to around 1.3 TeV. The combination of the 0, 1 and 2 leptons searches almost does not extend the exclusion limit for $m_{\tilde{\chi}_1^0} \leq 400$ GeV but extends it of ~ 100 GeV for more compressed mass spectra. For $m_{\tilde{\chi}_1^0} \gtrsim 700$ GeV, no model is excluded.

2.3.3 Slepton Production

Since they do not have a color charge, sleptons can only be produced by the electroweak force, thus reducing greatly their production cross section with respect to colored particles, as shown in Fig. 2.3. Their masses are therefore less constrained.

In addition to the missing transverse momentum, when directly decaying into the lightest neutralino, i.e. when $\tilde{\ell} \to \ell\tilde{\chi}_1^0$, the electrically charged sleptons of the first and second generations ($\tilde{e}$, $\tilde{\mu}$) are usually searched via final states containing two opposite-sign (OS) same-flavor (SF) leptons, while the $\tilde{\tau}$ is mainly searched for via final states containing hadronically decaying τ.

The summary of ATLAS exclusion limits on sleptons after Run 2 is shown in Fig. 2.18. The analyses searching for events with 2 ℓ and $\not{E}_T$, i.e. the orange, yellow and pink areas, constrain the pair production of $\tilde{e}$ and $\tilde{\mu}$. They exclude $m_{\tilde{\ell}}$ up to

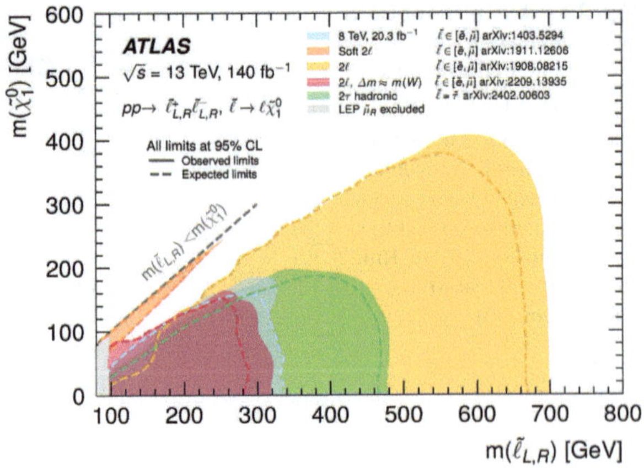

Fig. 2.18 Summary of ATLAS limits on direct slepton pair production, in the $m_{\tilde{\chi}^0_1}$ vs. $m_{\tilde{\ell}_{L,R}}$ plane, after LHC Run 2. The $\tilde{\ell}_L$ and $\tilde{\ell}_R$ are assumed to be mass degenerate. Taken from [34]

$\sim$ 700 GeV for 300 GeV $\gtrsim m_{\tilde{\chi}^0_1} \geq 0$ GeV when assuming the direct production of mass-degenerate $\tilde{\ell}_L$ and $\tilde{\ell}_R$. In the compressed mass region ($\Delta m(\tilde{\ell}, \tilde{\chi}^0_1) \sim 10$ GeV), they are excluded up to $\sim$ 250 GeV. The limits on the direct production of $\tilde{\tau}\tilde{\tau}$ are shown by the green area. Stau masses up to $\sim$ 480 GeV are excluded for 120 GeV $\gtrsim m_{\tilde{\chi}^0_1} \geq 0$ GeV. These limits reduce to $\sim$ 425 GeV ($\sim$ 350 GeV) when considering the production of a purely $\tilde{\tau}_L$ ($\tilde{\tau}_R$), due to a lower cross section.

Similarly to ATLAS, CMS excludes sleptons up to around 700 GeV for 300 GeV $\gtrsim m_{\tilde{\chi}^0_1} \geq 0$ GeV, and does not exclude any model for $m_{\tilde{\chi}^0_1} \geq$ 380 GeV [42]. In the compressed region, the limit goes up to around 215 GeV for $\Delta m(\tilde{\ell}, \tilde{\chi}^0_1) \lesssim 5$ GeV but quickly decreases for higher mass splittings. The CMS constraints on the pair production of $\tilde{\tau}$ are slightly below ATLAS ones [43].

2.3.4 *Electroweak-ino Production*

Regarding EW-inos, as described in Sect. 2.2.3, the final states can be fully hadronic, semileptonic, or fully leptonic, with jets and leptons possibly soft or hard, on top of having $\not{E}_T$.

The summary of ATLAS constraints on the production of the pure wino pair $\tilde{\chi}^0_2\tilde{\chi}^{\pm}_1$, with $\tilde{\chi}^0_2 \to Z\tilde{\chi}^0_1$ and $\tilde{\chi}^{\pm}_1 \to W^{\pm}\tilde{\chi}^0_1$, is shown in Fig. 2.19. Five statistically independent searches targeting different final states are combined, resulting in the exclusion of the area below the full red line. For 200 GeV $\gtrsim m_{\tilde{\chi}^0_1} \geq 0$ GeV, the $\tilde{\chi}^0_2$ and the $\tilde{\chi}^{\pm}_1$ are excluded up to $\sim$ 1 TeV, while for $m_{\tilde{\chi}^0_1} \gtrsim$ 400 GeV, no model is excluded. Similar results are obtained when considering the emission of a SM-like Higgs boson instead of a Z boson by the $\tilde{\chi}^0_2$. Regarding the compressed region, delimited by the two dashed grey lines, where $\Delta m(\tilde{\chi}^0_2, \tilde{\chi}^0_1) < m_Z$, the $\tilde{\chi}^0_2$ and the $\tilde{\chi}^{\pm}_1$ are excluded up to $\sim$ 290 GeV for $m_{\tilde{\chi}^0_1} \approx$ 260 GeV. The phenomenological

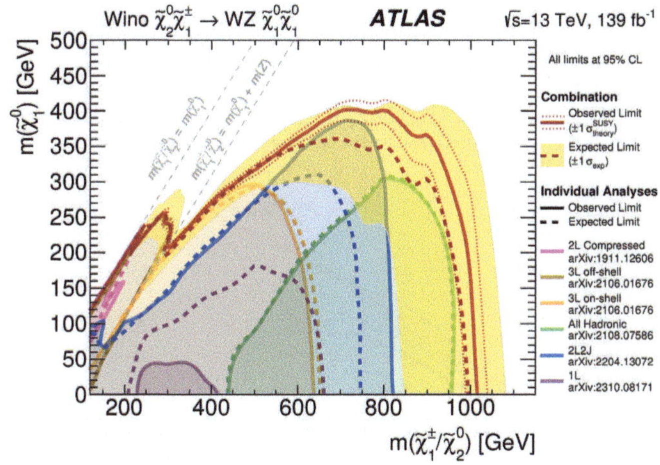

Fig. 2.19 Summary of ATLAS constraints on the direct associated production of purely wino $\tilde{\chi}_2^0$ and $\tilde{\chi}_1^\pm$, assuming $\tilde{\chi}_2^0 \to Z\tilde{\chi}_1^0$ and $\tilde{\chi}_1^\pm \to W^\pm\tilde{\chi}_1^0$, where the $\tilde{\chi}_1^0$ is a pure bino, in the $m_{\tilde{\chi}_1^0}$ vs. $m_{\tilde{\chi}_1^\pm/\tilde{\chi}_2^0}$ plane, after LHC Run 2. The combination includes the six individual analyses mentioned in the legend. Taken from [44]

study in Chap. 4 will make use of the results of all these searches, except for the compressed 2 ℓ search and the 1 ℓ search (the pink and violet lines, respectively), which could not be implemented in SModelS at the time of the study.

The combined CMS constraint is similar to the ATLAS one, while a bit lower for light $\tilde{\chi}_1^0$. It is shown in teal in Fig. 2.20. The figure also shows the limits when $\tilde{\chi}_2^0 \to h\tilde{\chi}_1^0$ (blue lines). In this case, the exclusion limit extends up to around 1 TeV for $m_{\tilde{\chi}_1^0} \approx 0$ GeV, but is weaker for higher $m_{\tilde{\chi}_1^0}$. The pink and red lines correspond to indirect wino decays into the $\tilde{\chi}_1^0$ via on-shell sleptons. In these models, the final states consist in $\not{E}_T$ and either 2 or 3 ℓ, for $\tilde{\chi}_1^\pm\tilde{\chi}_1^\mp$ or $\tilde{\chi}_2^0\tilde{\chi}_1^\pm$ productions, respectively. Here, models featuring a $\tilde{\tau}$ at a mass in between the LSP and the NLSP are excluded up to around 1 TeV for 300 GeV $\gtrsim m_{\tilde{\chi}_1^0} \geq 0$ GeV but are not constrained for $m_{\tilde{\chi}_1^0} \gtrsim 450$ GeV. Models featuring left-handed sleptons close to the LSP mass are excluded up to ~ 1.3 TeV for a nearly massless $\tilde{\chi}_1^0$ and up to ~ 1.35 TeV for $m_{\tilde{\chi}_1^0} \approx 800$ GeV.

ATLAS and CMS constraints on models where the $\tilde{\chi}_1^0$ is the NLSP and the $\tilde{G}$ is the LSP are shown in Fig. 2.21 in the BR($\tilde{\chi}_1^0 \to h\tilde{G}$) vs. pure higgsino mass plane. Both ATLAS and CMS performed a combination of analyses and they exclude pure higgsinos up to ~ 960 GeV and ~ 840 GeV for a low BR($\tilde{\chi}_1^0 \to h\tilde{G}$), respectively. For BR($\tilde{\chi}_1^0 \to h\tilde{G}$) ≈ 1, ATLAS and CMS exclusion limits go up to around 940 and 1000 GeV, respectively. However, both decrease for intermediate branching ratios, with a minimum at ~ 850 GeV for BR($\tilde{\chi}_1^0 \to h\tilde{G}$) ≈ 0.58 for ATLAS, and a minimum at ~ 750 GeV for BR($\tilde{\chi}_1^0 \to h\tilde{G}$) ≈ 0.40 for CMS.

Probably the most interesting results appear in the compressed region, where small excesses were recorded in different channels, resulting in around 2 σ deviation

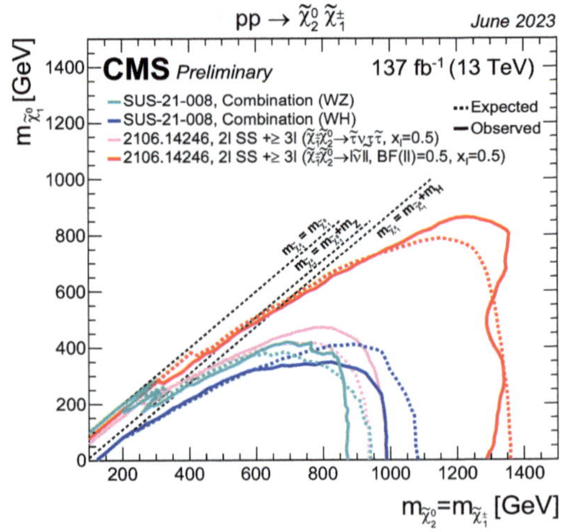

Fig. 2.20 Summary of CMS constraints on the direct associated production of purely wino $\tilde{\chi}_2^0$ and $\tilde{\chi}_1^\pm$, in the $m_{\tilde{\chi}_1^0}$ vs. $m_{\tilde{\chi}_1^\pm/\tilde{\chi}_2^0}$ plane, after LHC Run 2. Winos direct decays into the pure bino $\tilde{\chi}_1^0$ are assumed to happen via the emission of a W and a Z (teal lines) or via a W and a h (blue lines). Winos indirect decays are assumed to happen via the $\tilde{\tau}$, with $m_{\tilde{\tau}} = 0.5(m_{\tilde{\chi}_2^0} + m_{\tilde{\chi}_1^0})$, (pink lines), or via a flavor-democratic $\tilde{\ell}_\mathrm{L}$, with $m_{\tilde{\ell}_\mathrm{L}} = 0.05 m_{\tilde{\chi}_2^0} + 0.95 m_{\tilde{\chi}_1^0}$, (red lines). Taken from [35]

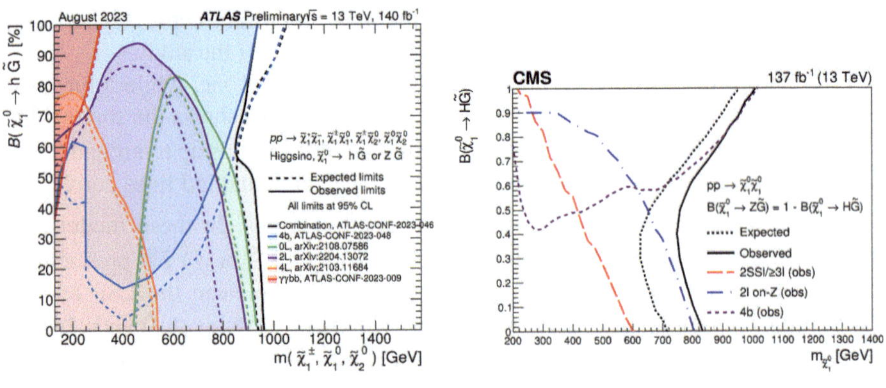

Fig. 2.21 Summary of ATLAS (left) [45] and CMS (right) [42] constraints on the BR($\tilde{\chi}_1^0 \to h\tilde{G}$) vs. pure higgsino mass plane, after LHC Run 2. Both collaborations perform a combination (black lines) of the analyses mentioned in the legend

from the SM when combined, both for ATLAS and CMS. The ATLAS combination of a 2 ℓ [46] and a 3 ℓ [47] search shows a small excess at $(m_{\tilde{\chi}_1^\pm}, \Delta m(\tilde{\chi}_1^\pm, \tilde{\chi}_1^0)) \approx$ (150 GeV, 10 GeV for models producing purely higgsino $\tilde{\chi}_2^0$, $\tilde{\chi}_1^\pm$ and $\tilde{\chi}_1^0$, as shown in the top panel of Fig. 2.22. Assuming a wino-bino scenario, the same excess shows up at $(m_{\tilde{\chi}_1^\pm}, \Delta m(\tilde{\chi}_1^\pm, \tilde{\chi}_1^0)) \approx$ (180 GeV, 20 GeV). The bottom panel shows the limits obtained from an analysis searching for a low-p_T, isolated, good-quality

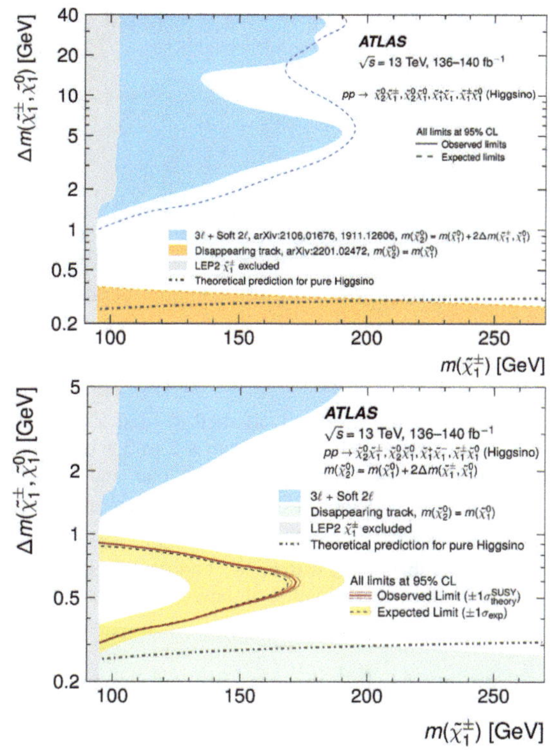

Fig. 2.22 Summary of ATLAS constraints, in the $\Delta m(\tilde{\chi}_1^\pm, \tilde{\chi}_1^0)$ vs. $m_{\tilde{\chi}_1^\pm}$ plane, on the production of pure higgsinos with off-shell decays of $\tilde{\chi}_2^0$ and $\tilde{\chi}_1^\pm$ into the LSP, after LHC Run 2. The dot-dashed grey line indicates the theoretically predicted mass-splitting in the pure higgsino scenario [48]. The full red and dashed black lines in the bottom panel are obtained by selecting final states with an energetic jet, missing transverse momentum, and a low-momentum track. Taken from [34] (top) and [49] (bottom)

track [49]. Mass splittings between 0.3 and 0.9 GeV are excluded for $m_{\tilde{\chi}_1^\pm}$ up to 170 GeV. Smaller mass splittings are excluded up to $m_{\tilde{\chi}_1^0} \approx 700$ GeV by a disappearing track search [50].

Analogous CMS results are shown in Fig. 2.23. The left panel shows the limits from a search for soft leptons assuming a higgsino-LSP scenario. A small excess shows up for $m_{\tilde{\chi}_2^0} \lesssim 180$ GeV and $\Delta m(\tilde{\chi}_2^0, \tilde{\chi}_1^0) \gtrsim 10$ GeV. It overlaps with the ATLAS small excess described above. The right panel shows the limits when the data is interpreted assuming a wino-bino scenario and is combined with a search for final states with 2, 3 or 4 ℓ and up to 2 hadronically decaying τ. A small excess shows up for $(m_{\tilde{\chi}_2^0}, \Delta m(\tilde{\chi}_2^0, \tilde{\chi}_1^0)) \approx (230$ GeV , 30 GeV). It also overlaps with the ATLAS small excess when interpreted in a wino-bino scenario, although the remaining place for the excess to live is reduced. The smallest mass splittings are also excluded by a search for disappearing tracks [53], up to $m_{\tilde{\chi}_1^\pm} \approx 950$ GeV when considering the higgsino-LSP scenario.

2.3.5 *Going Beyond Simplified Models*

So far, the presented results were derived assuming simplified models. Despite being useful to constrain phase-space regions with little theory assumption, their

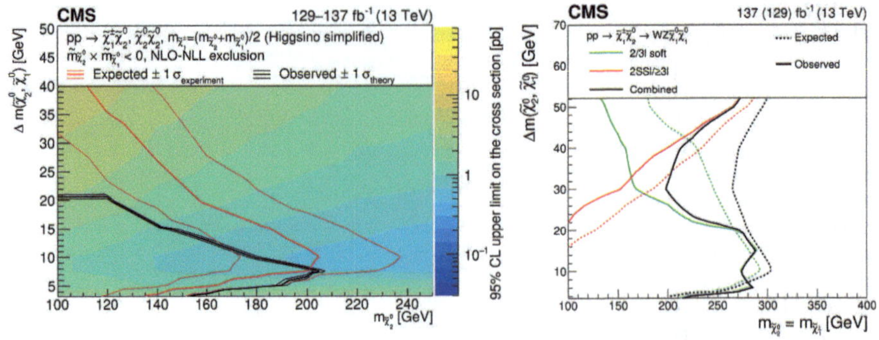

Fig. 2.23 CMS constraints, in the $\Delta m(\tilde{\chi}_2^0, \tilde{\chi}_1^0)$ vs. $m_{\tilde{\chi}_2^0}$ plane, on the production of EW-inos decaying into the LSP and off-shell W and Z bosons, using between 129 and 137 fb^{-1} of LHC Run 2 data. Left: results from a 2 to 3 e or μ plus $\not{E}_T$ search [51], interpreted assuming the production of a pure wino $\tilde{\chi}_2^0\tilde{\chi}_1^{\pm}$ pair and of a pure higgsino $\tilde{\chi}_2^0\tilde{\chi}_1^0$ pair, with $m_{\tilde{\chi}_1^{\pm}} = 0.5(m_{\tilde{\chi}_2^0} + m_{\tilde{\chi}_1^0})$. Right: results from the same analysis but interpreted assuming the production of a pure wino $\tilde{\chi}_2^0\tilde{\chi}_1^{\pm}$ pair only (green lines). Are also shown the results from an analysis looking for events with 3 or 4 ℓ, with up to 2 hadronically decaying τ, or 2 same-sign light ℓ (red lines) [52] and the resulting combination of these two analyses (black lines). Taken from [42]

constraints on full theories are not trivial. This is mainly because full theories comprise more SUSY particles and no pure state, which can reduce or enhance the production cross sections, and change the branching ratios. It is therefore necessary to quantify how well full theories are constrained by these analyses.

In particular, a recent ATLAS study used 8 ATLAS analyses, all performed with around 140 fb^{-1} of LHC Run 2 data, to constrain more realistic EW-ino models based on the RPC pMSSM with the $\tilde{\chi}_1^0$ as the LSP [54]. Gluinos, squarks and sleptons were considered decoupled. More precisely, 2×10^4 model points were produced by randomly sampling (with a flat prior) over the 19 parameters of the pMSSM. The M_1, M_2 and μ parameters were sampled between -2 and 2 TeV, while $\tan\beta$ was sampled between 1 and 60. LEP constraints on the chargino mass and a constraint on the SM Higgs boson's mass were then applied, as well as related constraints from previous electroweak, flavor and dark matter measurements. In addition to these "external" constraints, models featuring long-lived $\tilde{\chi}_1^{\pm}$ and $\tilde{\chi}_2^0$ (on the detector scale), or to which no analysis was sensitive, were removed, leaving a total of 2460 model points. Almost all models featuring a bino-like LSP produced too much DM relic density to fulfil experimental observations. Only remained the ones where the LSP had a higgsino component and a $\tilde{\chi}_1^0$ mass close to half of the Z or h masses (so that the $\tilde{\chi}_1^0$ self-annihilation is enhanced and the resulting relic density is at most equal to the observed value; this is called the "Z/h funnel").

Then, each model point was confronted to the 8 analyses independently (no combination was performed) and was excluded if at least one of them gave a CL_s below 0.05. The results are shown in Fig. 2.24. The left panel shows the fraction of model points, after applying the LEP and Higgs boson's mass constraints only, that are excluded by at least 1 of the 8 analyses. It also shows the simplified model exclusion envelope of 3 of the most sensitive analyses. It can be seen that the limits

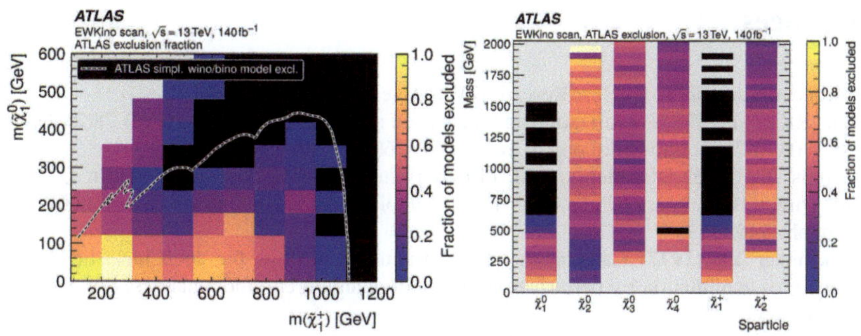

Fig. 2.24 Fraction of pMSSM models excluded by ATLAS. Left: for models satisfying the LEP and the Higgs boson's mass constraints, without further requirement, in the $m_{\tilde{\chi}_1^0}$ vs. $m_{\tilde{\chi}_1^\pm}$ plane. The dashed line represents the envelope of the exclusion limits set by the 3ℓ [47], $2\ell 2j$ [55] and fully hadronic [28] searches when assuming the direct production of purely wino $\tilde{\chi}_1^\pm \tilde{\chi}_2^0$, with $\text{BR}(\tilde{\chi}_1^\pm \to W^\pm \tilde{\chi}_1^0) = 1$ and $\text{B}(\tilde{\chi}_2^0 \to Z\tilde{\chi}1^0) = 1$. Right: for models that additionally satisfy electroweak, flavor and DM constraints, across the masses of each EW-ino. Taken from [54]

derived assuming simplified models are usually way stronger than the limits derived using more realistic models. Indeed, a large number of bins under the curve exclude less than 50% of the model points. If the electroweak, flavor and DM constraints are further applied (not shown here), almost all the model points with $m_{\tilde{\chi}_1^+} < 1.2$ TeV and $m_{\tilde{\chi}_1^0} < 600$ GeV are excluded, except those featuring a small $\Delta m(\tilde{\chi}_1^\pm, \tilde{\chi}_1^0)$.

The right panel shows the fraction of model points, among the one which passed the "external" constraints, that are excluded by at least 1 of the 8 analyses. It can be seen that for those remaining points, almost none of the EW-ino masses are completely excluded by the ATLAS analyses, except for very light $\tilde{\chi}_1^0$ or heavy $\tilde{\chi}_2^0$. The exclusion of the heaviest masses and of the heaviest EW-inos ($\tilde{\chi}_4^0$, $\tilde{\chi}_2^\pm$ and $\tilde{\chi}_3^0$) is mostly based on the exclusion of the lightest EW-inos of their models. This, combined with the fact that the pair production of bino-like $\tilde{\chi}_1^0$ is suppressed, and that no model point featuring a $m_{\tilde{\chi}_1^\pm} \gtrsim 600$ GeV is excluded by ATLAS while evading all the "external" constraints, explains why the $\tilde{\chi}_1^0$ and $\tilde{\chi}_1^\pm$ masses are not excluded above ~ 600 GeV but the masses of the other EW-inos are.

This exemplifies how limits derived using simplified models are overly constraining in comparison to the true constraints on full models (at least in the absence of any combination of analysis). Consequently, it is crucial to reinterpret experimental results in order to accurately evaluate how full models are constrained. Many reinterpretation software are designed to accomplish this, however, they all depend on the reinterpretation material published by the experimental collaborations. A detailed list of what material to publish and recommendations on how to do it can be found in the LHC Reinterpretation Forum report after LHC Run 2 [16]. Further information about the publication of full statistical models can also be found in [56]. Unfortunately, despite the efforts of ATLAS and CMS, the publication and the preservation of reinterpretation materials remains more exceptional than systematic.

References

1. European Council for Nuclear Research Web Page. https://home.cern/
2. Ellis RK et al (1978) Factorization and the parton model in QCD. Phys Lett B 78(2):281–284. ISSN: 0370-2693. https://doi.org/10.1016/0370-2693(78)90023-0
3. Collins JC, Soper DE, Sterman GF (1989) Factorization of hard processes in QCD. Adv Ser Direct High Energy Phys 5:1–91. https://doi.org/10.1142/9789814503266.0001. arXiv: hep-ph/0409313
4. Dokshitzer YL (1977) Calculation of the structure functions for deep inelastic scattering and e+ e- annihilation by perturbation theory in quantum chromodynamics. Sov Phys JETP 46:641–653
5. Gribov VN, Lipatov LN (1972) Deep inelastic e p scattering in perturbation theory. Sov J Nucl Phys 15:438–450
6. Altarelli G, Parisi G (1977) Asymptotic freedom in parton language. Nucl Phys B 126:298–318. https://doi.org/10.1016/0550-3213(77)90384-4
7. Wikipedia Contributors (2024) Large hadron collider — wikipedia, the free encyclopedia. https://en.wikipedia.org/wiki/Large_Hadron_Collider. Accessed 10 Aug 2024
8. Lippmann C (2012) Particle identification. Nucl Instrum Meth A 666:148–172. https://doi.org/10.1016/j.nima.2011.03.009. arXiv:1101.3276 [hep-ex]
9. Detector cross section of The Particle Adventure web page. https://www.particleadventure.org/end_view.html. Courtesy of Michael Barnett and the Particle Data Group
10. Lester CG, Summers DJ (1999) Measuring masses of semiinvisibly decaying particles pair produced at hadron colliders. Phys Lett B 463:99–103. https://doi.org/10.1016/S0370-2693(99)00945-4. arXiv:hep-ph/9906349
11. Barr A, Lester C, Stephens P (2003) A variable for measuring masses at hadron colliders when missing energy is expected; m_{T2}: the truth behind the glamour. J Phys G 29:2343–2363. https://doi.org/10.1088/0954-3899/29/10/304. arXiv:hep-ph/0304226
12. Arkani-Hamed N et al (2007) MARMOSET: the path from LHC data to the new standard model via on-shell effective theories. arXiv:hep-ph/0703088 [HEP-PH]
13. Alwall J, Schuster P, Toro N (2009) Simplified models for a first characterization of new physics at the LHC. Phys Rev D79:075020. https://doi.org/10.1103/PhysRevD.79.075020. arXiv:0810.3921 [hep-ph]
14. Alves D (2012) Simplified models for LHC new physics searches. J Phys G39:105005. https://doi.org/10.1088/0954-3899/39/10/105005. arXiv:1105.2838 [hep-ph]
15. Okawa H (2011) Interpretations of SUSY searches in ATLAS with simplified models. In: Particles and fields. Proceedings, meeting of the division of the american physical society, DPF 2011, Providence, August 9–13, 2011. arXiv:1110.0282 [hep-ex]
16. Abdallah W et al (2020) Reinterpretation of LHC results for new physics: status and recommendations after run 2. SciPost Phys 9(2):022. https://doi.org/10.21468/SciPostPhys.9.2.022. arXiv:2003.07868 [hep-ph]
17. Bailey S et al (2022) Data and analysis preservation, recasting, and reinterpretation. arXiv:2203.10057 [hep-ph]
18. Borschensky C et al (2023) LHC SUSY cross section working group. https://twiki.cern.ch/twiki/bin/view/LHCPhysics/SUSYCrossSections. Accessed 13 Dec 2023
19. Martin SP (1998) A supersymmetry primer. Adv Ser Direct High Energy Phys 18:1–98. https://doi.org/10.1142/97898128396570001. arXiv:hep-ph/9709356
20. Farrar GR, Fayet P (1978) Phenomenology of the production, decay, and detection of new hadronic states associated with supersymmetry. Phys Lett B 76:575–579. https://doi.org/10.1016/0370-2693(78)90858-4
21. Ellis JR, Kelley S, Nanopoulos DV (1991) Probing the desert using gauge coupling unification. Phys Lett B 260:131–137. https://doi.org/10.1016/0370-2693(91)90980-5

22. Amaldi U, de Boer W, Furstenau H (1991) Comparison of grand unified theories with electroweak and strong coupling constants measured at LEP. Phys Lett B 260:447–455. https://doi.org/10.1016/0370-2693(91)91641-8
23. Langacker P, Luo M-X (1991) Implications of precision electroweak experiments for M_t, ρ_0, $\sin^2\theta_W$ and grand unification. Phys Rev D 44:817–822. https://doi.org/10.1103/PhysRevD.44.817
24. Kim JE, Nilles HP (1984) The mu problem and the strong CP problem. Phys Lett B 138:150–154. https://doi.org/10.1016/0370-2693(84)91890-2
25. Giudice GF, Masiero A (1988) A natural solution to the mu problem in supergravity theories. Phys Lett B 206:480–484. https://doi.org/10.1016/0370-2693(88)91613-9
26. Jungman G, Kamionkowski M, Griest K (1996) Supersymmetric dark matter. Phys Rept 267:195–373. https://doi.org/10.1016/0370-1573(95)00058-5. arXiv:hep-ph/9506380
27. Altakach MM et al (2024) Global LHC constraints on electroweak-inos with SModelS v2.3. SciPost Phys 16:101. https://doi.org/10.21468/SciPostPhys.16.4.101. arXiv:2312.16635 [hep-ph]
28. Aad G et al (2021) Search for charginos and neutralinos in final states with two boosted hadronically decaying bosons and missing transverse momentum in pp collisions at $\sqrt{s} = 13$ TeV with the ATLAS detector. Phys Rev D 104(11):112010. https://doi.org/10.1103/PhysRevD.104.112010. arXiv:2108.07586 [hep-ex]
29. Tumasyan A et al (2023) Search for electroweak production of charginos and neutralinos at s =13TeV in final states containing hadronic decays of WW, WZ, or WH and missing transverse momentum. Phys Lett B 842:137460. https://doi.org/10.1016/j.physletb.2022.137460. arXiv:2205.09597 [hep-ex]
30. Haber HE, Kane GL (1985) The search for supersymmetry: probing physics beyond the standard model. Phys Rept 117:75–263. https://doi.org/10.1016/0370-1573(85)90051-1
31. Martin SP (2013) TASI 2011 lectures notes: two-component fermion notation and supersymmetry. In: Theoretical advanced study institute in elementary particle physics: the dark secrets of the terascale, pp 199–258. https://doi.org/10.1142/97898143901630005. arXiv:1205.4076 [hep-ph]
32. Read AL (2002) Presentation of search results: the CL(s) technique. J Phys G28:2693–2704. https://doi.org/10.1088/0954-3899/28/10/313
33. Cowan G et al (2011) Asymptotic formulae for likelihood-based tests of new physics. Eur Phys J C71:1554. https://doi.org/10.1140/epjc/s10052-011-1554-0. arXiv:1007.1727 [physics.data-an]. [Erratum: Eur Phys J C73:2501 (2013). https://doi.org/10.1140/epjc/s10052-013-2501-z]
34. Aad G et al (2024) The quest to discover supersymmetry at the ATLAS experiment. arXiv:2403.02455 [hep-ex]
35. CMS Collaboration (2024) CMS SUSY results. https://twiki.cern.ch/twiki/bin/view/CMSPublic/PhysicsResultsSUS
36. Aad G et al (2015) Summary of the ATLAS experiment's sensitivity to supersymmetry after LHC Run 1 interpreted in the phenomenological MSSM. J High Energy Phys 10:134. https://doi.org/10.1007/JHEP10(2015)134. arXiv:1508.06608 [hep-ex]
37. Khachatryan V et al (2016) Phenomenological MSSM interpretation of CMS searches in pp collisions at $\sqrt{s} = 7$ and 8 TeV. J High Energy Phys 10:129. https://doi.org/10.1007/JHEP10(2016)129. arXiv:1606.03577 [hep-ex]
38. Workman RL et al (2022) Review of particle physics. Progress Theor Exp Phys 2022:083C01. https://doi.org/10.1093/ptep/ptac097
39. The CMS Collaboration et al (2019) Search for supersymmetry in proton-proton collisions at 13 TeV in final states with jets and missing transverse momentum. J High Energy Phys 10:244. https://doi.org/10.1007/JHEP10(2019)244. arXiv:1908.04722 [hep-ex].
40. Sirunyan AM et al (2020) Search for physics beyond the standard model in events with jets and two same-sign or at least three charged leptons in proton-proton collisions at $\sqrt{s} = 13$ TeV. Eur Phys J C 80(8):752. https://doi.org/10.1140/epjc/s10052-020-8168-3. arXiv:2001.10086 [hep-ex]

41. Tumasyan A et al (2023) Search for supersymmetry in final states with a single electron or muon using angular correlations and heavy-object identification in proton-proton collisions at $\sqrt{s}$ = 13 TeV. J High Energy Phys 09:149. https://doi.org/10.1007/JHEP09(2023)149. arXiv:2211.08476 [hep-ex]
42. Hayrapetyan A et al (2024) Combined search for electroweak production of winos, binos, higgsinos, and sleptons in proton-proton collisions at $\sqrt{s}$ = 13 TeV. arXiv:2402.01888 [hep-ex]
43. Tumasyan A et al (2023) Search for direct pair production of supersymmetric partners of τ leptons in the final state with two hadronically decaying τ leptons and missing transverse momentum in proton-proton collisions at $\sqrt{s}$ = 13 TeV. Phys Rev D 108(1):012011. https://doi.org/10.1103/PhysRevD.108.012011. arXiv:2207.02254 [hep-ex]
44. Aad G et al (2024) A statistical combination of ATLAS Run 2 searches for charginos and neutralinos at the LHC. arXiv:2402.08347 [hep-ex]
45. ATLAS Collaboration (2023) SUSY August 2023 summary plot update. https://atlas.web.cern.ch/Atlas/GROUPS/PHYSICS/PUBNOTES/ATL-PHYS-PUB-2023-025/
46. Aad G et al (2020) Searches for electroweak production of supersymmetric particles with compressed mass spectra in $\sqrt{s}$ = 13 TeV *pp* collisions with the ATLAS detector. Phys Rev D 101(5):052005. https://doi.org/10.1103/PhysRevD.101.052005. arXiv:1911.12606 [hep-ex]
47. Aad G et al (2011) Search for chargino-neutralino pair production in final states with three leptons and missing transverse momentum in $\sqrt{s}$ = 13 TeV *pp* collisions with the ATLAS detector. Eur Phys J C 81(12):1118. https://doi.org/10.1140/epjc/s10052-021-09749-7. arXiv:2106.01676 [hep-ex]
48. Thomas SD, Wells JD (1998) Phenomenology of massive vectorlike doublet leptons. Phys Rev Lett 81:34–37. https://doi.org/10.1103/PhysRevLett.81.34. arXiv:hep-ph/9804359
49. Aad G et al (2024) Search for nearly mass-degenerate higgsinos using low-momentum mildly-displaced tracks in *pp* collisions at $\sqrt{s}$ = 13 TeV with the ATLAS detector. arXiv:2401.14046 [hep-ex]
50. Aad G et al (2022) Search for long-lived charginos based on a disappearing-track signature using 136 fb^{-1} of *pp* collisions at $\sqrt{s}$ = 13 TeV with the ATLAS detector. Eur Phys J C 82(7):606. https://doi.org/10.1140/epjc/s10052-022-10489-5. arXiv:2201.02472 [hep-ex]
51. Tumasyan A et al (2022) Search for supersymmetry in final states with two or three soft leptons and missing transverse momentum in proton-proton collisions at $\sqrt{s}$ = 13 TeV. J High Energy Phys 04:091. https://doi.org/10.1007/JHEP04(2022)091. arXiv:2111.06296 [hep-ex]
52. Tumasyan A et al (2022) Search for electroweak production of charginos and neutralinos in proton-proton collisions at $\sqrt{s}$ = 13 TeV. J High Energy Phys 04:147. https://doi.org/10.1007/JHEP04(2022)147. arXiv:2106.14246 [hep-ex]
53. Hayrapetyan A et al (2024) Search for supersymmetry in final states with disappearing tracks in proton-proton collisions at s =13 TeV. Phys Rev D 109(7):072007. https://doi.org/10.1103/PhysRevD.109.072007. arXiv:2309.16823 [hep-ex]
54. Aad G et al (2024) ATLAS run 2 searches for electroweak production of supersymmetric particles interpreted within the pMSSM. J High Energy Phys 05:106. https://doi.org/10.1007/JHEP05(2024)106. arXiv:2402.01392 [hep-ex]
55. Aad G et al (2023) Searches for new phenomena in events with two leptons, jets, and missing transverse momentum in 139 fb^{-1} of $\sqrt{s}$ = 13 TeV *pp* collisions with the ATLAS detector. Eur Phys J C 83(6):515. https://doi.org/10.1140/epjc/s10052-023-11434-w. arXiv:2204.13072 [hep-ex]
56. Cranmer K et al (2022) Publishing statistical models: getting the most out of particle physics experiments. SciPost Phys 12(1):037. https://doi.org/10.21468/SciPostPhys.12.1.037. arXiv:2109.04981 [hep-ph]

Chapter 3
Reinterpreting LHC Results: The SModelS Approach

Abstract Because no evidence of new physics has been found in the slew of Large Hadron Collider (LHC) data, and limits on simplified models (typically used to interpret data from LHC searches for supersymmetry) do not translate to actual limits on full models, it is necessary to systematically quantify how full models are constrained by the data. This led to the development of many reinterpretation software. The SModelS software package is one of them and is the subject of this chapter. The latter gives a general description of SModelS v2.3 (used to obtain the main results of this thesis), its working principle, its database structure and its outputs. It also presents the statistical inference and the different types of likelihoods used by SModelS, and introduces the recent SModelS developments, with an emphasis on the author's contributions.

Because no evidence of new physics has been found in the slew of LHC data, and limits on simplified models do not translate to actual limits on full models, it is necessary to systematically quantify how full models are constrained by the data. This led to the development of many reinterpretation software. So far, two approaches have been considered. The first one goes in the direction of reproducing the experimental analysis through a Monte Carlo simulation, thus requiring a lightweight version of the data treatment performed in the experimental analysis (e.g. signal and background selections). These "recasting" tools are usually computationally heavy. The second one aims at reusing the simplified model results and requires some simplified form of analysis selection, such as the efficiency and acceptance maps (corresponding to the proportion of BSM events passing the experimental cuts and that are successfully reconstructed by the detector), assuming that they do not change too much compared to the original model. The second approach is usually agnostic to kinematic variables but is faster than the first one. It is thus well suited for scanning a large set of model points. The interested reader can find a review of reinterpretation methods and tools in [1, 2]. The SModelS software package [3–8] belongs to the second category and will be the subject of this chapter. The current version of SModelS and its database is v3.0 [9, 10]. However, the main results of this thesis (presented in Chap. 4) were obtained using v2.3. Although the

T. Pascal, *Beyond the Standard Model at the Large Hadron Collider: From Simplified Model Constraints to Artificial Proto-Modelling for Dispersed Signals*, Springer Theses, https://doi.org/10.1007/978-3-032-09391-2_3

project presented in Chap. 5 operates with v3.0, only minor features of v3.0 (on top of the new database entries) are used. Therefore, unless stated otherwise, the following description and discussion will be about v2.3.

This chapter is organised as follows. Section 3.1 is dedicated to the general description of SModelS v2.3, its working principle, its database structure and its outputs. Section 3.2 presents the statistical inference and the different types of likelihoods used by SModelS. Finally, Sect. 3.3 is about the recent SModelS developments, with an emphasis on my contributions.

SModelS is public software distributed under the GNU General Public License v3 (GPLv3). It is available on GitHub and in the Python Package Index (PyPI). The SModelS database is available in text form on GitHub and as a binary pickle file on Zenodo [11].

3.1 SModelS Working Principle

This section describes the working principle of SModelS in its version 2.3 and is inspired by the rich online documentation available at smodels.readthedocs.io. SModelS is a public tool for the fast reinterpretation of LHC searches on the basis of simplified model results. The general procedure is to decompose realistic models with a $\mathbb{Z}_2$-like symmetry (such as SUSY R-parity) into simplified models and confront them against the results implemented in the SModelS database. Figure 3.1 schematises this procedure. It assumes that the experimental results used to constrain simplified models are model-independent enough to be used to constrain other models with similar decay products and quantum numbers, independently of the kinematics.

Therefore, one of the main SModelS features is to test if realistic models with a $\mathbb{Z}_2$-like symmetry are excluded, with a 95% CL, by using experimental constraints

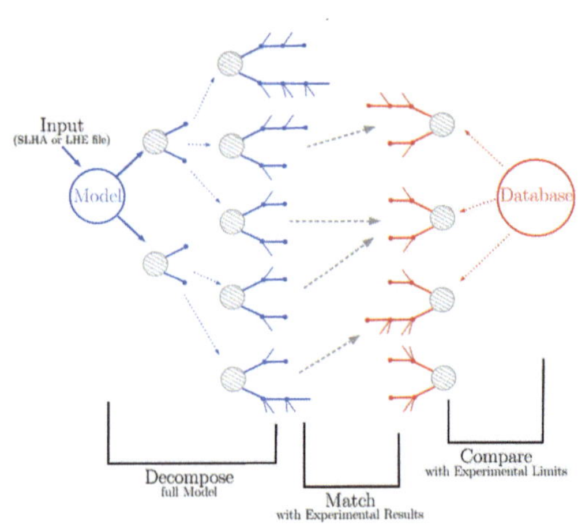

Fig. 3.1 SModelS working principle. The full model is decomposed into simplified ones, categorised by the widths of the primary produced BSM particles, their production cross-section, their branching ratios and the masses of the BSM particles. It is then confronted to the experimental results obtained for similar simplified models, provided that they are implemented in the SModelS database

on simplified models. It does so by means of an r-value:

$$r_{(\mathrm{exp})} = \frac{\sigma^{\mathrm{BSM}}}{\sigma_{95}^{(\mathrm{exp})}}, \tag{3.1}$$

where σ^{BSM} is the theoretical cross section of the tested model and $\sigma_{95}^{(\mathrm{exp})}$ is the 95% CL upper-limit on the observed (expected) cross section. The $\sigma_{95}^{(\mathrm{exp})}$ can be obtained either through upper limit (UL) maps on the observed (expected) cross Sect. [3], or by means of efficiency $\times$ acceptance maps (EMs) [5] published by the experimental collaborations or obtained through a recast, like in [12]. Details regarding the statistical inference through EMs will be provided in Sect. 3.2. While the observed limit is inferred from the number of events recorded by the experiment, the expected limit is the expected sensitivity to the tested model and can be obtained before doing the analysis. It indicates how much the BSM model can be constrained if the data fully agrees with the SM. When $\sigma_{95}^{(\mathrm{exp})}$ is obtained through UL maps, the theoretical prediction is simply the cross section multiplied by the branching ratios; when it is obtained through EMs, the theoretical prediction is multiplied by the experimental efficiency $\times$ acceptance of the model to get the fiducial cross section:

$$\sigma^{\mathrm{BSM}}_{(\mathrm{UL\text{-}type})} = \sum \sigma \prod \mathrm{BR} \quad \text{and} \quad \sigma^{\mathrm{BSM}}_{(\mathrm{EM\text{-}type})} = \sum \sigma \times \epsilon \times \mathcal{A} \prod \mathrm{BR}\,. \tag{3.2}$$

Here σ are the production cross sections and BR, ϵ and $\mathcal{A}$ are the corresponding branching ratios, efficiencies and acceptances, respectively. If $r_{(\mathrm{exp})} \geq 1$, then the model is (expected to be) excluded, otherwise it is not.

The input model can be given to SModelS either in the SUSY Les Houches Accord (SLHA) file format [13] or in the Les Houches event (LHE) file format [14]. It must contain all the relevant information to build simplified models, i.e. the masses of the particles, their production cross-sections, their widths and their branching ratios. From the input model, SModelS then builds simplified topologies. First, it gets all the allowed pairs of $\mathbb{Z}_2$-odd particles produced by the proton-proton collision. All the allowed decays of these two produced particles are then used to construct all the possible simplified topologies, as pictured in Fig. 3.2. Finally, these are compared to the experimental results implemented in the SModelS database.

Each simplified topology is given a shorthand notation, the TxName, reflecting the model's signature. For instance, Fig. 3.3 shows the TChiWZ and TChiWZoff topologies, whose characteristics are the production of on-shell and off-shell SM-like W and Z bosons from two different decay cascades. Even if the topology can be induced by the decay of neutralinos (hence the "Chi" in the TxName), the nature of the BSM particles has no impact on the simplified model, on its associated topology and therefore on its TxName. A complete list of TxNames can be found on smodels.github.io/docs/SmsDictionary.

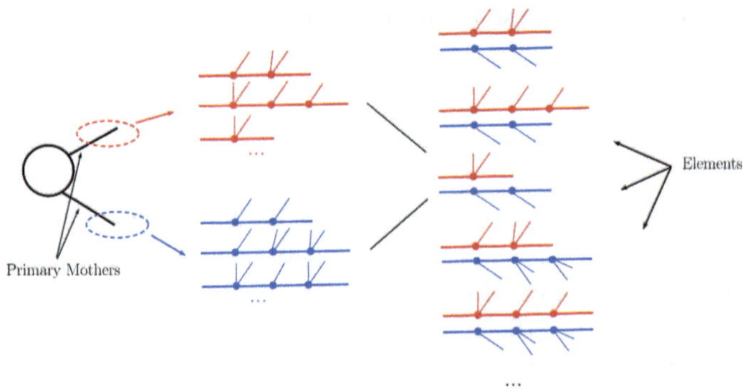

Fig. 3.2 SMODELS decomposition procedure. The possible decays of the produced BSM particles (primary mothers) are listed and combined to construct all the accessible simplified topologies. The latter, along with their widths, production cross-section, branching ratios and masses, form the basic elements that will be matched and compared to the experimental results implemented in the SMODELS database. By convention, the horizontal and tilted lines represent the produced BSM and SM particles, respectively

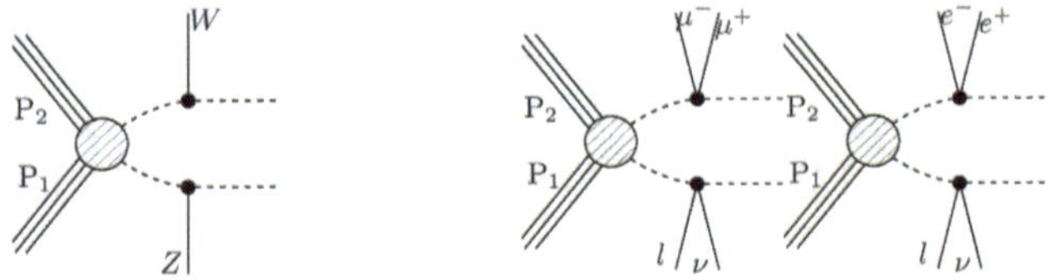

Fig. 3.3 TChiWZ (left) and TChiWZoff (right) topologies. P1 and P2 are the two colliding protons, the striped circles represent the hard processes, the dashed lines are the BSM particles, the black dots are the interacting vertices, and the full lines are the SM particles

To match an experimental result, the characteristics of the decomposed model and of the experimental result must be the same. It includes the widths of the primary mothers (above a certain width, the particles are considered to decay promptly), the produced SM particles, and the masses of the BSM particles. The latter must lie in the convex hull of the implemented UL maps or EMs, which map the masses (and, if relevant, the decay widths) of the BSM particles to the 95% CL upper limit on the cross-section or to the experimental efficiency × acceptance, respectively.

The structure of the database is depicted in Fig. 3.4. Two types of results can be implemented in the database: UL- or EM-type results, depending on the reinterpretation material published by the experimental collaborations. One analysis can have both UL- and EM-type results implemented, but it would then correspond to two different database entries (differentiated by the suffix "-eff", standing for efficiencies). Each entry has a `globalInfo.txt` file containing general information (e.g. its name, its corresponding LHC run and integrated luminosity), as well as one or multiple dataset folders. The latter contain a `dataInfo.txt` file indicating the dataset name and type, and either upper-limit or efficiency × acceptance maps, one for each simplified topology. For EM-type results, the datasets correspond to the SRs

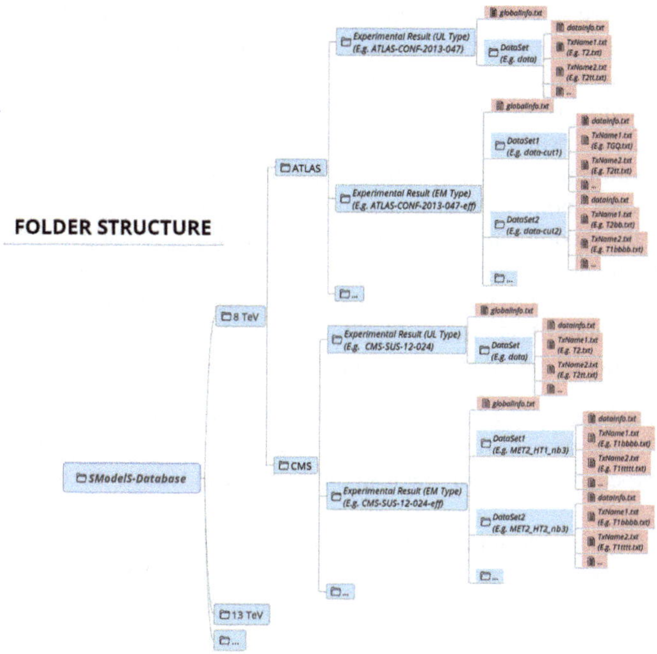

Fig. 3.4 SModelS database structure. The entries are categorised according to the LHC Run (so far, only 8 or 13 TeV) and the experiment (ATLAS or CMS) of their analyses, as well as the type of map they contain (UL- or EM-type results)

designed by the experimental collaboration, and the `dataInfo.txt` file additionally contains the number of expected SM events, its uncertainty, and the actual number of events observed, as well as the 95% CL on the expected and on the observed cross-section upper limits. These upper limits are obtained through a hypothesis test detailed in Sect. 3.2. The full list of analyses implemented in SModelS database can be found in Appendix B.

Each simplified model matching a database entry is compared against the corresponding experimental results. This is mainly done through the computation of the r-values defined in Eq. (3.1), confronting the model theoretical prediction with the experimentally-derived 95% CL upper-limit on this quantity. The results are reported in the SModelS outputs. Available outputs are screen-type (stdout), log (same as screen but redirected to a .log file), Python, XML, and SLHA. When running on multiple files, a summary file is also printed, which is a simplified output with a summary of the results for each input file. For each simplified model, the outputs typically print the following quantities for each matched database entry: the name of the database entry, the SR constraining the model ("upper limit" for UL-type results; "combined" if the SRs are combined (see Sect. 3.2)), the TxName(s) of the constrained topology(ies), the energy in the centre-of-mass frame, the theoretical prediction (i.e. the BSM cross section as defined in Eq. (3.2)), the observed experimental limit on the theoretical prediction, the observed and expected r-values (no expected r-value is computed if the dataset is of UL-type but no UL map on the expected cross section is implemented). For EM-type results, the likelihoods under the SM and BSM hypothesis, as well as the maximum likelihood are also printed. If a list of analyses to combine is provided (see below, Sect. 3.2), the aforementioned quantities are also displayed for the said combination. Moreover, details about unconstrained simplified models also appear in the outputs. It comprises the theoretical prediction and the simplified topology of the simplified models for which no match was found in the database (both for prompt and displaced decays), or for which a match was found but the particle masses were outside the UL and/or EM grid(s). Finally, are also listed the parameters used to run SMODELS. Some of the most relevant parameters are the following:

- `combineSRs`: set to `True` to combine the SRs when possible (see below, Sect. 3.2), otherwise only the result of the most sensitive SR is returned, i.e. the SR that expects to constrain the tested model the most, i.e. the one that gives the highest r_{exp}.
- `combineAnas`: an optional comma-separated list of analyses to combine. Only the combination of approximately uncorrelated analyses can give a sensitive result (see below, Sect. 3.2). Knowing which analyses can be safely combined is left to the user.
- `reportAllSRs`: set to `True` to report the results of all matched SRs, otherwise the result to return is dictated by `combineSRs`.
- `model`: path to the model file defining the BSM states and their quantum numbers. It can either be a Python or a SLHA file. By default, the MSSM is used.
- `promptWidth`: all particles with widths above this value are considered prompt. Default value is 10^{-11} GeV.
- `stableWidth`: all particles with widths below this value are considered stable. Default value is 10^{-25} GeV.
- `sigmacut`: simplified topologies with $\sigma \prod \text{BR}$ below this value are not considered. Default value is 5×10^{-3} fb.
- `minmassgap`: give the minimum mass gap for mass compression. Each time the decay products within a decay chain have an energy below this value, a similar topology without these decays is created and is also compared against the database entries. The default value is 5 GeV. More information is provided in Sect. 4.2.
- `path`: path to the database. It can be a path to the `smodels-database` folder, the path to a pickle file or a URL path. The available databases can be seen on the smodels-database-release page on Github. More details can be found in Appendix C.
- `analyses`: reduce the database entries to compare the tested model against to this list of analyses (separated by a comma). Set to `all` to use all the analyses included in the database.

- `txnames`: only the TxNames in this list will be considered. Set to `all` to use all TxNames included in the database.
- `dataselector`: set to `all` to use all upper limit and efficiency $\times$ acceptance maps results in the database. Set to `upperLimit` (`efficiencyMap`) to use only UL-type (EM-type) results.
- `outputType`: Define the output formats, e.g. Python, XML, SLHA.

These parameters can be modified through the `parameters.ini` file given to SModelS in the execution line. The basic command line to run SModelS is

```
./runSModelS.py -f FILENAME [-p PARAMETERFILE] [-o OUTPUTDIR] [-T TIMEOUT].
```

Here, `FILENAME` is the path to the file containing the model point to test (the SLHA or LHE file(s)), it can be a single file or a directory; `PARAMETERFILE` is the path to the `parameters.ini` file containing the values of the parameters mentioned above; `OUTPUTDIR` is the path to the directory where SModelS' output(s) will be printed; and `TIMEOUT` defines a limit on the running time per file (in seconds). The last three arguments are optional. If not provided, default ones will be used. More information can be obtained by executing the command `./runSModelS.py -h`.

3.1.1 Comments on the SModelS *Assumptions*

For the SModelS results to be valid, the tested model should be kinematically close to the experimental model used to constrain it, so that the UL- and EM-type results can be safely applied to it. Several works showed that this assumption can hold for complex models close to their simplified counterparts, see e.g. [15–18], and that the models not covered by simplified ones are likely a small fraction or kinematically similar to those that are [15]. Other works showed that when this assumption does not hold, the derived limits of the new model can nonetheless be similar to the original ones [19].

Nevertheless, these assumptions can be a source of errors that are pointed out here through the example of ATLAS-SUSY-2018-16 [20], a search for the EW production of sparticles with compressed mass spectra. This analysis searches for final states with missing transverse momentum and two same-flavor, oppositely-charged, low transverse momentum electrons or muons. The distributions of the dilepton invariant mass are shown in Fig. 3.5 for wino-bino and higgsino-LSP scenarios. The differences in the distribution of the invariant dilepton mass impact the efficiencies. Therefore, the efficiencies used to compute the fiducial cross sections depend on the sign of the product of the mass eigenstates $m_{\tilde{\chi}_2^0} \times m_{\tilde{\chi}_1^0}$, a quantity to which SModelS simplified models are agnostic to. This also translates into the exclusion limits set by the experimental collaboration, as shown in Fig. 3.6. One model seems to record stronger over- and under-fluctuations in the background than the other. This is partly due to the difference in efficiencies but also to the production cross-section which varies according to the model.

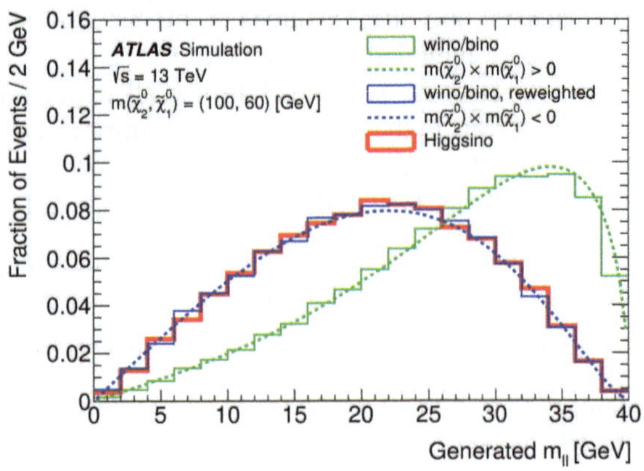

Fig. 3.5 Dilepton invariant mass for higgsino-LSP and wino-bino simplified models. Results from simulations (histograms) are compared with analytic calculations of the expected lineshape (dashed lines) presented in [21]. The product of the signed mass eigenvalues ($m_{\tilde{\chi}_2^0} \times m_{\tilde{\chi}_1^0}$) is negative for the higgsino-LSP model and can be either negative or positive for wino-bino scenarios. Taken from [20]

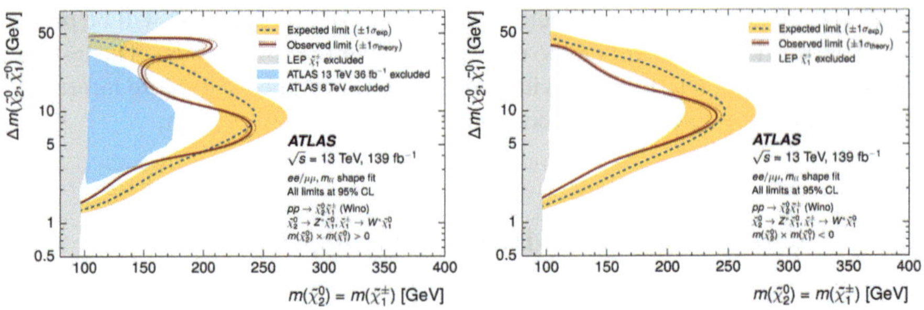

Fig. 3.6 Exclusion limits from ATLAS-SUSY-2018-16, a search for the production of EW-inos with compressed mass spectra. The limits are derived assuming a wino-bino simplified model, for $m_{\tilde{\chi}_2^0} \times m_{\tilde{\chi}_1^0} > 0$ (right) and $m_{\tilde{\chi}_2^0} \times m_{\tilde{\chi}_1^0} < 0$ (left). Taken from [20]

By default, the efficiencies used in SMODELS are from the $m_{\tilde{\chi}_2^0} \times m_{\tilde{\chi}_1^0} > 0$ model when $\tilde{\chi}_1^\pm$ and $\tilde{\chi}_2^0$ are mass degenerate, while they are from the $m_{\tilde{\chi}_2^0} \times m_{\tilde{\chi}_1^0} < 0$ model when $m_{\tilde{\chi}_1^\pm} \approx (m_{\tilde{\chi}_2^0} + m_{\tilde{\chi}_1^0})/2$, which is more in adequacy with the higgsino-LSP scenario. The impact of this on the SMODELS exclusion lines is shown in Fig. 3.7.

3.2 Statistical Inference

To exclude a model point, SMODELS needs to compare the BSM signal to its upper limit value. It can be done in two ways: by using UL maps if available, and through a hypothesis test if EMs are available. If EMs are implemented, not only SMODELS can compute both observed and expected r-values, which mainly hold binary information, but it can also compute a likelihood for the tested signal. This section explains how this is done and is based on [8].

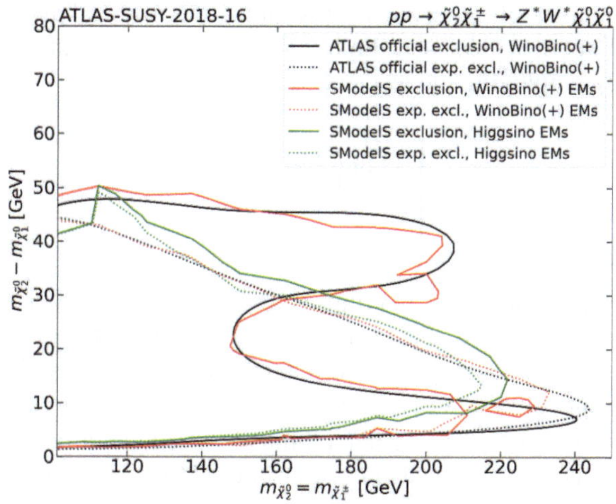

Fig. 3.7 Exclusion limits, obtained with SModelS, assuming a wino-bino simplified model with $m_{\tilde{\chi}_2^0} \times m_{\tilde{\chi}_1^0} > 0$, for different set of EMs. "WinoBino(+) EMs" corresponds to EMs obtained assuming wino-bino models with $m_{\tilde{\chi}_2^0} \times m_{\tilde{\chi}_1^0} > 0$. "Higgsino EMs" corresponds to EMs obtained assuming a model with higgsino-like $\tilde{\chi}_1^\pm$, $\tilde{\chi}_2^0$ and $\tilde{\chi}_1^0$ with $m_{\tilde{\chi}_2^0} \times m_{\tilde{\chi}_1^0} < 0$. For the wino-bino scenarios, the SRs are combined using a full statistical model, while for the higgsino scenario, only constraints from the most sensitive inclusive SR (i.e. giving the highest r_{exp}) can be computed. The official limits are only shown for completeness

The likelihood describes the plausibility of a signal strength μ given the data D:

$$\mathcal{L}(\mu, \theta|D) = P(D|\mu s + b + \theta)\, p(\theta)\,. \tag{3.3}$$

Here, s and b are the numbers of predicted signal and background events, respectively, while θ denotes the nuisance parameters that describe the variations in the signal and background contributions due to systematic effects, with $p(\theta)$ being the probability distribution of the nuisances.

3.2.1 Likelihood for a Single Signal Region

The simplest case is to compute the likelihood for a single SR (by setting `combineSRs=False` in the SModelS `parameters.ini` file). This assumes $p(\theta)$ to follow a Gaussian distribution centred around zero with a variance of δ^2, whereas $P(D)$ corresponds to a counting variable and is thus described by a Poissonian. The likelihood for each SR thus takes the form [5]

$$\mathcal{L}(\mu, \theta|D) \propto \frac{(\mu s + b + \theta)^{n_{\text{obs}}} e^{-(\mu s + b + \theta)}}{n_{\text{obs}}!} \exp\left(-\frac{\theta^2}{2\delta^2}\right) \tag{3.4}$$

with n_{obs} the number of observed events and δ the 1σ background uncertainty. Given the likelihood, a 95% CL limit on μ, μ_{95}, is computed using the $\mathrm{CL_s}$ prescription [22], employing the test statistic q_μ [23]:

$$q_\mu = \begin{cases} -2\ln\left(\frac{\mathcal{L}(\mu,\hat{\hat{\theta}})}{\mathcal{L}(\hat{\mu},\hat{\theta})}\right) & \hat{\mu} \leq \mu\ , \\ 0 & \hat{\mu} > \mu\ , \end{cases} \tag{3.5}$$

where $\hat{\hat{\theta}}$ is the value of θ that maximises $\mathcal{L}$ for the specified μ ($\mathcal{L}(\mu, \hat{\hat{\theta}})$ is the profiled likelihood), while $\hat{\mu}$ and $\hat{\theta}$ are the maximum likelihood estimators. The upper limit on the cross section is then obtained by multiplying the fiducial cross section with the upper limit on μ: $\sigma_{95} = \mu_{95}\sigma^{\mathrm{BSM}}$. Equivalently, μ_{95}^{exp}, and therefore $\sigma_{95}^{\mathrm{exp}}$, can be obtained by assuming that all the observed yields are coming from the SM background, i.e. $n_{\mathrm{obs}} = b$. Unless `reportAllSRs = True`, SModelS then reports the r-values ($r_{(\mathrm{exp})} = \sigma^{\mathrm{BSM}}/\sigma_{95}^{(\mathrm{exp})} = 1/\mu_{95}^{(\mathrm{exp})}$), as well as the values for the observed $\mathcal{L}_{\mathrm{BSM}} \equiv \mathcal{L}(\mu = 1)$, $\mathcal{L}_{\mathrm{SM}} \equiv \mathcal{L}(\mu = 0)$ and $\mathcal{L}_{\max} \equiv \mathcal{L}(\hat{\mu})$ for the most sensitive (a.k.a. "best") SR of each analysis.

If information on the background correlations across SRs is provided by the experimental collaboration, either in the form of a correlation or covariance matrix, or –better– in the form of a full statistical model, SModelS can go a significant step further and compute the likelihood for the entire analysis, combining its SRs. To this end, one has to set `combineSRs=True` in the SModelS parameters settings. Three different approaches are now available in SModelS as detailed below in Sect. 3.2.2.

Moreover, independent of whether or not SRs are combined, SModelS now offers the possibility to combine likelihoods from different analyses [8]. This important new feature will be described in Sect. 3.3.1.

3.2.2 Combination of Signal Regions Within an Analysis

3.2.2.1 Simplified Likelihood Version 1: Gaussian Uncertainties

In this framework, initially introduced in [24] and available in SModelS since v1.1.3 [6], all nuisance parameters are consolidated into a single distribution using a multivariate Gaussian. Concurrently, Poissonians are utilised to accommodate the statistical behaviour arising from counting events in individual signal regions. The likelihood for N SRs takes the form

$$\mathcal{L}(\mu, \theta|D) \propto \prod_{i=1}^{N} \mathrm{Pois}\left(n_{\mathrm{obs}}^{i}|\mu s_i + b_i + \theta_i\right)\, \exp\left(-\frac{1}{2}\vec{\theta}^{T} V^{-1}\vec{\theta}\right), \tag{3.6}$$

where μ is the overall signal strength and V represents the covariance matrix.[1] Signal uncertainties are neglected.

Here referred to as SLv1, this simplified likelihood approach holds the distinction of being the first technique that enabled the combination of signal regions in a non-trivial manner for phenomenologists whenever a correlation or covariance matrix is available for an analysis. The SLv1 has demonstrated satisfactory performance as long as the Gaussian approximation for the nuisances is valid. However, it may not be a good approximation in case of very small expected event yields.

3.2.2.2 Simplified Likelihood Version 2: Gaussian with a Skew

A possible solution to account for non-Gaussian effects in the nuisances is incorporating a skewness term in the Gaussian distribution as proposed in [25]. In the formalism of [25], again assuming a Poisson statistics for the observed event counts, the likelihood takes the form

$$\mathcal{L}(\mu, \theta | D) \propto \prod_{i=1}^{N} \text{Pois}\left(n^i_{\text{obs}} | \mu s_i + \alpha_i + \beta_i \theta_i + \gamma_i \theta_i^2\right) \exp\left(-\frac{1}{2}\vec{\theta}^T \rho^{-1} \vec{\theta}\right), \tag{3.7}$$

which is here referred to as SLv2. Note that here the b_i from Eq. (3.6) has been relabelled α_i and that the nuisances have been reparametrised as $\theta_i \rightarrow \beta_i \theta_i$. The coefficients α_i, β_i and γ_i can be related to the first three statistical moments. Specifically, the first moment is the mean, while the second moment is the covariance $V_{ij} = \beta_i \beta_j \rho_{ij} + 2\gamma_i \gamma_j \rho_{ij}^2$; the diagonal element of the third moment is $m_{3,i} = 6\beta_i^2 \gamma_i + 8\gamma_i^3$.[2] In the end, all that is effectively needed to extend the SLv1 to SLv2 is the third moment m_3, which, given asymmetric background uncertainties δ_- and δ_+, may be computed from a bifurcated Gaussian as [26]

$$m_3 = \frac{2}{\delta_- + \delta_+} \left[\delta_- \int_{-\infty}^{0} x^3 \,\text{No}\left(x; 0, \delta_-^2\right) dx + \delta_+ \int_{0}^{\infty} x^3 \,\text{No}\left(x; 0, \delta_+^2\right) dx\right], \tag{3.8}$$

where No refers to the normal distribution. The SLv2 has been technically available in SModelS for a while, but was not used due to a lack of experimental information. The CMS-SUS-20-004 analysis [27] (see Sect. 4.1) is the first analysis

[1] The correlation matrix ρ and the covariance matrix V are related by $V_{ij} = \rho_{ij}\delta_i\delta_j$.

[2] In the words of [26], α_i is the central value of the background prediction; β_i corresponds to the effective sigma of the background uncertainty, with $\beta_i = \sqrt{V_{ii}}$ in the limit of symmetric uncertainties; and γ_i describes the asymmetry of the background uncertainty.

to provide the required information. With this, SModelS is the first reinterpretation tool to make actual use of the formalism of [25].

3.2.2.3 HistFactory Statistical Models

ATLAS searches are often based on HistFactory [28] for their statistical modelling. Following [29], the collaboration has started to provide JSON serialisations of the full HistFactory workspaces for results with full Run 2 luminosity on HEPData. Thus, the full set of nuisance parameters, changes under systematic variations, and observed data counts are provided at the same fidelity as used in the experiment.

SModelS supports the usage of these JSON-serialised statistical models since v1.2.4 [30] via an interface to the pyhf package [31], a pure-Python implementation of the HistFactory family of statistical models. This means that with `combineSRs=True`, whenever a HistFactory statistical model is available in the database, the evaluation of the likelihood is relegated to pyhf (where, internally, the calculation is again based on the asymptotic formulas of [23]). More technically, the HistFactory likelihood takes the form:

$$\mathcal{L}(\eta, \chi|n, a) = \prod_{c\,\in\,\text{channels}} \prod_{b\,\in\,\text{bins}} \text{Pois}(n_{cb}|\nu_{cb}(\eta, \chi)) \prod_{\chi} c_\chi(a_\chi|\chi)\,, \tag{3.9}$$

where η, χ, n and a are the free parameters, the constrained parameters, the number of observed yields, and the additional global observables from auxiliary measurements, respectively. The first two products contain the simultaneous measurement of multiple channels (defined as disjoint binned distributions), while the third product constrains the constrained parameters χ with a constraint term c based on auxiliary data a. For each bin of each channel, the overall expected event rate is the sum of the number of events predicted by each physics process (called sample):

$$\begin{aligned}\nu_{cb}(\eta, \chi) &= \sum_{s\,\in\,\text{samples}} \nu_{scb}(\eta, \chi) \\ &= \sum_{s\,\in\,\text{samples}} \underbrace{\left(\sum_{\kappa} \kappa_{scb}(\eta, \chi)\right)}_{\text{multiplicative modifiers}} \left(\nu^0_{scb}(\eta, \chi) + \underbrace{\sum_{\Delta} \Delta_{scb}(\eta, \chi)}_{\text{additive modifiers}}\right).\end{aligned} \tag{3.10}$$

The global event rate is determined from the nominal event rate ν^0_{scb} and a set of multiplicative and additive multipliers, which are functions of the constrained and unconstrained parameters. The list of modifiers and their associated constraint term (for constrained parameters) are listed in Fig. 3.8. Here $\eta, \chi \in \{\gamma, \alpha, \lambda, \mu\}$, and σ_b and δ_b are the relative uncertainty of the event rate and the event rate uncertainty

Modifiers and Constraints

Description	Modification	Constraint Term c_χ	Input
Uncorrelated Shape	$\kappa_{scb}(\gamma_b) = \gamma_b$	$\prod_b \text{Pois}\left(r_b = \sigma_b^{-2} \middle| \rho_b = \sigma_b^{-2}\gamma_b\right)$	σ_b
Correlated Shape	$\Delta_{scb}(\alpha) = f_p\left(\alpha \middle| \Delta_{scb,\alpha=-1}, \Delta_{scb,\alpha=1}\right)$	$\text{Gaus}\left(a = 0 \middle| \alpha, \sigma = 1\right)$	$\Delta_{scb,\alpha=\pm1}$
Normalisation Unc.	$\kappa_{scb}(\alpha) = g_p\left(\alpha \middle| \kappa_{scb,\alpha=-1}, \kappa_{scb,\alpha=1}\right)$	$\text{Gaus}\left(a = 0 \middle| \alpha, \sigma = 1\right)$	$\kappa_{scb,\alpha=\pm1}$
MC Stat. Uncertainty	$\kappa_{scb}(\gamma_b) = \gamma_b$	$\prod_b \text{Gaus}\left(a_{\gamma_b} = 1 \middle| \gamma_b, \delta_b\right)$	$\delta_b^2 = \sum_s \delta_{sb}^2$
Luminosity	$\kappa_{scb}(\lambda) = \lambda$	$\text{Gaus}\left(l = \lambda_0 \middle| \lambda, \sigma_\lambda\right)$	$\lambda_0, \sigma_\lambda$
Normalisation	$\kappa_{scb}(\mu_b) = \mu_b$		
Data-driven Shape	$\kappa_{scb}(\gamma_b) = \gamma_b$		

Fig. 3.8 HistFactory modifiers and constraints, from pyhf.readthedocs.io

of the sample relative to the total event rate $\nu_b = \sum_s \nu_{sb}^0$, respectively. f_p and g_p are interpolating functions constructed from a small number of evaluations of the expected rate at fixed values of the parameter α. Usually, $f_p(\alpha = 0) = 0$ and $g_p(\alpha = 0) = 1$.

It has to be noted here that the evaluation of full HistFactory models, which can have hundreds of nuisance parameters ($\theta \in \{\eta, \chi\}$), can be very computationally intensive, in particular when combining analyses (see Sect. 3.3.1). For this reason, the official SModelS database contains mostly simplified HistFactory models [32], which were derived from the full ones by means of the simplify [33] tool.[3] The currently only exceptions are the ATLAS-SUSY-2018-16 [20] and the ATLAS-SUSY-2019-08 [34] analyses, for which the simplify'ed statistical models do not reproduce the experimental limits well enough. For these two analyses, the full statistical models are used by default. The failure of the simplified statistical model therefore often comes from the fact that the systematics can not be approximated by a Gaussian. In any case, when CPU performance is not an issue, the simplify'ed statistical models in the database can be replaced by the full ones through a "`full_llhds`" database add-on as explained in Appendix C. Figure 3.9 shows one successful and one unsuccessful attempt at simplifying the statistical model. One can also see how relevant the combination of SRs can be, and how statistical models can help accurately reproducing the experimental limits.

3.3 Recent SModelS Developments

This section lists some of the recent SModelS developments to which I contributed regarding likelihood computation and statistical inference.

[3] This simplify method collapses all the modifiers into a single, additive, one, constrained by a Gaussian (labelled "correlated shape" in Fig. 3.8).

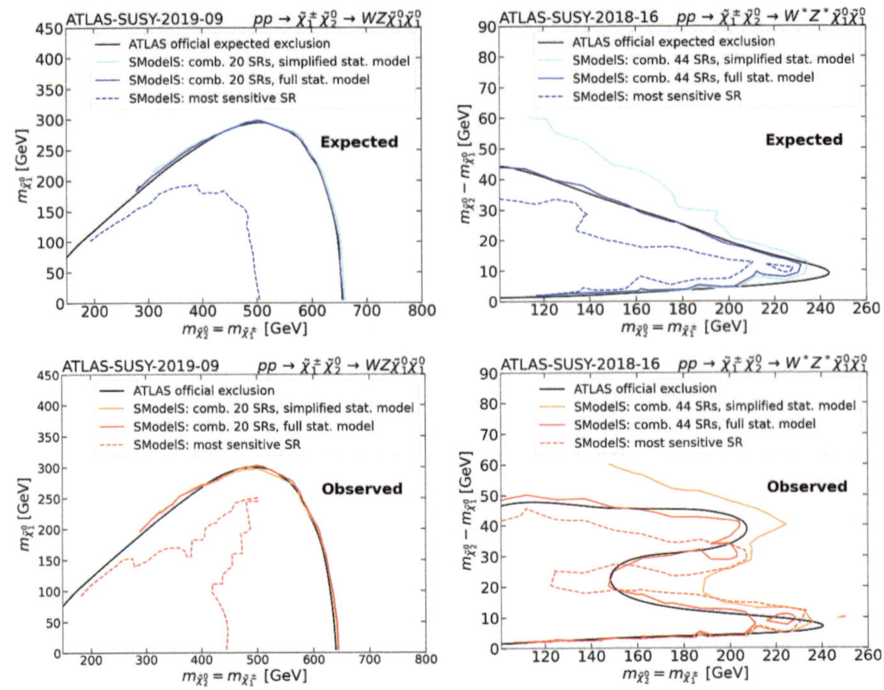

Fig. 3.9 Capability of SModelS to reproduce the expected (top) and observed (bottom) experimental limits using the most sensitive SR, the full statistical model and the simplified one, for ATLAS-SUSY-2019-09 [35] (left) and ATLAS-SUSY-2018-16 [20] (right)

3.3.1 Combination of Likelihoods of Orthogonal Analyses

SModelS now provides the possibility to combine likelihoods from different analyses under the assumption that they are approximately uncorrelated. Here, approximately uncorrelated means that SRs do not overlap and inter-analyses correlations of systematic uncertainties (stemming from, e.g., luminosity measurements) can be neglected.[4] The combined likelihood $\mathcal{L}_C$ is then simply the product of the likelihoods $\mathcal{L}_i$ of the individual analyses. Furthermore, a common signal strength μ is assumed for all analyses. Thus

$$\mathcal{L}_C(\mu) = \prod_{i=1} \mathcal{L}_i(\mu\, s^i)\,. \tag{3.11}$$

[4] Overlaps of SRs of one analysis with the control regions of another analysis in the combination can in principle induce correlations of systematic uncertainties and therefore should also be checked. However, the effect is generally expected to be negligible compared to other uncertainties in SModelS.

The individual likelihoods can correspond to best signal region likelihoods and/or combined signal region likelihoods from any of the three approaches explained above (turned on/off with the `combineSRs=True/False` switch). For the determination of the maximum likelihood, or more precisely the minimum negative log-likelihood $-\log \mathcal{L}_{\max} = -\log \mathcal{L}_C(\hat{\mu})$, scipy.optimize.minimize is used with the BFGS method.

As of now, the information of which analyses should be combined has to be provided by the user. Generally, results from different experiments (ATLAS and CMS), different LHC runs (8 and 13 TeV), as well as fully hadronic and fully leptonic analyses can be regarded as approximately uncorrelated [36], at least within the approximations of SModelS. For more sophisticated combinations, deeper scrutiny of the signal (and control) region definitions is needed; see [37] for an approach based on Monte Carlo simulation.

The combination of analyses is interesting for two reasons. First, the signal of a particular BSM scenario may be manifest in different final states, which are constrained by different analyses. Combining them uses more of the available data and thus provides more robust, and usually stronger, constraints. Second, experimental analyses can always be subject to over- or under-fluctuations of the backgrounds. In the former case, the observed limit is weaker, in the latter case stronger, than the expected limit. Again, the combination of different, approximately independent analyses can mitigate this effect and provide more robust constraints.

This is illustrated in Fig. 3.10 for a sample point from the EW-ino scan used in Sect. 4.3, which features a bino-like $\tilde{\chi}_1^0$ with a mass of 257 GeV and wino-like $\tilde{\chi}_1^\pm$ and $\tilde{\chi}_2^0$ with masses of 617 GeV. The $\tilde{\chi}_1^\pm$ decays to 100% into $\tilde{\chi}_1^0 W^\pm$, while the $\tilde{\chi}_2^0$ decays to 96% into $\tilde{\chi}_1^0 h$ and to 4% into $\tilde{\chi}_1^0 Z$. The strongest constraints for this scenario come from the fully hadronic EW-ino searches using boosted W, Z and Higgs bosons, ATLAS-SUSY-2018-41 [38] and CMS-SUS-21-002 [39]. Plotted in Fig. 3.10 are the likelihoods as a function of the signal strength μ for these two analyses and their combination, on the left for the expected and on the right for the observed data. As can be seen from the left panel, the CMS-SUS-21-002 analysis

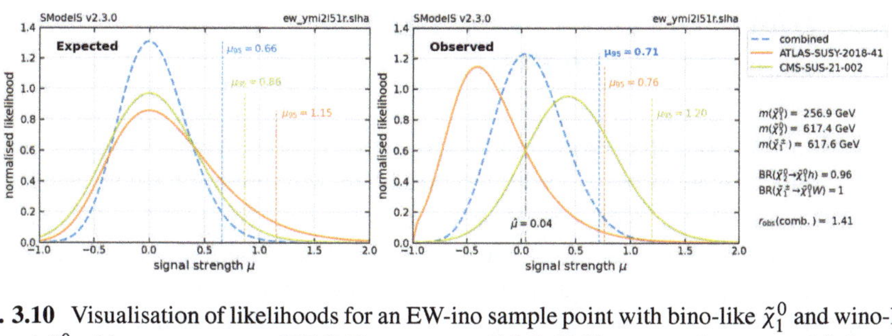

Fig. 3.10 Visualisation of likelihoods for an EW-ino sample point with bino-like $\tilde{\chi}_1^0$ and wino-like $\tilde{\chi}_1^\pm$ and $\tilde{\chi}_2^0$ with masses $m_{\tilde{\chi}_1^0} = 257$ GeV, $m_{\tilde{\chi}_1^\pm,\tilde{\chi}_2^0} = 617$ GeV. Shown are the expected (left panel) and observed (right panel) likelihoods as a function of the signal strength μ for the two hadronic EW-ino searches from ATLAS and CMS and their combination

has the highest sensitivity (lowest μ_{95}^{exp}) and is expected to exclude the point with $r_{\text{exp}} = 1.16$ if there is no new physics in the data (recall that a point is expected to be excluded if $\mu_{95}^{\text{exp}} > 1$ or equivalently $r_{\text{exp}} \equiv 1/\mu_{95}^{\text{exp}} < 1$). The ATLAS analysis has slightly less sensitivity and is not expected to individually exclude the point. Combining the likelihoods from both analyses (dashed blue curve in the plot) gives an expected exclusion of $r_{\text{exp}}(\text{combined}) = 1.52$, which illustrates the gain in sensitivity.

With the observed data, however, shown in the plot on the right, the strongest constraint comes from the ATLAS-SUSY-2018-41 analysis, excluding the point with an $r_{\text{obs}} = 1.32$. The CMS-SUS-21-002 analysis only gives an r_{obs} value of 0.84. The reason is that the former analysis observed a deficit of events, while the latter observed a small excess. In an analysis-by-analysis approach, one could exclude the point based on the highest observed r-value, or conclude that it is still allowed because the most sensitive analysis does not allow to exclude it. With the combined likelihood, however, one arrives at $r_{\text{obs}}(\text{combined}) = 1.41$, thus robustly excluding the point.

It is worth noting that here the expected limit always becomes stronger when combining. For the observed limit, however, the effect can go in either direction. The observed limit can become stronger, like in the example in Fig. 3.10. In the same way, it is possible that a point is excluded only by the combination of analyses, but not by any of the individual ones. But, the opposite can also happen, namely that a point excluded by some analysis becomes "unexcluded" by the combination with one or more other analyses. Indeed, all these cases will occur in the physics application in Chap. 4.

An example code for visualising likelihoods and their combinations as in Fig. 3.10 is given in the How To's section of the online manual. Moreover, there are code examples showing how to compute the confidence level of an exclusion, and how to define and use a "combinability matrix" that describes which analyses and/or SRs are approximately orthogonal and thus combinable.

3.3.2 *Trick to Get Positive Total Yields for Likelihood Computation with* PYHF

It has been observed that the computation of likelihoods with PYHF was failing for low values of negative μ. While no problem was reported when using each statistical model in its usual operating mode, i.e. when computing $\mu_{95}^{(\text{exp})}$, $\mathcal{L}_{\text{BSM}}$, $\mathcal{L}_{\text{SM}}$ and $\mathcal{L}_{\text{max}}$, having undefined value for a likelihood could potentially lead into issues when computing such quantities for a combination of them.

The problem arises when at least one SR has a negative total yield, i.e. when $\nu_{cb} < 0$. Of course, ν_{scb}^{0} is always positive, but ν_{cb} can be negative if $\mu < 0$ with the other modifiers and the background yields not pulling enough towards positive values. This situation can occur when initialising the fit of the profiled likelihood.

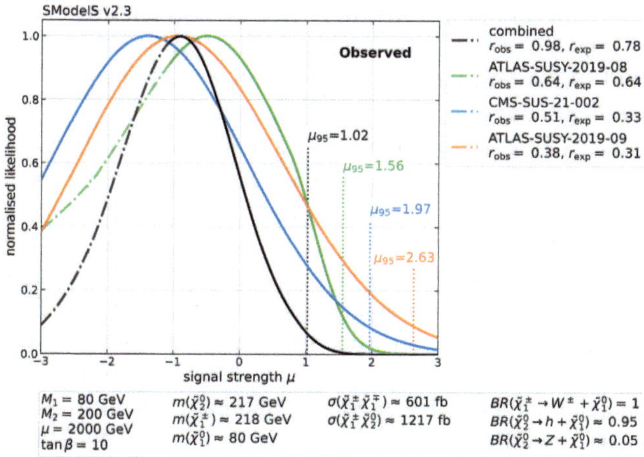

Fig. 3.11 Visualisation of the normalized likelihoods vs. the signal strength μ of three analyses searching for EW-inos and their combination. The full lines (dot-dashed lines) correspond to the likelihood obtained without (with) the trick explained in the text. Other information such as the r-values and the model parameters are displayed only for completeness. Here, all the other SUSY particles are decoupled and only shown are the relevant cross sections

Indeed, the modifiers have an initial value that can lead to negative total yields and prevent the profiling of the nuisances parameters from starting for too negative values of μ. This can be circumvented to some extent by modifying the initial values of these modifiers, while keeping them in their respective boundaries (encoded in the statistical model), to make each total yield positive. Only are affected the initial values of the nuisance parameters. They are free to vary within the subsequent fit to find their values that maximise the likelihood for the given value of μ. An illustration of the impact of this likelihood continuation is shown in Fig. 3.11. Without this trick, the likelihood of ATLAS-SUSY-2019-08 [40] would stop abruptly at $\mu \approx -1.1$, and so would be the combined one, potentially impacting the computation of likelihood-based quantities. With this trick, all the likelihoods can be computed to at least $\mu = -3$.

3.3.3 *Control Regions and Signal Uncertainty with* PYHF

This subsection describes some improvements regarding the SMODELS interface to PYHF originally presented in [30]. First of all, unlike what is described in [30], the `datasetOrder` field from the `globalInfo.txt` file in the database entry is not a mandatory argument anymore. Moreover, some progress has been achieved in the implementation of the full statistical model in order to reproduce as closely as possible the statistical treatment performed in the experiments. It is now possible to not remove the CRs from the HISTFACTORY statistical model by adding `includeCRs = True` in the `globalInfo.txt` file. So far, this option is used for three analyses: ATLAS-SUSY-2018-16 [20], ATLAS-SUSY-2018-32 [41], ATLAS-SUSY-2019-09 [35]. For all the other analyses that make use of a HISTFACTORY statistical model, `includeCRs = False` by default.

On top of that, if EMs are available for the CRs (most of the time they can be extracted from the corresponding patchset) SMODELS v3.0 can also emulate signal leaking into the CRs. In that case, SRs and CRs are treated on the same footing within the statistical model.

On the same note, an uncertainty on the computed signal can now be patched into the statistical model along the signal yields (this also is a new feature of v3.0). More precisely, when a value is provided in the `signalUncertainty` field of the `globalInfo.txt` file, each nominal signal of each bin of each channel (CRs as well as SRs) can be modified through an additive modifier constrained by a Gaussian (a "correlated shape" modifier in Fig. 3.8). For instance, if `signalUncertainty = 0.22`, the signal of the bin b of the channel c will be $s_{cb} = s^0_{cb} + f_p(\alpha)$, with s^0_{cb} the nominal signal of the bin b of the channel c and f_p an interpolation function defined by $f_p(0) = 0$, $f_p(-1) = -0.22\ s^0_{cb}$ and $f_p(1) = 0.22\ s^0_{cb}$. This modifier is constrained by a Gaus($a = 0|\alpha$, $\sigma = 1$) multiplied to the likelihood. This can help to get one step closer to the statistical treatment performed by the experimental collaborations. So far, only ATLAS-SUSY-2018-16 [20] is implemented in the SMODELS database with a signal uncertainty. As shown in Fig. 3.12, this can be necessary to reproduce the experimental limits as closely as possible. These choices are done at the level of analysis implementation and are not directly accessible to the user.

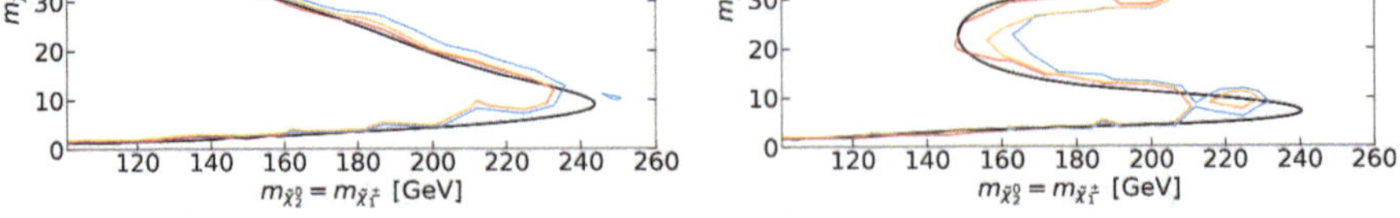

Fig. 3.12 SMODELS attempts at reproducing the expected (left) and observed (right) exclusion limits of ATLAS-SUSY-2018-16 [20] for different statistical model treatments. The full statistical model available on HEPDATA is used [42], "sig. in CRs" means signal leaking into the CRs, and δs indicates the `signalUncertainty` value provided in the `globalInfo.txt` file. The official SMODELS implementation allows for signal leaking into the CRs and has `signalUncertainty = 0.22` (red lines). The computation of the exclusion limits with the CRs removed from the statistical model failed, with and without uncertainty on the signal

3.3.4 *Towards a Delegation of the Statistical Inference to* SPEY

Following the recent publication of a new Python-based package allowing for the computation of statistical quantities from different likelihood prescriptions through a plug-in system, SPEY [43], a SMODELS interface to it has been developed. Delegating the statistical inference to SPEY has multiple advantages. First, it provides a clearer, more compartmentalised and centralised code for statistical computations; second, this centralisation allows all the reinterpretation tools to be interfaced to the same well-tested framework; and third, it provides an easy way to implement new statistical backends (the simplified likelihood framework and PYHF are examples of implemented backends). The SMODELS code has been refactored to be interfaced to SPEY, with good results, but some issues persist and this update has not been released yet.

3.3.5 SMODELS *Graph-Based Topologies*

Finally, the version 3.0 of SMODELS modifies the element decomposition described in Sect. 3.1 into a graph-based approach. An example is shown in Fig. 3.13. The root node represents the hard scattering collision ($pp \to$ produced particles) and is labelled "PV" (primary vertex). All the particles appearing in the simplified model topology correspond to graph nodes, while the decays are represented by edges connecting the parent particles to their daughters. This graph-based description is very flexible. Specifically, it allows SMODELS to go beyond the two-branch structure (from pair-production of new particles followed by cascade decays).

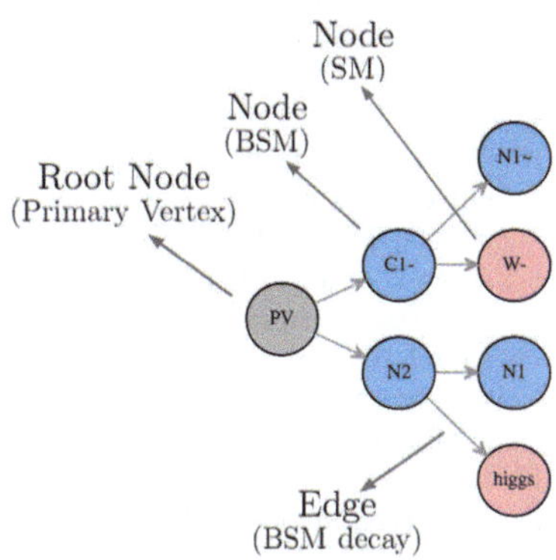

Fig. 3.13 Graph representation of a simplified model topology and of its elements. Taken from [9]

References

1. Abdallah W et al (2020) Reinterpretation of LHC results for new physics: status and recommendations after run 2. SciPost Phys 9(2):022. https://doi.org/10.21468/SciPostPhys.9.2.022. arXiv:2003.07868 [hep-ph]
2. Bailey S et al (2022) Data and analysis preservation, recasting, and reinterpretation. arXiv:2203.10057 [hep-ph]
3. Kraml S et al (2014) SModelS: a tool for interpreting simplified-model results from the LHC and its application to supersymmetry. Eur Phys J C 74:2868. https://doi.org/10.1140/epjc/s10052-014-2868-5. arXiv:1312.4175 [hep-ph]
4. Kraml S et al (2014) SModelS v1.0: a short user guide. arXiv:1412 1745 [hep-ph]
5. Ambrogi F et al (2018) SModelS v1.1 user manual: improving simplified model constraints with efficiency maps. Comput Phys Commun 227:72–98. https://doi.org/10.1016/j.cpc.2018.02.007. arXiv:1701.06586 [hep-ph]
6. Ambrogi F et al (2020) SModelS v1.2: long-lived particles, combination of signal regions, and other novelties. Comput Phys Commun 251:106848. https://doi.org/10.1016/j.cpc.2019.07.013. arXiv:1811.10624 [hep-ph]
7. Alguero G et al (2022) Constraining new physics with SModelS version 2. J High Energy Phys 08:068. https://doi.org/10.1007/JHEP08(2022)068. arXiv:2112.00769 [hep-ph]
8. Altakach MM et al (2023) SModelS v2.3: enabling global likelihood analyses. SciPost Phys 15(5):185. https://doi.org/10.21468/SciPostPhys.15.5.185. arXiv: 2306.17676 [hep-ph]
9. Altakach MM et al (2024) SModelS v3: going beyond Z_2 topologies. J High Energy Phys 11:074. https://doi.org/10.1007/JHEP11(2024)074. arXiv:2409.12942 [hep-ph]
10. SModelS Collaboration (2024) SModelS database v3.0.0. Zenodo. https://doi.org/10.5281/zenodo.13354582
11. SModelS Collaboration (2023) SModelS database v2.3.0. Zenodo. https://doi.org/10.5281/zenodo.7961638
12. Khosa CK et al (2020) SModelS database update v1.2.3. Lett High Energy Phys 2020:158. https://doi.org/10.31526/lhep.2020.158. arXiv:2005.00555 [hep-ph]
13. Skands PZ et al (2004) SUSY Les Houches accord: interfacing SUSY spectrum calculators, decay packages, and event generators. J High Energy Phys 0407:036. https://doi.org/10.1088/1126-6708/2004/07/036. arXiv:hep-ph/0311123 [hep-ph]
14. Alwall J et al (2007) A standard format for les houches event files. Comput Phys Commun 176:300–304. https://doi.org/10.1016/j.cpc.2006.11.010. arXiv: hep-ph/0609017 [hep-ph]
15. Gutschow C, Marshall Z (2012) Setting limits on supersymmetry using simplified models. arXiv:1202.2662 [hep-ex]
16. Cacciapaglia G et al (2013) LHC missing-transverse-energy constraints on models with universal extra dimensions. Phys Rev D87(7):075006. https://doi.org/10.1103/PhysRevD.87.075006. arXiv:1302.4750 [hep-ph]
17. Arina C et al (2015) Constraints on sneutrino dark matter from LHC Run 1. J High Energy Phys 05:142. https://doi.org/10.1007/JHEP05(2015)142. arXiv:1503.02960 [hep-ph]
18. Kraml S et al (2016) Scalar versus fermionic top partner interpretations of $t\bar{t} + E_T^{\mathrm{miss}}$ searches at the LHC. J High Energy Phys 11:107. https://doi.org/10.1007/JHEP11(2016)107. arXiv:1607.02050 [hep-ph]
19. Edelhäuser L, Krämer M, Sonneveld J (2015) Simplified models for same-spin new physics scenarios. J High Energy Phys 04:146. https://doi.org/10.1007/JHEP04(2015)146. arXiv:1501.03942 [hep-ph]
20. Aad G et al (2020) Searches for electroweak production of supersymmetric particles with compressed mass spectra in $\sqrt{s} = 13$ TeV pp collisions with the ATLAS detector. Phys Rev D 101(5):052005. https://doi.org/10.1103/PhysRevD.101.052005. arXiv:1911.12606 [hep-ex]
21. De Sanctis U et al (2007) Perspectives for the detection and measurement of Supersymmetry in the focus point region of mSUGRA models with the ATLAS detector at LHC. Eur Phys J C 52:743–758. https://doi.org/10.1140/epjc/s10052-007-0415-3. arXiv:0704.2515[hep-ex]

22. Read AL (2002) Presentation of search results: the CL(s) technique. J Phys G28:2693–2704. https://doi.org/10.1088/0954-3899/28/10/313
23. Cowan G et al (2011) Asymptotic formulae for likelihood-based tests of new physics. Eur Phys J C71:1554. [Erratum: Eur Phys J C73:2501 (2013)]. https://doi.org/10.1140/epjc/s10052-013-2501-z. arXiv:1007.1727 [physics.data-an]
24. CMS Collaboration (2017) Simplified likelihood for the re-interpretation of public CMS results. Tech. Rep. CMS-NOTE-2017-001. CERN, Geneva. https://cds.cern.ch/record/2242860
25. Buckley A et al (2019) The simplified likelihood framework. J High Energy Phys 04:064. https://doi.org/10.1007/JHEP04(2019)064. arXiv:1809.05548 [hep-ph]
26. Ford W (2022) Reproducing a CMS higgsino search from public data. https://indico.cern.ch/event/1197680/contributions/5164689/. Talk at the 7th Workshop of the LHC Reinterpretation Forum, 12–15 Dec 2022 at CERN
27. Tumasyan A et al (2022) Search for higgsinos decaying to two Higgs bosons and missing transverse momentum in proton-proton collisions at $\sqrt{s} = 13$ TeV. J High Energy Phys 05:014. https://doi.org/10.1007/JHEP05(2022)014. arXiv: 2201.04206 [hep-ex]
28. Cranmer K et al (2012) HistFactory: a tool for creating statistical models for use with RooFit and RooStats. CERN-OPEN-2012-016
29. ATLAS Collaboration (2019) Reproducing searches for new physics with the ATLAS experiment through publication of full statistical likelihoods Tech. Rep. ATL-PHYS-PUB-2019-029 CERN, Geneva
30. Alguero G, Kraml S, Waltenberger W (2021) A SModelS interface for pyhf likelihoods. Comput Phys Commun 264:107909. https://doi.org/10.1016/j.cpc.2021. 107909. arXiv:2009.01809 [hep-ph]
31. Heinrich L et al (2021) pyhf: pure-python implementation of HistFactory statistical models. J Open Source Softw 6(58):2823. https://doi.org/10.21105/joss.02823
32. ATLAS Collaboration (2021) Implementation of simplified likelihoods in HistFactory for searches for supersymmetry. Tech. Rep. ATL-PHYS-PUB-2021-038
33. Schanet E simplify: v0.1.6. Version 0.1.6. https://github.com/eschanet/simplify
34. Aad G et al (2020) Search for direct production of electroweakinos in final states with one lepton, missing transverse momentum and a Higgs boson decaying into two b-jets in pp collisions at $\sqrt{s} = 13$ TeV with the ATLAS detector. Eur Phys J C80(8):691. https://doi.org/10.1140/epjc/s10052-020-8050-3. arXiv:1909.09226 [hep-cx]
35. Aad G et al (2021) Search for chargino-neutralino pair production in final states with three leptons and missing transverse momentum in $\sqrt{s} = 13$ TeV pp collisions with the ATLAS detector. Eur Phys J C 81(12):1118. https://doi.org/10.1140/epjc/s10052-021-09749-7. arXiv:2106.01676 [hep-ex]
36. Waltenberger W, Lessa A, Kraml S (2021) Artificial proto-modelling: build-ing precursors of a next standard model from simplified model results. J High Energy Phys 03:207. https://doi.org/10.1007/JHEP03(2021)207. arXiv:2012.12246 [hep-ph]
37. Araz JY et al (2023) Strength in numbers: optimal and scalable combination of LHC new-physics searches. SciPost Phys 14(4):077. https://doi.org/10.21468/SciPostPhys.14.4.077. arXiv:2209.00025 [hep-ph]
38. Aad G et al (2021) Search for charginos and neutralinos in final states with two boosted hadronically decaying bosons and missing transverse momentum in pp collisions at $\sqrt{s} = 13$ TeV with the ATLAS detector. Phys Rev D 104(11):112010. https://doi.org/10.1103/PhysRevD.104.112010. arXiv:2108.07586 [hep-ex]
39. Tumasyan A et al (2023) Search for electroweak production of charginos and neutralinos at s =13TeV in final states containing hadronic decays of WW, WZ, or WH and missing transverse momentum. Phys Lett B 842:137460. https://doi.org/10.1016/j.physletb.2022.137460. arXiv:2205.09597 [hep-ex]
40. Aad G et al (2020) Search for direct production of electroweakinos in final states with one lepton, missing transverse momentum and a Higgs boson decaying into two b-jets in pp collisions at $\sqrt{s} = 13$ TeV with the ATLAS detector. Eur Phys J C 80(8):691. https://doi.org/10.1140/epjc/s10052-020-8050-3. arXiv:1909.09226 [hep-ex]

41. Aad G et al (2020) Search for electroweak production of charginos and sleptons decaying into final states with two leptons and missing transverse momentum in $\sqrt{s} = 13$ TeV pp collisions using the ATLAS detector. Eur Phys J C80(2):123. https://doi.org/10.1140/epjc/s10052-019-7594-6. arXiv:1908.08215 [hep-ex]
42. ATLAS Collaboration (2022) statistical models.tar.gz of Searches for electroweak production of supersymmetric particles with compressed mass spectra in $\sqrt{s} = 13$ TeV pp collisions with the ATLAS detector (Version 5). HEPData (Other). https://doi.org/10.17182/hepdata.91374.v5/r6
43. Araz JY (2024) Spey: smooth inference for reinterpretation studies. SciPost Phys 16(1):032. https://doi.org/10.21468/SciPostPhys.16.1.032. arXiv: 2307.06996 [hep-ph]

Chapter 4
Constraining the EW-ino Sector of the MSSM Through a Global Likelihood

Abstract The richness of the supersymmetry phenomenology constrains Large Hadron Collider experiments to follow a channel-by-channel approach and to interpret their data through specific simplified models, despite often being unrealistic. Therefore, global reinterpretations of the wealth of experimental results become more and more important in order to understand which scenarios are really excluded and where new physics may still be hiding. The work of this chapter goes in this direction, by using the reinterpretation tool SModelS to combine different ATLAS and CMS searches for electroweak-inos within a global likelihood. It presents the 16 relevant electroweak-ino searches in the SModelS database v2.3, focuses on the impact of the newly implemented analyses compared to v2.1, as well as the impact of a fixed combination of two analyses. Then, a general study dynamically finding the most sensitive combination of uncorrelated analyses and assessing its impact is presented.

The richness of the SUSY phenomenology constrains LHC experiments to follow a channel-by-channel approach and to interpret their data through specific simplified models, despite often being unrealistic. Therefore, global reinterpretations of the wealth of experimental results become more and more important in order to understand which scenarios are really excluded and where new physics may still be hiding. The work of this chapter goes in this direction, by combining individual analyses within a global likelihood. A global likelihood from the combination of different individual analyses is relevant for two reasons. First, the signal of a particular BSM scenario may be manifest in different final states, which are constrained by different analyses. Combining them uses more of the available data and thus provides more robust (and usually stronger) constraints. In any case, the sensitivity of a combination of relevant analyses is always bigger than the sensitivity of any single analysis taken alone. Second, experimental analyses can always be subject to over- or under-fluctuations in the data. In the former case, the observed limit is weaker, in the latter case stronger, than the expected limit. Again, the combination of different, approximately independent analyses can mitigate this effect and provide more robust constraints.

T. Pascal, *Beyond the Standard Model at the Large Hadron Collider: From Simplified Model Constraints to Artificial Proto-Modelling for Dispersed Signals*, Springer Theses, https://doi.org/10.1007/978-3-032-09391-2_4

The gain in exclusion power relative to single-analysis limits was recently demonstrated in [1], which introduced the TACO code, for models with varying degrees of complexity. Such global approach is also attempted by, e.g., [2], which uses the code TACO [1] to constrain the production of squarks and lightest neutralinos through the dynamical combination of SRs from 4 analyses, and by the GAMBIT collaboration: in [3] for the case of a neutralino as the LSP, and in [4] for the case of a gravitino as the LSP. The GAMBIT approach is different from the one developed here in several ways. In particular, GAMBIT aims at maximizing the profile likelihood in a global fit, taking into account also other constraints than those from LHC searches. Regarding LHC searches, GAMBIT performs a simulation-based reinterpretation. Regarding the set of experimental analysis, [3] studied the impact of ATLAS and CMS EW-ino searches for 36 fb^{-1} of 13 TeV LHC data. In contrast, [4] considered 15 ATLAS and 12 CMS searches at 13 TeV, along with a number of ATLAS and CMS measurements of SM signatures.

In this chapter, a different objective is pursued. Instead of asking which region of the EW-ino parameter space best fits the data, it is investigated in depth how the current EW-ino search results constrain this sector of the MSSM. To this end, the combination of analyses introduced in SModelS v2.2 is used to constrain the EW-ino sector of the MSSM by means of global likelihoods. More precisely, is studied which combinations maximise the sensitivity in different regions of parameter space, how fluctuations in the data observed in individual analyses influence the global likelihood, and what is the resulting exclusion power of the combination compared to the analysis-by-analysis approach.

The SModelS version used for this study is v2.3. This means that the "v3.0" analyses listed in Appendix B are not included, except for the Run 2 $1\ell + h(bb) + \not{E}_T$ search (ATLAS-SUSY-2019-08) [5], for which the SModelS database v2.3 has EMs. In particular, the Run 2 soft dilepton + $\not{E}_T$ search (ATLAS-SUSY-2018-16) [6], pointed out in Sect. 2.3.4 as one analysis having recorded a small excess, is not part of the SModelS database v2.3 and is therefore not part of this study.

Finally, the motivation to evaluate constraints on the EW-ino sector of the MSSM has already been discussed in Sect. 2.2.3. However, it is worth adding that the distribution of the SR significances under the SM hypothesis, i.e. the observed deviation from the SM prediction in units of error on the SM prediction, is slightly more skewed in the EW-ino sector compared to the full set of EM-type results validated in SModelS v2.3, as shown in Fig. 4.1. Regarding only 8 TeV results (green histograms), the EW-ino sector has a lower mean significance than the full set of EM-type results (-0.20 versus -0.10, respectively). However, regarding only the 13 TeV results, it is the opposite (0.37 versus 0.15 for an integrated luminosity below 100 fb^{-1}, and 0.14 versus 0.07 for an integrated luminosity above 100 fb^{-1}). This indicates that slightly more excesses are recorded in the EW-ino sector than in the rest of the implemented results, hence motivating the study of this sector.

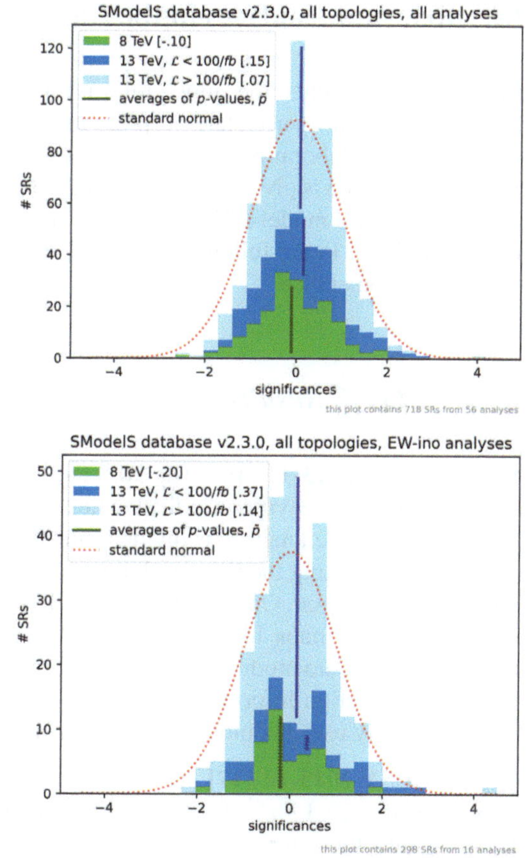

Fig. 4.1 Distribution of the SR significances under the SM hypothesis, for all the validated analyses with EM-type results in the SModelS database v2.3 (top), and for the EW-ino searches only (bottom). A standard normal distribution is expected if no new physics is in the data. New physics would manifest itself as an overabundance of large (positive) significances

This chapter is organised as follows. Section 4.1 describes the relevant EW-ino searches implemented in the v2.3 of the SModelS database. Section 4.2 presents the set of points used in this study. Section 4.3 focuses on the impact of the newly implemented analyses in the SModelS database v2.3 compared to v2.1 [7], as well as the impact of a fixed combination of two analyses. This section is based on [8]. Section 4.4 generalises the approach by dynamically finding the most sensitive combination of uncorrelated analyses (i.e. the combination that is the most likely to reject the BSM hypothesis). This section is based on [9]. Finally, Sect. 4.5 summarises and concludes.

My contribution to this work appears in every section. More specifically, it includes finding the relevant analyses for the study, setting up the data, assessing the combinability of the analyses between each other, designing and implementing the combination strategy, running the code, analysing the outputs, producing the figures and writing up the results.

The complete dataset (input and output files) from the scans used in Sects. 4.3 and 4.4 are made available on Zenodo here [10] and here [11], respectively, ensuring full reproducibility of the results presented.

4.1 Relevant Analyses in the Database

This section presents the most relevant EW-inos searches in the SModelS database v2.3. In particular, it presents which material is provided by the experimental collaborations and how it is used in SModelS. For EM-type results, particular attention is given to the combination of SRs through full [12] or simplified [13, 14] statistical models. The overlap of their SRs among the different analyses is also highlighted. In the following, 'lepton' refers to electrons or muons ($\ell = e, \mu$) unless stated otherwise.

4.1.1 ATLAS Results

ATLAS-SUSY-2013-11 [15] The 20.3 fb^{-1} Run 1 search for charginos, neutralinos and sleptons decaying into final states with two opposite-sign leptons, 0 or $\geq$ 2 jets, and $\not{E}_T$. The analysis is divided into 13 SRs, six targeting the production of sleptons, six targeting the pair-production of charginos and one targeting chargino-neutralino production. Only the latter considers jets. Implemented in the SModelS database are the maps for the observed UL on the cross sections and EMs for slepton and chargino production produced through the MadAnalysis 5 recast [16]. Since no correlation information was provided by the experimental collaboration, SModelS only uses the most sensitive (a.k.a. "best") SR.

ATLAS-SUSY-2013-12 [17] The EMs of this analysis are new in the database v2.3. It is the 20.3 fb^{-1} Run 1 search for chargino-neutralino pair production decaying into three leptons (e, μ or τ) and $\not{E}_T$. At least one electron or muon is required among the three leptons. The ATLAS collaboration published on HEPData the observed ULs on the cross sections and the EMs of 4 SRs [18], each of them targeting a specific (WZ, Wh, $\tilde{e}/\tilde{\mu}$ or $\tilde{\tau}$-mediated) scenario. These 4 EMs are implemented in the SModelS database; no correlation information being available, only the best SR is used.

The SRs of the two analyses above do not overlap, since one requires exactly 2 leptons and the other one exactly 3.

ATLAS-SUSY-2016-24 [19] A search in final states with exactly two or three leptons plus $\not{E}_T$, targeting the EW production of charginos, neutralinos and sleptons. The analysis is based on 36.1 fb^{-1} of data from Run 2. Its three search regions (2ℓ + 0 jets, 2ℓ+jets and 3ℓ), are split into a total of 37 SRs, 28 exclusive and 9 inclusive ones. Acceptance and efficiency maps are available on HEPData [20] for the 9 inclusive 2ℓ SRs and the 11 exclusive 3ℓ SRs and implemented in SModelS. Relevant for EW-inos decaying via SM gauge bosons (as opposed to decays via light sleptons) are 9 SRs: the 3 inclusive 2ℓ+jets SRs and 6 of the exclusive 3ℓ SRs. No correlation information is provided for the exclusive SRs. However, the most sensitive SR reproduces quite well the official limit for $pp \to \tilde{\chi}_1^{\pm}\tilde{\chi}_2^0 \to WZ + \not{E}_T$

from this analysis. Observed ULs on the cross sections are also implemented. The observed limit in this analysis is about 1σ stronger than the expected one.

ATLAS-SUSY-2017-03 [21] Another search in two-lepton and three-lepton final states based on 36.1 fb^{-1} of Run 2 data. The main difference to ATLAS-SUSY-2016-24 above is that it uses a recursive jigsaw reconstruction. Moreover, it targets only chargino-neutralino pair production with decays via W/Z bosons. It considers two types of search regions, one with two leptons and at least two jets (2ℓ category), and one with three leptons and up to three jets (3ℓ category). In total, the analysis has 8 SRs, for all of which EMs for the $pp \to \tilde{\chi}_1^{\pm}\tilde{\chi}_2^0 \to WZ + \not{E}_T$ simplified model are available on HEPDATA [22] and implemented in SMODELS. Lacking an explicit statistical model, SMODELS uses only the best SR for the statistical interpretation, which leads to a slight under-exclusion compared to the official ATLAS result, which combines the 2ℓ and 3ℓ SRs of the same type. The observed ULs on the cross section are also implemented. The observed limit is about 1σ weaker than the expected one in the low mass region, and about 1σ stronger for LSP masses around 300 GeV.

The SRs not being orthogonal, ATLAS-SUSY-2016-24 and ATLAS-SUSY-2017-03 are not combinable in this study; only one of them can enter the global combination for any given parameter point.

ATLAS-SUSY-2018-05 [23] This analysis is new in the database v2.3. It is a search in final states with an e^+e^- or $\mu^+\mu^-$ pair, jets, and $\not{E}_T$. This and the following ATLAS analyses are based on 139 fb^{-1} of Run 2 data. The analysis is done in two parts, one targeting $pp \to \tilde{\chi}_1^{\pm}\tilde{\chi}_2^0 \to (W\tilde{\chi}_1^0)\,(Z\tilde{\chi}_1^0) \to (q\bar{q}'\tilde{\chi}_1^0)\,(\ell\bar{\ell}\tilde{\chi}_1^0)$, the other targeting pair production of colored SUSY particles (squarks or gluinos) decaying through the next-to-lightest neutralino. Since these two parts are completely distinct, the EM-type results are implemented in SMODELS with different analysis IDs: ATLAS-SUSY-2018-05-ewk (comprising EMs for 13 SRs) and ATLAS-SUSY-2018-05-strong (comprising EMs for 30 SRs). The analysis provides extensive material on HEPDATA [24], including observed ULs on the cross sections, acceptance and efficiency values for all SRs, and the full HISTFACTORY statistical models in JSON format [25].

The EW implementation in SMODELS contains EMs for the $WZ^{(*)} + \not{E}_T$ signature in 13 SRs. These EMs were extracted from the `ewk_signal_patchset.json` file contained in [25], instead of the HEPDATA tables. As can be seen in Fig. 4.2 (top row), the combination of SRs is important to reproduce the ATLAS limit. The full and SIMPLIFY'edstatistical models give very similar results, so the latter is used by default. Control regions are ignored, as including them does not improve the agreement with the official limits from ATLAS. The observed limit is about 1σ stronger than the expected one.
The SRs of this analysis overlap with those of ATLAS-SUSY-2016-24 and ATLAS-SUSY-2017-03.

ATLAS-SUSY-2018-06 [26] This is an EW-ino search in the three-leptons plus $\not{E}_T$ final state by means of the recursive jigsaw reconstruction technique [27, 28]. It specifically targets the channel $pp \to \tilde{\chi}_1^{\pm}\tilde{\chi}_2^0 \to (W\tilde{\chi}_1^0)\,(Z\tilde{\chi}_1^0) \to (\ell\nu\tilde{\chi}_1^0)\,(\ell\ell\tilde{\chi}_1^0)$. The analysis has two SRs, one vetoing jets, and one requiring 1–3 jets from initial-state radiation. The auxiliary material provided on HEPDATA [29] includes observed and expected ULs on the cross section, EMs for the $WZ^{(*)} + \not{E}_T$ signature in the 2 SRs as well as the full HISTFACTORY statistical model. A complication arises from the fact that the `BkgOnly.json` model [30] leads to issues[1] which so far could not be clarified. While its SIMPLIFY'ed version works and yields reasonable results, the best SR is also a good approximation and much faster, so only the best SR results will be used for this analysis. The observed limit is about 1σ weaker than the expected one.

The SRs of this analysis overlap with those of ATLAS-SUSY-2016-24 and ATLAS-SUSY-2017-03.

ATLAS-SUSY-2018-32 [31] This analysis is new in the database v2.3. It is a search for EW production of charginos or sleptons decaying into final states with two leptons and $\not{E}_T$. It considers events with 0–1 jets, but vetoes b-tagged jets. The analysis is made of 36 exclusive SRs (binned in m_{T2}) and 16 inclusive SRs (overlapping in m_{T2}). The HEPDATA entry [32] includes observed ULs on the cross sections, acceptance and efficiency values of the $pp \to \tilde{\chi}^+\tilde{\chi}^- \to W^+W^- + \not{E}_T$ simplified model for all SRs, as well as the full HISTFACTORY statistical models (`bkgonly.json` and signal patchsets).

Implemented in SMODELS are $WW + \not{E}_T$ EMs for the 36 binned SRs. These were extracted from the `C1C1WW` signal patchsets contained in [33]. The SIMPLIFY'ed statistical model reproduces well the official observed limit but somewhat overestimates the expected limit, see Fig. 4.2 (middle). It is nonetheless used in order to save CPU time. The best SR does not give a sensible limit. This analysis relies on a combined fit of signal and control regions, so the latter must not be pruned when patching the signal counts in the SRs (this means setting `includeCRs:True` in the analysis' `globalInfo.txt` file, see [34]).
It is noteworthy that the SRs of this analysis do not overlap with the SRs of ATLAS-SUSY-2018-05. Indeed, this analysis requires at most 1 jet, while ATLAS-SUSY-2018-05 requires at least 2 jets, except for 1 SR which requires only 1 jet. However, this SR looks for opposite-sign same-flavor (OSSF) leptons with $m_{\ell\ell} \in [71, 111]$ GeV, while ATLAS-SUSY-2018-32 requires the dilepton mass of the OSSF lepton pair to be higher than 121.2 GeV. Moreover, ATLAS-SUSY-2016-24 and ATLAS-SUSY-2017-03 SRs targeting direct EW-ino decays into the LSP with 2 leptons require at least 2 jets. They hence do not overlap with those of ATLAS-SUSY-2018-32.

[1] See PYHFissue #1320 for details.

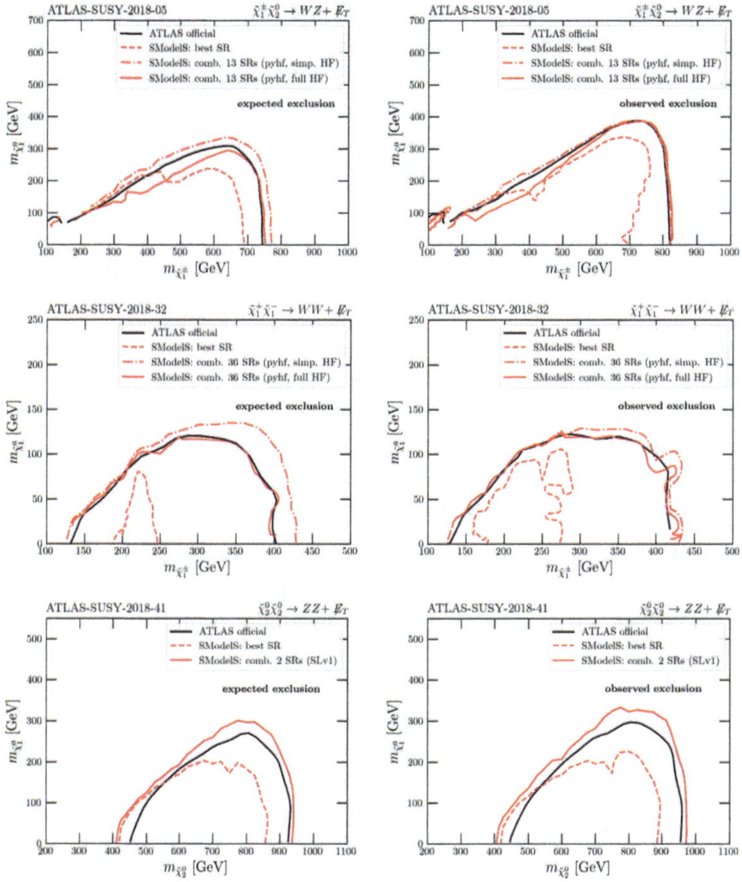

Fig. 4.2 Examples for the validations of the ATLAS-SUSY-2018-05 (top), ATLAS-SUSY-2018-32 (middle) and ATLAS-SUSY-2018-41 (bottom) EMs in SMODELS v2.3

ATLAS-SUSY-2018-41 [35] This analysis is new in the database v2.2 and the EMs were updated in v2.3. It is a search for charginos and neutralinos in fully hadronic final states, using large-radius jets and jet substructure information to identify high-p_T W, Z or Higgs bosons. Two orthogonal SR categories, 4Q and 2B2Q, are defined according to the $qqqq$ and $bbqq$ final states, respectively; events with leptons are vetoed. Multiple SRs are defined in each category to target final states from different combinations of SM bosons (WW, WZ, Wh, ZZ, Zh, hh). Moreover, 3 SRs are inclusive in $V = W, Z$: SR-2B2Q-VZ, SR-2B2Q-Vh and SR-4Q-VV. Acceptance and efficiency tables are provided on HEPDATA [36] for the three inclusive $V = W, Z$ SRs for different signal hypotheses; their implementation in SMODELS covers WW, WZ, Wh, ZZ, Zh, hh $(+\not{E}_T)$ from chargino/neutralino production, provided the masses of the produced particles are not too different. Are also implemented the observed and expected UL-type results.

Since the SR-2B2Q-VZ, SR-2B2Q-Vh and SR-4Q-VV signal regions are described as statistically independent in [35], SModelS combines them by means of a trivial (diagonal) correlation matrix. This means the covariance matrix is given by the background uncertainty squared on the diagonal, $(\text{cov})_{ii} = (\delta b_i)^2$, and zero otherwise.[2] An example of the validation is shown in Fig. 4.2 (bottom). For most scenarios (with the exception of the $WW + \not{E}_T$ channel) the analysis poses a stronger limit than expected. The effect is only about 1σ but, as we will see, has a strong influence in the combination.
Since leptons are vetoed, the SRs of this analysis do not overlap with any of the other analyses.

ATLAS-SUSY-2019-02 [37] This analysis is new in the database v2.3. It is another search in final states with two leptons plus $\not{E}_T$. Like ATLAS-SUSY-2018-32 above, it considers events with opposite-sign di-leptons (of same or different flavor) with 0 or 1 non-b-tagged jets. The difference is that ATLAS-SUSY-2019-02 specifically targets the kinematic region where $m_{\text{mother}} - m_{\text{LSP}}$ is close to the W-boson mass. Specific SRs are defined for targeting slepton or chargino production.

Observed ULs on the cross sections, as well as acceptance and efficiency tables are provided on HEPData [38] for all SRs and implemented in the SModelS database. However, no statistical model is available. For the slepton simplified model, the best SR provides a fairly good sensitivity. This is, however, not the case for the chargino simplified model: here, SR combination is essential, as the best SR alone has no sensitivity. Lacking more information, it was tried, among others, the assumption that the exclusive different-flavor (DF) SRs are correlated and the exclusive SF SRs are correlated, but DF and SF are not correlated with each other. This indeed allowed SModelS to reproduce the official ATLAS limit on $pp \to \tilde{\chi}_1^+ \tilde{\chi}_1^- \to W^+W^- + \not{E}_T$. Observed and expected limits agree in this case.
The analysis overlaps with ATLAS-SUSY-2018-32, but, since the SRs targeting EW-ino production veto jets, they do not overlap with SRs of other analyses.

ATLAS-SUSY-2019-08 [39] A search for chargino-neutralino pairs in final states with $W \to \ell\nu$ and $h \to b\bar{b}$. The signal selection thus requires one lepton, a pair of b-tagged jets consistent with the decay of a Higgs boson, plus $\not{E}_T$. Three sets of SRs target 'low mass', 'medium mass' and 'high mass' scenarios through cuts on the transverse mass variable m_{T}. Each of the three m_{T} regions is further binned in three m_{CT} regions, resulting in 9 exclusive SRs. In addition, there are 3 inclusive SRs with only lower cuts on m_{T} and m_{CT}.

The analysis provides acceptance and efficiency values for the $Wh + \not{E}_T$ simplified model for all 9 exclusive SRs on HEPData [40], together with the full HistFactory statistical model and observed ULs on the cross section. Combining

[2] In the meanwhile, ATLAS provided also the full statistical model and extensive patchsets for all SRs on HEPData; their implementation and validation in SModelS is ongoing.

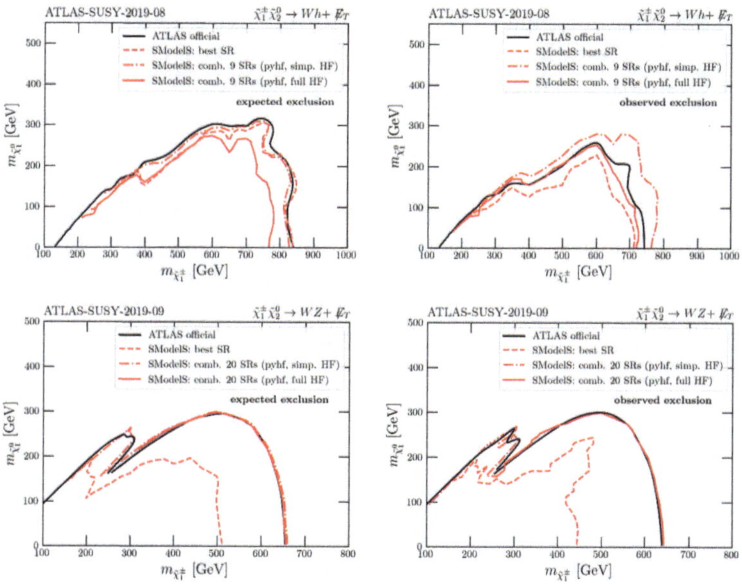

Fig. 4.3 Validation plots for the ATLAS-SUSY-2019-08 (top) and ATLAS-SUSY-2019-09 (bottom) EMs in SModelS v2.3

the 9 SRs by means of the SIMPLIFY'ed statistical model results in a small over-exclusion, so SModelS uses the full one, although it requires significantly more CPU time (several minutes instead of $\mathcal{O}(1)$ sec). The validation is shown in Fig. 4.3 (top). The observed limit from this analysis is about 1σ lower than the expected limit.

Given the $1\ell + 2b$ requirement, the SRs of this analysis do not overlap with any other analysis.

ATLAS-SUSY-2019-09 [41] This analysis is new in the database v2.2. It is another search for chargino-neutralino pairs in three-lepton final states with $\not{E}_T$. Events are classified by jet multiplicity (0 or ≥ 1 jets) with a veto on b-tagged jets. The analysis considers leptonically decaying W, Z and SM Higgs bosons with SRs optimised for on-shell WZ, off-shell WZ or Wh selections. In total, there are 20 SRs for on-shell WZ, 31 SRs for off-shell WZ and 21 SRs for Wh selections.

The HEPData record [42] provides observed and expected ULs on the cross section, as well as the full HistFactory statistical models together with patches for the on- and off-shell WZ signal models considered in the paper.[3] Truth-level acceptances and reconstruction efficiencies are also provided on HEPData [42] but only for inclusive SRs, which do not allow one to reproduce the official limits

[3] It is noteworthy that [43] also includes recipes of how to do a combined fit of the on-shell and the off-shell channels, and how to combine this analysis with the "2-lepton compressed" analysis [6].

from ATLAS. The EMs implemented in SModelS have therefore been extracted from the JSON patchsets. By default, the SIMPLIFY'ed statistical model is used, which gives almost identical results to the full one, see Fig. 4.3 (bottom). As for ATLAS-SUSY-2018-32, it is important to include the background yields in the control regions in the statistical evaluation.

For the Wh selection, however, no JSON patches are available; this part of the analysis could not be validated and is therefore not included in the SModelS database. This is a pity, as a small excess (between 1–2σ) was observed in this case. A small excess was also reported in the off-shell WZ channel for compressed spectra.
The SRs overlap with ATLAS-SUSY-2016-24, ATLAS-SUSY-2017-03 and ATLAS-SUSY-2018-06.

4.1.2 CMS Results

CMS-SUS-13-006 [44] This is a 19.5 fb^{-1} Run 1 search for EW-inos in signatures with leptons (e, μ and/or τ) along with W, Z, and Higgs bosons and $\not{E}_T$ final states. Only the ULs on the observed cross sections are published and therefore implemented in the database.

CMS-SUS-13-012 [45] An inclusive search for new physics in multijet events with large $\not{E}_T$ from Run 1. The analysis was designed as a generic search for gluinos and squarks and has 36 SRs characterised by jet multiplicity, the scalar sum of jet transverse momenta, and the $\not{E}_T$. Observed and expected ULs on the cross sections are implemented, and besides some EMs available on the analysis wiki page, the SModelS database contains a large number of EMs maps which were obtained through a MadAnalysis 5 recast [46]; these include also WW, WZ and ZZ EMs for chargino/neutralino production. Only the best SR is used for limit setting.

CMS-SUS-16-039 [47] The EMs of this analysis are new in the database v2.2. It is a search for charginos and neutralinos in multilepton final states from Run 2 with 36 fb^{-1}. Considered are final states with 2 same-sign leptons as well as final states with 3–4 leptons including up to 2 τ. The observed ULs on the cross sections are implemented in the database. Regarding the EMs, the analysis is made of 11 search categories, each subdivided into up to 44 SRs. CMS provided EMs for a set of 8 "super signal regions" defined by simpler selections, but these turned out to have very little sensitivity. The EMs used in SModelS were therefore obtained through a MadAnalysis 5 recast [48]. These are for the "category A" SRs (3 e/μ's that form at least one OSSF pair). To save CPU time, the 43 SRs of this category are aggregated to 11 in the SModelS database. The covariance matrix provided by CMS is used to combine the SRs. The combination of the 11 aggregated SRs reproduces fairly well the observed limit from CMS but underestimates the expected exclusion, see Fig. 4.4 (top). That the analysis saw a slight excess in $3\ell + \not{E}_T$ events, so the observed limit is slightly weaker than expected.

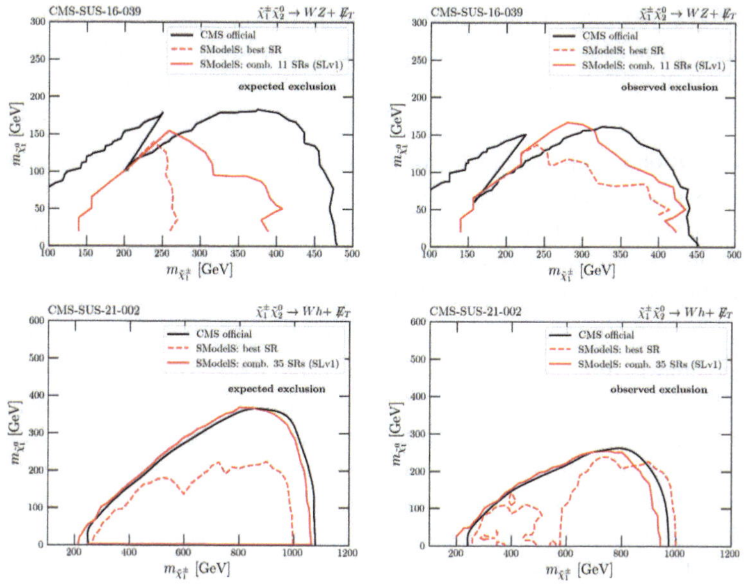

Fig. 4.4 Validation plots for the CMS-SUS-16-039 (top) and CMS-SUS-21-002 (bottom) EMs in SModelS v2.3

CMS-SUS-16-048 [49] This analysis is new in the database v2.2. This is a 36 fb^{-1} Run 2 search in final states with two soft leptons of opposite electric charge and of same or different flavor, originating from off-shell W and Z boson decays, and $\not{E}_T$. To reduce the SM background, the 2 leptons are required to be produced along with at least one hard, non-b-tagged jet, which is supposed to come from ISR. The search is then sensitive to lepton transverse momenta between 5 and 30 GeV.

This way the analysis targets compressed spectra of EW-inos or stops, with 12 SRs optimised for $\tilde{\chi}_1^{\pm}\tilde{\chi}_2^0$ and 9 SRs optimised for stops. CMS supplied a covariance matrix for the 21 SRs; the corresponding efficiency×acceptance values for EW-inos or stops were, however, not made available. The EMs implemented in the SModelS database were therefore obtained through a MadAnalysis 5 recast [50]. For mass differences around 20 GeV, the observed limit is about 1σ stronger than the expected one.
The CMS-SUS-16-048 analysis and CMS-SUS-16-039 are orthogonal to each other.

CMS-SUS-17-004 [51] This is a statistical combination of several searches for EW-inos using 35.9 fb^{-1} of data from Run 2 [47, 49, 52–55]. All of them, except [54], are implemented in the SModelS database. In addition, the analysis performs a search for EW-inos decaying into three or more electrically charged leptons + $\not{E}_T$. Only the observed ULs on the cross sections are implemented in the database.

CMS-SUS-18-004 [56] This analysis is new in the database v2.2. It is a 137 fb^{-1} Run 2 search for EW-inos and top squarks with a low mass difference with the lightest neutralino. Selected events contain two or three low-momentum leptons and $\not{E}_T$. Only observed and expected ULs on the cross sections are implemented.

CMS-SUS-20-004 [57] This analysis is new in the database v2.3. It is a search for new physics in channels with two Higgs bosons, each decaying into $b\bar{b}$, and large $\not{E}_T$, using 137 fb^{-1} of Run 2 data. Events with leptons are vetoed. It comprises 22 SRs with 3–4 b-jets, which are binned in p_T; 16 of these SRs target a resolved signature and 6 target a boosted signature.

CMS provides observed and expected cross section UL maps, as well as a covariance matrix for the 22 SRs together with EMs for a $pp \to \tilde{\chi}_2^0 \tilde{\chi}_2^0 \to (h\tilde{\chi}_1^0)(h\tilde{\chi}_1^0)$ simplified model on HEPDATA [58], which are implemented in the SMODELS database. Since the reported uncertainties are asymmetric, the SLv2 (Gaussian with a skew) [59] approach is used for SR combination. The necessity of going beyond the Gaussian approximation for this analysis was discussed in detail in [60]. It is interesting to note that this analysis sees deviations from the SM of the level of about 1–2σ in several SRs.
Given the multi b requirement, the SRs of this analysis do not overlap with any other analysis.

CMS-SUS-21-002 [61] This analysis is new in the database v2.3. It is the CMS all-hadronic search for EW-inos with 137 fb^{-1} of Run 2 data. It considers final states with large $\not{E}_T$ and pairs of hadronically decaying SM bosons (W, Z or Higgs), which are identified using novel algorithms. In total 35 SRs are defined in one b-veto search region (0 b-tagged jets) and three b-tag search regions ($\geq$ 1 b-tagged jets). Events with leptons are vetoed.

Both observed and expected cross section UL maps are available on HEPDATA [62] and are implemented in the database. Efficiency maps for Wh, WZ and $WW + \not{E}_T$ simplified models as well as a covariance matrix for all SRs are available as ROOT files on the analysis wiki page (but not on HEPDATA). A validation example illustrating the importance of the SR combination is shown in Fig. 4.4 (bottom). The observed limit is weaker than the expected one, which signals an excess in observed events. As for the hadronic ATLAS search (which saw a deficit), the effect is of the order of 1σ but given the high sensitivity of the analysis it will play an important role in the global likelihood.
CMS-SUS-20-004 and CMS-SUS-21-002 overlap but are orthogonal to the other CMS analyses.

4.2 Setup of the Numerical Analysis

To set global constraints on the EW-ino sector of the MSSM, the EW-ino scan points from [7] are reused. This not only saves CPU time, it also allows for direct comparison with the earlier publications. In Sect. 4.2 of [7], M_1, M_2, μ, and $\tan\beta$ were randomly scanned over within the following ranges:[4]

$$
\begin{aligned}
10\,\text{GeV} < &\; M_1 \; < 3\,\text{TeV} \\
100\,\text{GeV} < &\; M_2 \; < 3\,\text{TeV} \\
100\,\text{GeV} < &\; \mu \; < 3\,\text{TeV} \\
5 < &\tan\beta < 50
\end{aligned}
$$

All other SUSY breaking parameters were fixed to 10 TeV, assuming that the stop-sector parameters can always be adjusted such that $m_h \simeq 125$ GeV without influencing the EW-ino sector. The lower limits on M_2 and μ were chosen so as to avoid the LEP constraints on light charginos, while the bounds on $\tan\beta$ were chosen to avoid Yukawa couplings from becoming non-perturbatively large. The mass spectra and decay tables were computed with SOFTSUSY 4.1.11, setting $m_h = 125$ GeV for consistency of the decay calculations. Following the SLHA conventions [63], in the absence of CP violation the neutralino and chargino mixing matrices are taken to be real, allowing for *signed* (negative) mass eigenvalues for the neutralinos. All points have a neutralino LSP.

From the close to 100k points of the complete scan in [7], is selected the subset of points with only prompt decays (no long-lived particles, all decay widths $\Gamma_{\text{tot}} > 10^{-11}$ GeV).[5] Moreover, $m_{\tilde{\chi}_1^0} < 500$ GeV and $m_{\tilde{\chi}_1^\pm} < 1200$ GeV are required in order to focus on the region to which the current prompt EW-ino searches are sensitive. Moreover, a few points with erroneous branching ratios in SOFTSUSY 4.1.11[6] were removed. This leaves a total of 18544 points, which are analysed with SModelS v2.3.

When running SModelS, a `sigmacut` of 10^{-3} fb is used; this parameter sets the minimum cross section value considered in SLHA decomposition. Moreover, a `minmassgap` of 5 GeV (the default SModelS value) will first be used in Sect. 4.3. A different value will be used in Sect. 4.4. This parameter defines the minimum

---

[4] While in principle there can be a relative phase between the EW-ino mass parameters, for simplicity only non-negative values are considered in this study ($M_1, M_2, \mu \geq 0$).

[5] While searches for long-lived particles are included in the SModelS database, they are very different from the analyses searching for promptly decaying EW-inos and, if relevant, their constraints are completely dominant. Therefore, models featuring long-lived EW-inos, i.e. models with a pure wino LSP will not be considered here.

[6] Due to a bug in the transition from 2-body to 3-body neutralino decays, which was in the meanwhile corrected, see https://ballanach.github.io/softsusy/.

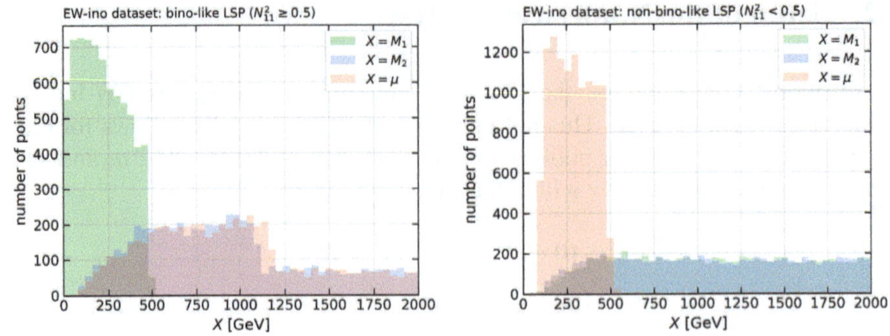

Fig. 4.5 Distributions of M_1, M_2 and μ in the set of EW-ino scan points used in this study; on the left for points with a bino-like LSP, on the right for points with a non-bino-like LSP. Only values up to 2 TeV are shown

mass difference for mass compression in SModelS:[7] if BSM particle P_2 decays into BSM particle P_1 but the mass difference $m_{P_2} - m_{P_1} <$ `minmassgap`, the decay products are assumed to be potentially too soft to be visible in a typical LHC analysis. Two signal topologies are hence compared with the simplified model results in the database: the compressed one, where the decay $P_2 \to P_1 + X$, X being any SM decay product(s), is ignored, and the non-compressed one, where the decay is kept. This is relevant in particular if one or more SUSY particles are very close in mass to the LSP, as is the case for non-bino-like LSP points.

In what follows, it is often relevant to distinguish between scenarios with a bino-like LSP and those with non-bino-like LSP. Model points are considered to feature a bino-like LSP when the $\tilde{\chi}_1^0$ has at least 50% bino component, i.e. $N_{11}^2 \geq 0.5$, where N is the neutralino mixing matrix defined in Eq. (1.86). In turn, non-bino-like LSP points ($N_{11}^2 < 0.5$) feature a mostly higgsino-like, mostly wino-like, or strongly mixed $\tilde{\chi}_1^0$. Figure 4.5 shows the distributions of M_1, M_2 and μ values in the dataset; in the left plot for points with a bino-like LSP and in the right plot for points with a non-bino-like LSP. The cutoff at $M_1 \approx 500$ GeV in the plot on the left comes from the requirement $m_{\tilde{\chi}_1^0} < 500$ GeV, while the edges around M_2, $\mu \approx 1.2$ TeV come from $m_{\tilde{\chi}_1^\pm} < 1200$ GeV. In the plot on the right, $m_{\tilde{\chi}_1^0} < 500$ GeV together with the requirement of promptly decaying $\tilde{\chi}_1^\pm$ leads to the edge at $\mu \approx 500$ GeV (higher values of μ being hardly visible in the plot), while M_1 and M_2 can range from small to very large values. That the M_2 distribution is suppressed towards small values, instead of showing an edge around 500 GeV, is due to the fact that light winos are generally long-lived; as mentioned, these cases were removed from the scan. Also, there is no edge around 1.2 TeV, because for non-bino-like LSP points the $\tilde{\chi}_1^\pm$ is always close in mass to the $\tilde{\chi}_1^0$ and thus $m_{\tilde{\chi}_1^\pm} < 540$ GeV.

[7] A detailed description of the mass compression procedure is available in the SModelS online documentation: https://smodels.readthedocs.io/en/latest/Decomposition.html#masscomp.

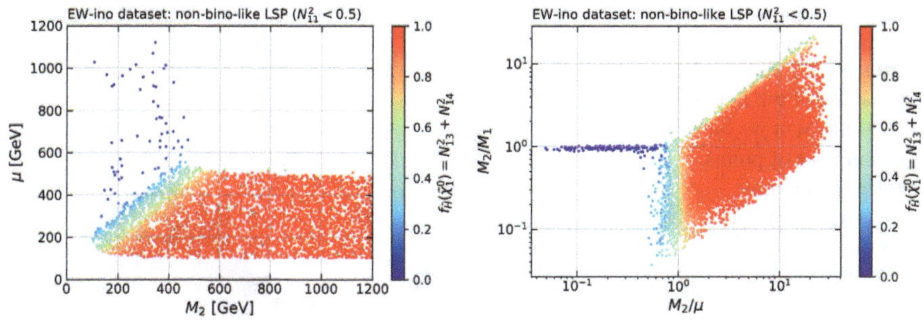

Fig. 4.6 Non-bino-like LSP points, on the left in the μ vs. M_2 plane and on the right in the M_2/M_1 vs. M_2/μ plane. The colour code shows higgsino content of the LSP. In the left plot, the x- and y-axes go only up to 1.2 TeV for better visibility of the mixed region, but in both directions scan points extend up to 3 TeV

The properties of non-bino-like LSP points can be understood from Fig. 4.6, which shows these points in the μ vs. M_2 plane (left) and the M_2/M_1 vs. M_2/μ plane (right). The colour code indicates the higgsino content of the LSP: for red points the $\tilde{\chi}_1^0$ is mostly higgsino, while for purple points it is mostly wino. In between, from yellow to green to turquoise, the $\tilde{\chi}_1^0$ is a strongly mixed state of higgsino and wino ($\mu \approx M_2$), higgsino and bino ($\mu \approx M_1$) or wino and bino ($M_2 \approx M_1$). The sparsity of points at low M_2 (visible in the left plot) is due to the lifetime constraint, as points with long-lived charginos have been removed from the scan. In fact, for all the remaining points in the region $M_2 < 500$ GeV and $\mu \gtrsim M_2 + 100$ GeV in Fig. 4.6, the bino mass parameter M_1 is close to M_2, roughly $M_1 \in [1, 1.2] \times M_2$. This increases the splitting between the $\tilde{\chi}_1^0$ and the $\tilde{\chi}_1^\pm$, such that the latter is no longer long-lived.

Each point in the EW-ino dataset is confronted against the experimental results of the analyses detailed in Sect. 4.1. Often, a point is considered excluded if the highest observed r-value is greater or equal to 1: $r_{\mathrm{obs}}^{\mathrm{max}} \geq 1$. This corresponds to an exclusion by the *most constraining* analysis. The statistically more sound approach, however, is to base the exclusion on the *most sensitive* (or "best") analysis, i.e. the analysis with the highest r_{exp}. A point is then considered as excluded if the corresponding observed r-value is greater or equal to 1: $r_{\mathrm{obs}}^{\mathrm{best}} \geq 1$. As we will see, either approach is quite sensitive to statistical fluctuations, which is a motivation to attempt a global combination.

4.3 Combination of Analyses: A Simple Example

In this section, the impact of the new EW-ino analyses implemented in the SMODELS database v2.3 is evaluated by comparing its results on the dataset described above to the results obtained using the SMODELS database v2.1 originally

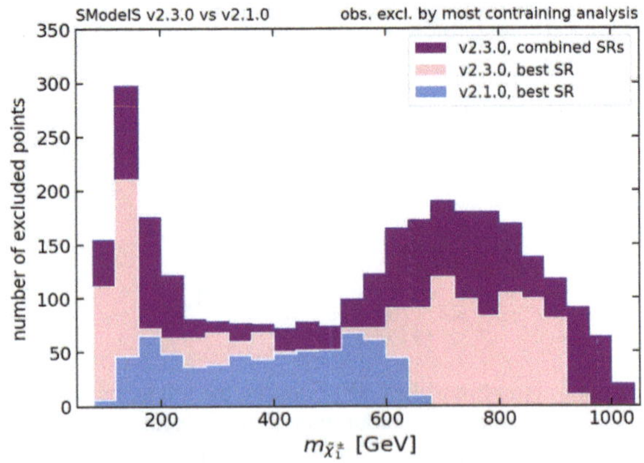

Fig. 4.7 Comparison of exclusion power of SModelS v2.3 versus v2.1 as a function of the lighter chargino mass, $m_{\tilde{\chi}_1^\pm}$. Shown are the number of points excluded by the most constraining analysis, i.e. the number of points with $r_{\max} \equiv \max(r_{\text{obs}}) \geq 1$

presented in [7]. In addition, a simple combination of the two ATLAS and CMS hadronic searches will be presented (since they belong to two different experiments, they are here assumed to be approximately uncorrelated).

4.3.1 Comparison of v2.3 with v2.1

The increase of exclusion power as compared to [7] is illustrated in Fig. 4.7, which shows the number of excluded points as a function of the $\tilde{\chi}_1^\pm$ mass. The plot compares v2.1 to v2.3 with and without SR combination.[8] Different analyses are not combined at this stage. The new experimental results in v2.3 extend the reach in chargino mass by about 300 GeV in the best SR approach (i.e., when SRs are not combined). Combination of SRs extends this reach by another 100 GeV, but mostly it increases the number of excluded points in the $m_{\tilde{\chi}_1^\pm} \approx$ 500–1000 GeV mass range; this concerns primarily scenarios with a bino-like LSP. The effect comes mostly from the ATLAS-SUSY-2018-41 analysis, where the combination of SRs allows to simultaneously take into account the signal contributions to VV, Vh and hh final states ($V = W^\pm, Z$), thus increasing the constraining power of the search. Finally, there is also a significant increase in the number of excluded points at low $m_{\tilde{\chi}_1^\pm} \lesssim 200$ GeV; these are to large extent points with a higgsino-like LSP. Overall, the number of excluded points increases from 661 with v2.1 to 2974 (1787) with v2.3 when SR combination is turned on (off).

Given Fig. 4.7, it is interesting to ask which experimental results are driving the exclusion in different regions of the parameter space. To answer this question, Fig. 4.8 shows the points excluded by the LHC searches in the SModelS v2.3

[8] The combination of SRs is turned on/off by setting `combineSRs=True/False` in the `parameters.ini` file.

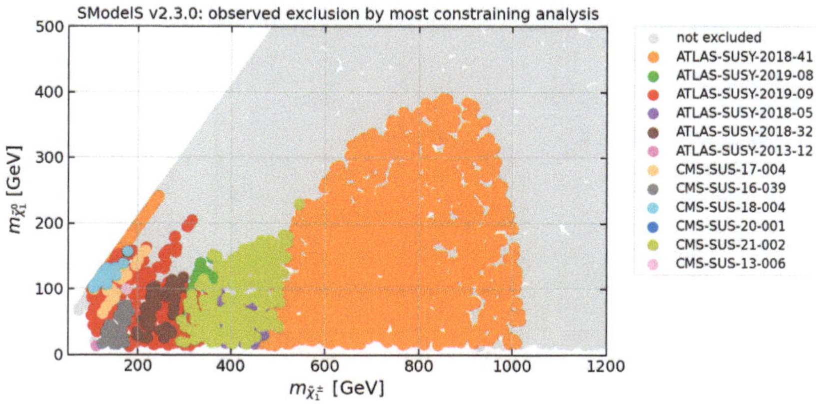

Fig. 4.8 Scan points excluded by the most constraining analysis in SModelS v2.3, with combination of SRs turned on. The colour denotes the analysis that gives the highest r-value (see legend). Grey points are not excluded

database in the $m_{\tilde{\chi}_1^\pm}$ versus $m_{\tilde{\chi}_1^0}$ plane. The color of each excluded point denotes the most constraining analysis, that is the analysis giving the highest observed r-value. As can be seen, in the low mass range, $m_{\tilde{\chi}_1^\pm} \lesssim 500$ GeV, the constraints come from a variety of different analyses, while the high mass range, $m_{\tilde{\chi}_1^\pm} \gtrsim 500$ GeV is completely dominated by the ATLAS-SUSY-2018-41 analysis. The reason for this is that the ATLAS-SUSY-2018-41 analysis observed less events than expected (at least in the three super signal regions for which EMs are available) and therefore sets stronger limits than expected, and also stronger limits than the equivalent analysis from CMS, CMS-SUS-21-002, which saw a small excess in events.

This brings us to the issue that the most constraining analysis is not necessarily also the most sensitive one. From the statistics point of view, however, as long as limits are set on an analysis-by-analysis basis, using only the most sensitive result is the more rigorous approach in order to stay at the 95% confidence level. Indeed, as can be seen in Fig. 4.9, the picture changes quite significantly when considering the exclusion from the most sensitive analysis only. In particular when the most sensitive analysis is either the ATLAS or the CMS search in fully hadronic final states, the excluded region shrinks from $(m_{\tilde{\chi}_1^\pm}, m_{\tilde{\chi}_1^0}) \lesssim (1000,\ 400)$ GeV in Fig. 4.8 to $(m_{\tilde{\chi}_1^\pm}, m_{\tilde{\chi}_1^0}) \lesssim (850,\ 250)$ GeV in Fig. 4.9.

4.3.2 Combination of the Two Hadronic Searches

The assessment of the excluded parameter space can be improved by statistically combining the relevant analyses, as already demonstrated in Fig. 3.10 for a specific benchmark point. The results in Figs. 4.8 and 4.9 indeed motivate a combination of

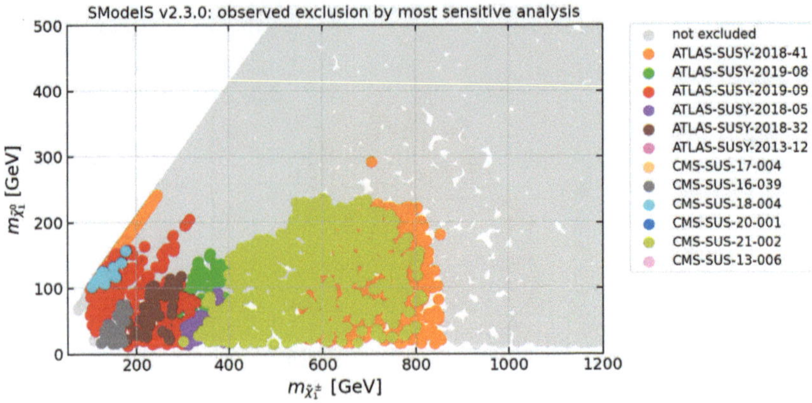

Fig. 4.9 As Fig. 4.8 but for points excluded by the most sensitive analysis

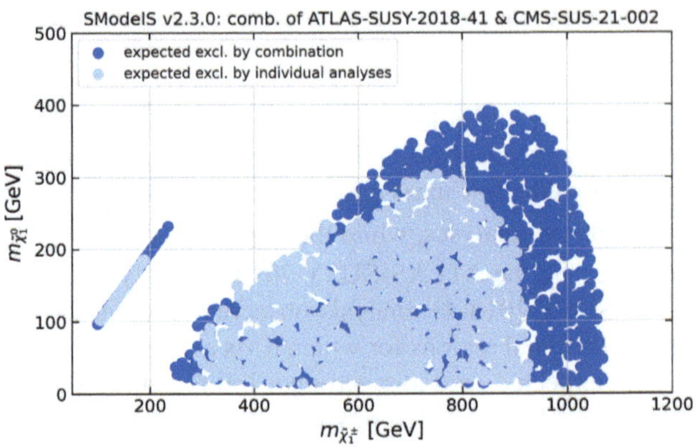

Fig. 4.10 Effect of statistically combining the hadronic EW-ino searches from ATLAS and CMS, ATLAS-SUSY-2018-41 and CMS-SUS-21-002 on the **expected reach**: in light blue the expected exclusion of the individual ATLAS or CMS analyses, in dark blue the expected exclusion of the combination

the two hadronic EW-ino searches, ATLAS-SUSY-2018-41 and CMS-SUS-21-002, as they are the most sensitive and/or most constraining ones in the high $m_{\tilde{\chi}_1^\pm}$ range (roughly for $m_{\tilde{\chi}_1^\pm} \gtrsim 400$ GeV). Figure 4.10 shows how the combination of these two analyses improves the sensitivity to EW-ino signals as compared to the single analysis approach: the expected limits are extended by about 200 GeV in chargino mass and by up to 100 GeV in LSP mass. The effect on the observed exclusion is shown in Fig. 4.11. Here, the khaki (light yellowish) and dark red coloured points are excluded by the combination, with the dark red points being those which are excluded only by the combination but not by either the ATLAS or the CMS analysis alone. In contrast, the light blue points would be excluded by one analysis, but are not excluded any more in the combination.

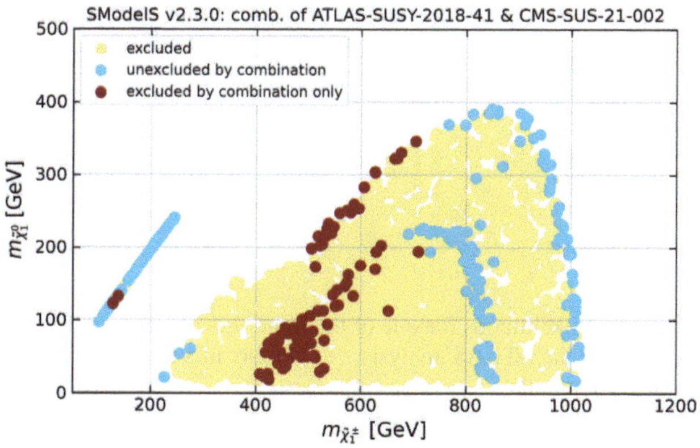

Fig. 4.11 Effect of statistically combining the hadronic EW-ino searches from ATLAS and CMS, ATLAS-SUSY-2018-41 and CMS-SUS-21-002 on the **observed limit**. The khaki (light yellowish) and dark red coloured points are excluded at 95% confidence level by the combination, with the dark red points being those which are excluded only by the combination but not by one of the two analyses alone. The light blue points would be excluded by either ATLAS-SUSY-2018-41 or CMS-SUS-21-002, but are not excluded any more in the combination

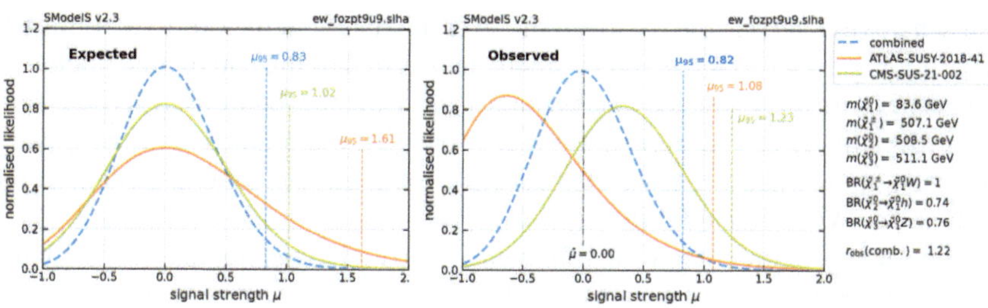

Fig. 4.12 Visualisation of likelihoods for one of the dark red points from Fig. 4.11, i.e. a point which is only excluded by the combination of the considered ATLAS and CMS analyses, but not by any of the two analyses alone. See text for details

Concrete examples for a dark red and a light blue point from Fig. 4.11 are given in Figs. 4.12 and 4.13, respectively. The sample point in Fig. 4.12 lies at $(m_{\tilde{\chi}_1^\pm}, m_{\tilde{\chi}_1^0}) =$ (507, 84) GeV, the point in Fig. 4.13 at $(m_{\tilde{\chi}_1^\pm}, m_{\tilde{\chi}_1^0}) =$ (834, 36) GeV. Both feature a bino-like $\tilde{\chi}_1^0$ and higgsino-like $\tilde{\chi}_1^\pm$, $\tilde{\chi}_{2,3}^0$. Moreover, in both cases neither the ATLAS nor the CMS hadronic analysis is expected to exclude the point; only the combination of both analyses gives high enough sensitivity. For the point in Fig. 4.12, this behaviour is replicated also with the observed data: indeed, $r_{\rm obs} =$ 0.81 and 0.93 for the CMS and ATLAS analysis, respectively, so neither analysis individually excludes the point (recall that $r = 1/\mu_{95}$). In the combination, this moves to $r_{\rm obs}^{\rm comb} = 1.22$, resulting in a solid exclusion. In contrast, in the example in Fig. 4.13, the behaviour of the observed limits is quite different: the ATLAS analysis excludes the point with an $r_{\rm obs}$ of 1.07, while the CMS analysis, although having

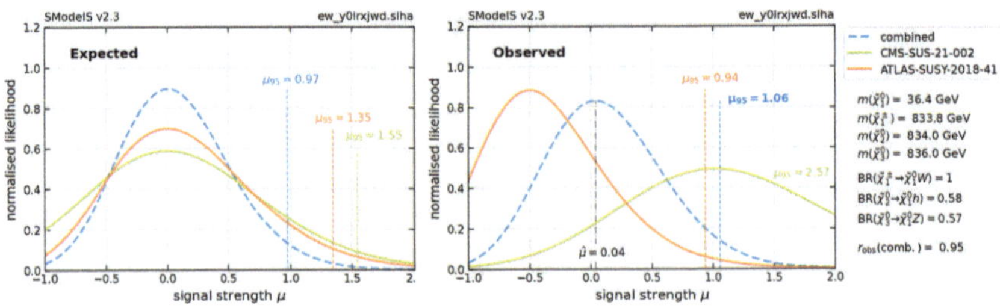

Fig. 4.13 Visualisation of likelihoods for one of the light blue points from Fig. 4.11, concretely a point which is excluded by the ATLAS analysis, but not any more so when combining it with the CMS analysis. See text for details

very similar sensitivity, gives only $r_{\text{obs}} = 0.39$. Combining the two analyses results in $r_{\text{obs}}^{\text{comb}} = 0.95$, which means the point is not excluded any more.

Altogether, the combination of the ATLAS and CMS hadronic EW-ino searches excludes 2258 of the scan points. Of the remaining points, further 614 are excluded by some other EW-ino search. This gives a total of 2872 excluded points, as compared to 2974 (2048) when considering the limit from the most constraining (most sensitive) analysis only. Indeed the overall change in the observed limit when comparing Figs. 4.11 to 4.8 is rather small. This is due to the data under-fluctuations (with respect to the expected background) seen by the hadronic ATLAS search and the over-fluctuations seen by the CMS one. The combined limit is, however, statistically more reliable, since it makes use of a larger amount of data coming from both the ATLAS and CMS searches.

A comment is in order regarding the bino, wino and higgsino mixing. The bulk of the excluded points at $m_{\tilde{\chi}_1^\pm} \approx 200$–$1000$ GeV in Fig. 4.11 has a (mostly) bino-like LSP, with the next heavier states being mostly higgsino- or wino-like; highly mixed charginos and neutralinos make only a fraction of the scan points, due to volume effects. As a consequence, the dark red and light blue points form two "arcs" in the $m_{\tilde{\chi}_1^\pm}$ versus $m_{\tilde{\chi}_1^0}$ plane, the lower one featuring higgsino-like $\tilde{\chi}_1^\pm$ and $\tilde{\chi}_{2,3}^0$ and the higher one featuring wino-like $\tilde{\chi}_1^\pm$ and $\tilde{\chi}_2^0$. The few dark red/light blue points scattered away from these arcs have significant wino-higgsino mixing. The isolated diagonal line of points at $m_{\tilde{\chi}_1^\pm} \approx 100$–$250$ GeV, on the other hand, is characterised by a higgsino-like LSP, with the signal coming from the production of wino-like $\tilde{\chi}_2^\pm$ and $\tilde{\chi}_3^0$.

This is illustrated by the projections of the results in terms of gaugino and higgsino mass parameters M_1, M_2 and μ, giving more insights into the mass hierarchies and mixings at play. Figure 4.14 shows the scan points excluded by the most sensitive or the most constraining analysis, equivalent to Figs. 4.8 and 4.9. The three cases considered are:

- top row: $M_1 < M_2 < \mu$, leading to a bino-like $\tilde{\chi}_1^0$ and wino-like $\tilde{\chi}_2^0$ and $\tilde{\chi}_1^\pm$;

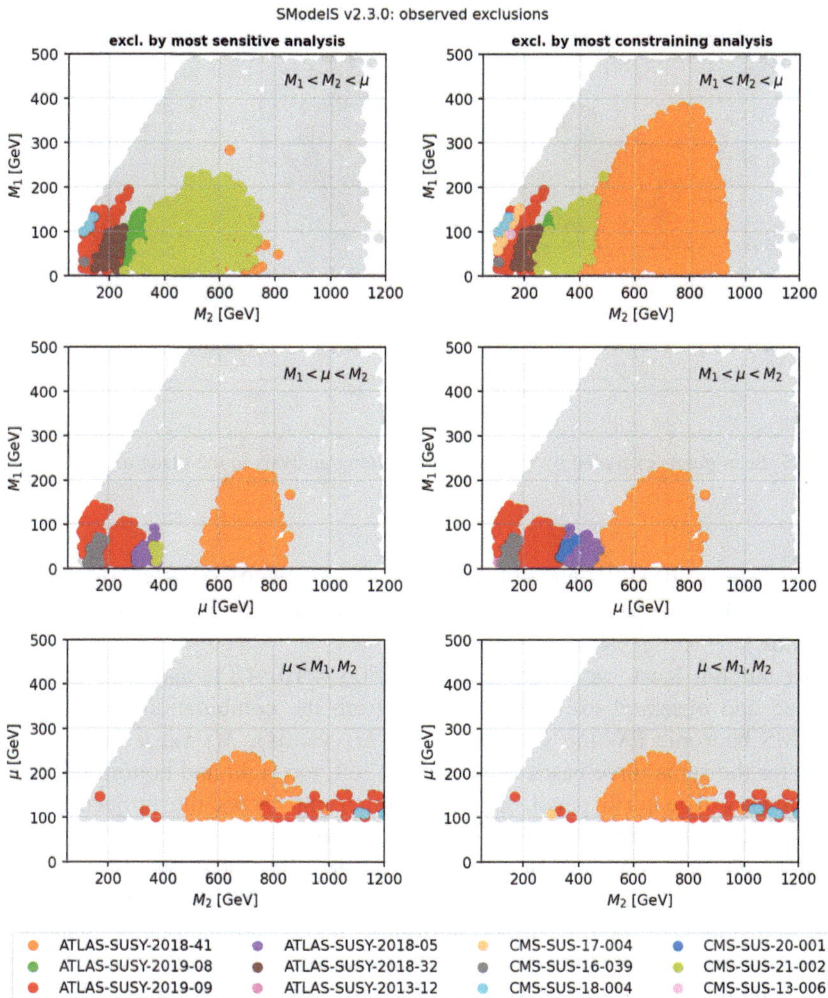

Fig. 4.14 Scan points excluded by the most sensitive analysis (left panels) or the most constraining analysis (right panels) in terms of M_1, M_2 and μ. Three different hierarchies are considered: $M_1 < M_2 < \mu$ (top row), $M_1 < \mu < M_2$ (middle row) and $\mu < M_1,\ M_2$ (bottom row)

- middle row: $M_1 < \mu < M_2$, leading to a bino-like $\tilde{\chi}_1^0$, higgsino-like $\tilde{\chi}_{2,3}^0$ and $\tilde{\chi}_1^{\pm}$; and
- bottom row: $\mu < M_1,\ M_2$, leading to higgsino-like and near mass-degenerate $\tilde{\chi}_{1,2}^0$ and $\tilde{\chi}_1^{\pm}$.

The gaugino/higgsino mixing is further illustrated in Fig. 4.15, which shows the scan points excluded by the most constraining analysis in the plane of M_2/M_1 versus M_2/μ. The points below the dashed line are in the compressed region

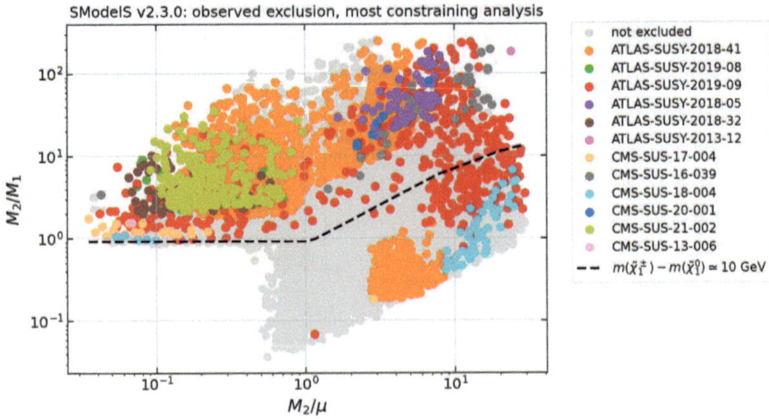

Fig. 4.15 Scan points excluded by the most constraining analysis in the plane of M_2/M_1 versus M_2/μ

($m_{\tilde{\chi}_1^\pm} - m_{\tilde{\chi}_1^0} < 10$ GeV) and correspond to the higgsino-LSP scenario. Note that the wino-LSP scenario ($M_2 < M_1, \mu$) is not considered, since it leads to long-lived charginos.

Last but not least, as a supplement to Figs. 4.10, 4.11, and 4.16 shows the expected and observed exclusion obtained from the combination of the ATLAS and CMS hadronic EW-ino searches in the M_1 vs. M_2, M_1 vs. μ and μ vs. M_2 planes for the same three cases as above (top left, top right and bottom left panels, respectively). The forth panel on the bottom right shows the projection of the excluded points onto the M_2/M_1 versus M_2/μ plane.

All in all, the combination of analyses, here illustrated by the example of combining the fully hadronic EW-ino searches from ATLAS and CMS, leads to better and statistically more robust constraints: first, combinations of analyses use more of the available data thus increasing the sensitivity to the BSM signal; second, while individual analyses can lead to an over(under)-aggressive exclusion of the parameter space due to under(over)-fluctuations of the data with respect to the expected background, the combination of analyses reduces the impact of such fluctuations.

4.4 Global Study

We will now turn to the dynamical combination of analyses. While in the previous section only two analyses, one from ATLAS and one from CMS, were combined, a significant step further is taken to dynamically determine the best (i.e., the most sensitive) combination from a set of 16 analyses (11 from ATLAS and 5 from CMS), for which likelihoods for EW-inos can be computed. Since the aim is to build global

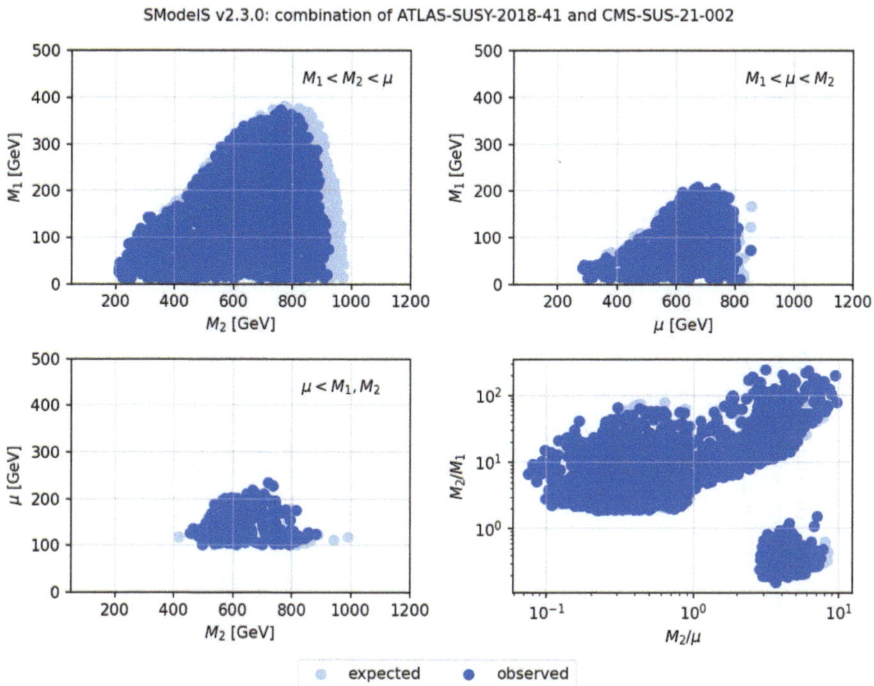

Fig. 4.16 Expected and observed exclusion by the combination of the ATLAS and CMS hadronic EW-ino searches in terms of M_1, M_2 and μ. Note that, except for a small strip at high M_2, the expected and observed exclusions are almost the same

likelihoods from these data, are only considered EM-type results [64]. The relevant analyses implemented in the SModelS database v2.3 are summarised in Table 4.1. The procedure relies on the combiner algorithm developed in [65] for the search for possible dispersed signals in the LHC results.

The parameter scan described in Sect. 4.2 is again used, but with some improvements. An important difference is that the production cross sections have now been computed at NLO with Resummino 3.1.2 [66, 67] if the corresponding LO cross section was above 7×10^{-4} fb. The interface to Resummino is described in Appendix A of [9]. Moreover, few points with $m_{\tilde{\chi}_1^\pm} < 103$ GeV as well as some points for which the NLO computation failed[9] were removed, which leaves a total of 18247 points as the EW-ino dataset for this dynamical combination study. On top of that, a non-default `minmassgap` of 10 GeV is used. The choice of `minmassgap` and its influence on the results is explained in Sect. 4.4.3.

[9] Most of these points possibly failed because of the way the computation was setup on the server running the computation. If done locally, the computation would likely have succeeded. These failing points were not recomputed for the sake of time and the small 'unessential' subset they represent.

Table 4.1 List of the EW-ino analyses from LHC Run 1 ($\sqrt{s} = 8$ TeV) and Run 2 (13 TeV) considered in this study. The column "Lumi." gives the integrated luminosity in fb^{-1}. The column "Comb." specifies whether and how SRs are combined: "pyhf" means a full or SIMPLIFY'ed HISTFACTORY model is used through interface with PYHF; "SLv1" ("SLv2") means that a covariance matrix is used in the Simplified Likelihood scheme of [14] ([59]); and a "–" means that only the most sensitive SR is used. See Sect. 4.1 for more details

ID	Run	Lumi.	Final state ($+ \not{E}_T$)	EMs ($+ \not{E}_T$)	SRs	Comb.
ATLAS-SUSY-2013-11 [15]	1	20.3	2 lept., 0 or ≥ 2 jets, 0b	$WW^{(*)}$	13	–
ATLAS-SUSY-2013-12 [17]	1	20.3	3 lept. (0–2 τ's), 0b	$WZ^{(*)}$, Wh	2	–
ATLAS-SUSY-2016-24 [19]	2	36.1	2–3 lept., 0 or ≥ 2 jets, 0b	WZ	9	–
ATLAS-SUSY-2017-03 [21]	2	36.1	2–3 lept., 0 or ≥ 1 jets, 0b	WZ	8	–
ATLAS-SUSY-2018-05 [23]	2	139	2 lept., ≥ 1 jets	$WZ^{(*)}$	13	pyhf
ATLAS-SUSY-2018-06 [26]	2	139	3 lept., 0 or 1–3 jets, 0b	$WZ^{(*)}$	2	–
ATLAS-SUSY-2018-32 [31]	2	139	2 lept., 0 or 1 jets, 0b	WW	36	pyhf
ATLAS-SUSY-2018-41 [35]	2	139	4 jets or $2b + 2$ jets, 0 lept.	WW, WZ, Wh, Zh, ZZ, hh	3	SLv1
ATLAS-SUSY-2019-02 [37]	2	139	2 lept., 0 or 1 jets, 0b	WW	24	SLv1
ATLAS-SUSY-2019-08 [39]	2	139	1 lept., $(h \rightarrow)b\bar{b}$	Wh	9	pyhf
ATLAS-SUSY-2019-09 [41]	2	139	3 lept., 0 or ≥ 1 jets, 0b	$WZ^{(*)}$	20+31	pyhf
CMS-SUS-13-012 [45]	1	19.5	0 lept., ≥ 3 jets (q or b)	WW, WZ, ZZ	36	–
CMS-SUS-16-039 [47]	2	35.9	2+ lept., 0–2 hadr. τ's, 0b	$WZ^{(*)}$	11	SLv1
CMS-SUS-16-048 [49]	2	35.9	2 soft lept., ≥ 1 jets, 0b	WZ^{*}	12+9	SLv1
CMS-SUS-20-004 [57]	2	137	0 lept., $2\,h(\rightarrow b\bar{b})$	hh	22	SLv2
CMS-SUS-21-002 [61]	2	137	≥ 2 AK8 jets, 0 or $\geq 1b$'s 0 lept.	WW, WZ, Wh	35	SLv1

4.4.1 Combination Strategy

The next step is to build a global likelihood from the individual analysis likelihoods. Since inter-analyses correlations are unknown, only analyses that are approximately uncorrelated may enter the combination. The combined likelihood $\mathcal{L}_C$ is then simply the product of the likelihoods $\mathcal{L}_i$ of the individual analyses. This raises two questions: (i) which analyses can be treated as uncorrelated, and (ii) how to choose the best combination from all possible combinations.

Regarding (i), analyses from different LHC runs, and from distinct experiments (ATLAS or CMS) are assumed to be uncorrelated. Furthermore, analyses which do not share any event in their SRs are also treated as approximately uncorrelated. In [65], the latter aspect was limited to clearly different final states (e.g., fully hadronic final states vs. final states with leptons). In this work, the SR definitions of all analyses under consideration (see Sect. 4.1) are carefully scrutinised to determine whether they are orthogonal or they overlap. With this procedure, the combinability matrix shown in Fig. 4.17 can be built.

It should be noted that this approximation assumes that inter-analyses correlations of systematic uncertainties (stemming from, e.g., luminosity measurements) can be neglected. Moreover, it neglects possible correlations of systematic uncer-

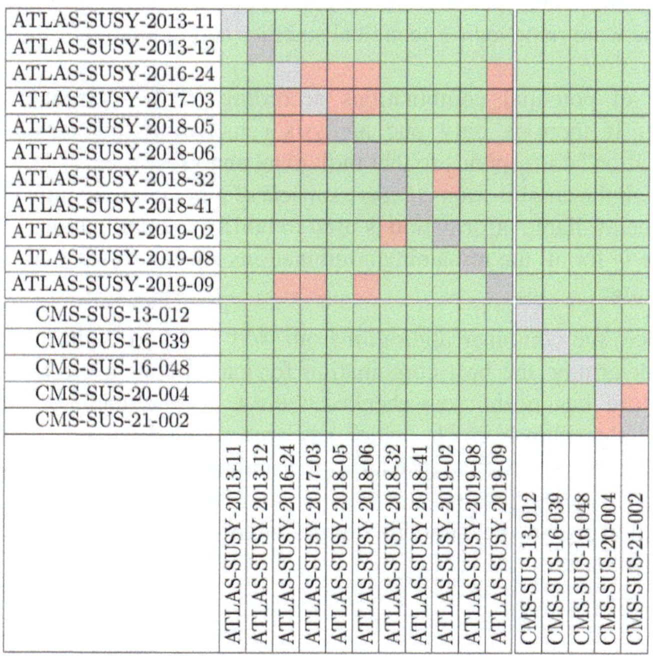

Fig. 4.17 Matrix displaying the combinability of the searches considered in this study. Green means the two analyses are assumed to be approximately uncorrelated and can be combined, while red means that they are not and shouldn't be combined

tainties due to overlaps of SRs of one analysis with the control regions of another analysis. Such effects can only be checked in full recasts including signal *and* control regions, which is beyond the present state-of-the-art. However, it is expected that the effects are small compared to other uncertainties in SModelS and in reinterpretation studies in general.

With respect to (ii), since not all the analyses can enter the same combination, many different combinations can be built based on this combinability matrix. In this work, the "best combination" is defined as the most sensitive one, i.e. the one with the lowest expected limit on the signal strength μ_{95}^{exp} (or, equivalently, highest r_{exp}). Technically, since the computation of μ_{95}^{exp} can be time-consuming, it is assumed that the combination that minimizes

$$w = \frac{\mathcal{L}^{\text{exp}}(\text{BSM})}{\mathcal{L}^{\text{exp}}(\text{SM})} \tag{4.1}$$

is the one with the lowest μ_{95}^{exp}. Here, $\mathcal{L}^{\text{exp}}(\text{H})$ is the expected combined likelihood under the H hypothesis. For a given model point, the series of steps to find the best combination of analyses is the following:

1. Compute μ_{95}^{exp} for every analysis and keep for the following steps only the sensitive ones. Sensitive here means $\mu_{95}^{\text{exp}} \leq 10$. Even though the insensitive analyses could have been used in the combination, they do not impact the result and hence were removed for technical stability and reduction of the computation time.
2. Identify all potential combinations according to the combinability matrix, Fig. 4.17. If there is only one analysis sensitive to the tested model, the "combination" corresponds to that individual analysis.
3. Remove all the combinations that are subsets of other combinations. This ensures that all the available information is used to build the combination.
4. Compute w for all the remaining combinations and select the one that gives the lowest value.

In practice, the combiner algorithm from [65] was used with minimal modifications to determine the best combination for each point in the EW-ino dataset. The procedure has been cross-checked with a second method, based on the "pathfinder" algorithm from [1], which has been adapted for this study to find the best combination of analyses, instead of the optimal SR combination in [1]. Both methods gave identical results.

In the following, the best combination of analyses will simply be referred to as "the combination".

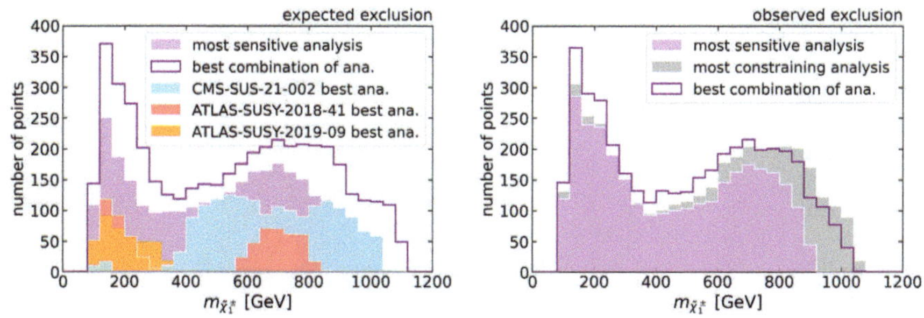

Fig. 4.18 Number of points expected to be excluded (left) and number of points actually excluded (right) as determined by the most sensitive analysis (pink area) and the best combination of analyses (purple line). In addition, the left panel shows the impact of the two hadronic searches: CMS-SUS-21-002 (light blue) and ATLAS-SUSY-2018-41 (light red), along with the $3\ell + \not{E}_T$ search ATLAS-SUSY-2019-09 (orange). On the other hand, the right panel also displays the observed exclusion with respect to the most constraining analysis (grey)

4.4.2 Results

Using the parameter scan and combination strategy discussed in the previous sections, the impact of global analysis combination on the EW-ino sector of the MSSM is now investigated. The gain in expected reach due to the combination can be seen in the left plot of Fig. 4.18. A point is expected to be excluded if an analysis gives $r_{\text{exp}} \geq 1$. The expected exclusion is mostly driven by three analyses: the $3\ell + \not{E}_T$ search ATLAS-SUSY-2019-09 and the all-hadronic searches CMS-SUS-21-002 and ATLAS-SUSY-2018-41. The points for which the most sensitive analysis is one of these three analyses are displayed by the orange, light blue and light red histograms. As anticipated, the expected reach is enhanced due to the accumulation of statistics within the combination, resulting in an increase of 48% in the expected number of excluded points (from 3081 to 4549 points).

The impact of the combination on the observed number of excluded points with respect to the most sensitive and most constraining analyses, i.e. the analyses that give the highest r_{exp} and r_{obs} respectively, is shown on the right plot of Fig. 4.18. It must be pointed out that using the largest r_{obs} is not statistically sound, since it gives preference to the analyses which have observed under-fluctuations in the data, thus artificially increasing the constraining power. The histogram for the most constraining analyses in Fig. 4.18 is shown only for reference and comparison to previous studies. Nonetheless, it is interesting to note that the combination excludes more points than the most constraining analysis up to $m_{\tilde{\chi}_1^\pm} \approx 700$ GeV. For higher masses, the under-fluctuations recorded by the hadronic ATLAS analysis result in a more aggressive exclusion when considering the most constraining analysis. All in all, the most sensitive analysis, the most constraining analysis and the combination exclude 3046, 3949 and 4124 points, respectively. Regarding the importance of

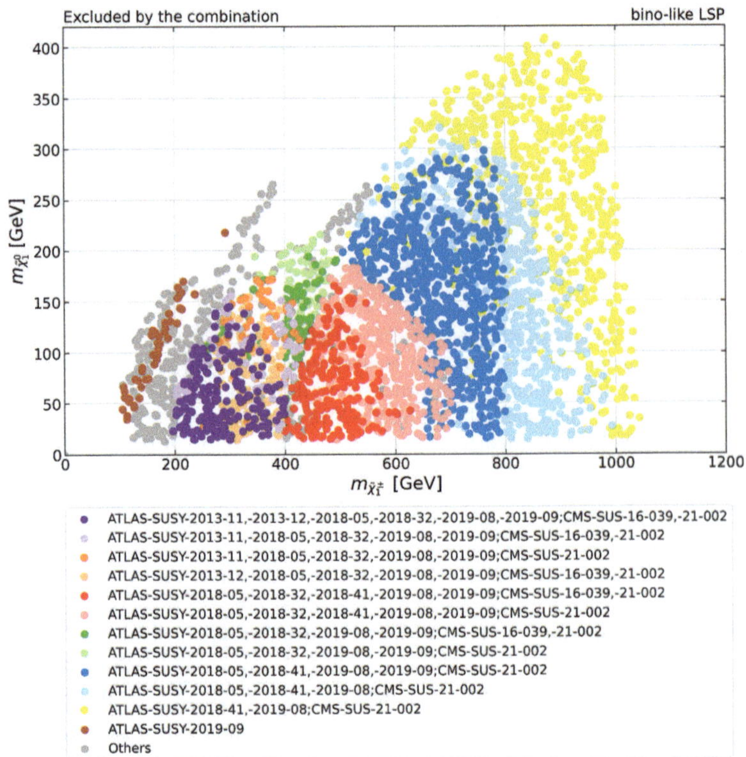

Fig. 4.19 Most sensitive combinations in the $m_{\tilde{\chi}_1^0}$ vs. $m_{\tilde{\chi}_1^\pm}$ plane for points excluded by the combination and featuring a bino-like LSP. Only the combinations appearing at least 25 times are shown. The combinations that appeared less than 25 times fall into the "Others" category

using NLO cross sections, 611 points ($\approx$ 15%) would not have been excluded by the combination if LO cross sections were used instead of NLO.

It is also relevant to verify which analyses are entering the combination as we move around the parameter space. In particular, it is interesting to check the stability of the combinations and the number of analyses contributing to each combination. Is first shown in Fig. 4.19 the combinations for the excluded points featuring a bino-like LSP, in the plane of $m_{\tilde{\chi}_1^0}$ vs. $m_{\tilde{\chi}_1^\pm}$. Two observations can be made. First, the combinations are clustered in different regions of the parameter space, which proves the stability of the procedure. Second, the higher the mass of $\tilde{\chi}_1^\pm$ or $\tilde{\chi}_1^0$, the fewer analyses enter the combination. This is because the majority of them are sensitive to low masses, while only ATLAS-SUSY-2018-05, ATLAS-SUSY-2018-41, ATLAS-SUSY-2019-08 and CMS-SUS-21-002 are sensitive to high masses. Indeed, moving from one colour patch to another can be understood as an analysis becoming (in)sensitive to the tested signal when $m_{\tilde{\chi}_1^0}$ or $m_{\tilde{\chi}_1^\pm}$ (increases) decreases. The (small) overlapping regions are due to the composition of the produced particles:

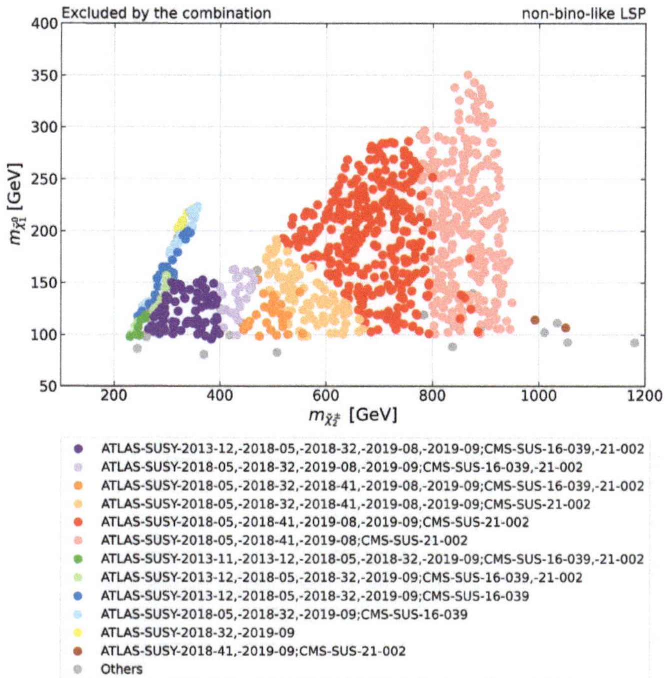

Fig. 4.20 Most sensitive combinations in the $m_{\tilde{\chi}_1^0}$ vs. $m_{\tilde{\chi}_2^\pm}$ plane for points excluded by the combination and featuring a non-bino-like LSP. Only the combinations appearing at least 2 times are shown. The other combinations that appeared 2 times or less fall into the "Others" category

in the wino-bino scenario the production cross sections are larger and the sensitivity of the analyses is extended, while for the higgsino-bino scenario, the sensitivity is suppressed. Finally, the remaining grey points are made of many different combinations and are concentrated at low masses due to a large number of analyses being sensitive only in this region.

Figure 4.20 displays the same information, but for points with a non-bino-like LSP (i.e. points with $N_{11}^2 < 0.5$). Since points featuring a long-lived $\tilde{\chi}_1^\pm$ were removed, the majority of points have a higgsino-like LSP. The number of points excluded by the combination is highly reduced for these scenarios. Indeed, as discussed in Sect. 2.2.3, the higgsino-like LSP scenario produces a large number of final state topologies, many of which have no available LHC results. As already seen in Fig. 4.19, the low mass region is populated by combinations made up of many analyses, while only four analyses are sensitive enough to enter the combination at high masses. The grey points at low LSP masses contain a sufficiently large $m_{\tilde{\chi}_1^\pm} - m_{\tilde{\chi}_1^0}$ mass gap and are excluded by higgsinos ($\tilde{\chi}_1^\pm$ and $\tilde{\chi}_2^0$) decays to off-shell gauge bosons. Finally, the two brown points at large $m_{\tilde{\chi}_2^\pm}$ are constrained by a combination of analyses sensitive to off-shell higgsino decays and on-shell wino decays.

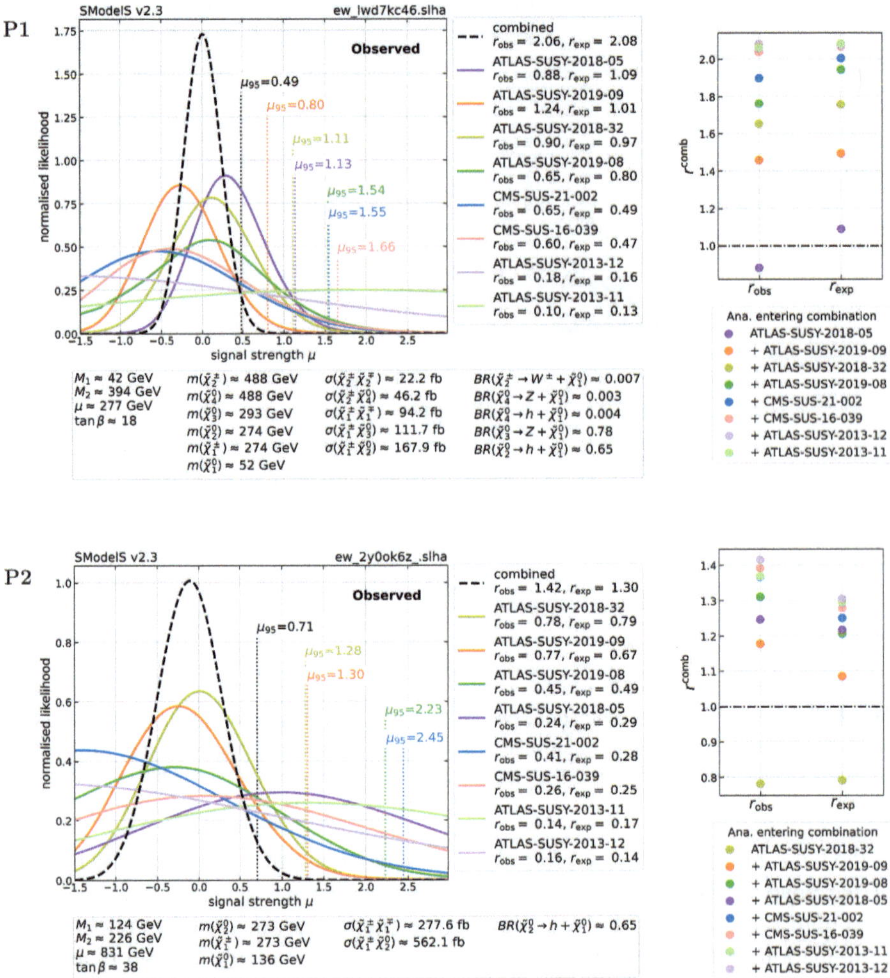

Fig. 4.21 Normalised likelihood versus the signal strength μ (left plots) for 2 different points of the random scan, denoted as P1 (top) and P2 (bottom). Shown are the combined likelihood (black dashed line) together with the likelihoods of every analysis entering the combination (coloured full lines). The smaller panels to the right show the evolution of the combined observed and expected r-values when the analyses entering the combination are added one by one in order of decreasing sensitivity. The boxes below the likelihood plots provide further details on the parameter points; $\tilde{\chi}_1^{\pm}$ decays are not specified as the only available decay mode is into $W + \tilde{\chi}_1^0$. Both points are excluded by the combination

The impact of analysis combination can be better understood once we consider the individual likelihoods for the analyses entering the combination and the combined likelihood. The top plots of Fig. 4.21 display the combination of 8 analyses for a sample point (P1) featuring a bino-like LSP with similar wino and higgsino mass scales. The left plot shows the individual and combined likelihoods, while the right plot shows the evolution of the combined r-values, when analyses are sequentially added to the combination. The most sensitive analysis is expected to exclude the point with $r_{\text{exp}} = 1.09$, but due to an over-fluctuation seen by the ATLAS search, the observed r-value is below 1. However, once 8 analyses are combined, the effect of over and under-fluctuations are drastically reduced and the point is robustly excluded with an observed r-value very close to the expected value. This can be seen on the left plot, where the analyses that saw a small excess ($r_{\text{obs}} < r_{\text{exp}}$) pull the combined likelihood towards positive values of the signal strength μ, while the analyses that saw an under-fluctuation ($r_{\text{obs}} > r_{\text{exp}}$) pull the combined likelihood in the opposite direction. In this case, the resulting likelihood is centred around 0, i.e. its observed r-value is close to its expected one, and its width is narrow, meaning that the observed r-value is high. One can also see on the right plot that the small excess recorded by the most sensitive analysis is already compensated by the second most sensitive analysis, which recorded an under-fluctuation, when the combination is only made of these two analyses.

The bottom plots of Fig. 4.21 show the results for a point (P2) representing the wino-bino scenario. This point is neither expected to be excluded nor excluded by any individual analysis entering the combination. However, the constraining power is considerably increased once the 8 shown analyses are combined, excluding the point with $r_{\text{obs}} = 1.42$. It is also interesting to note that, while the combination of the two most sensitive analyses is already sufficient to exclude the point, the inclusion of the remaining analyses significantly increases the sensitivity of the combination.

If the number of sensitive analyses entering the combination is not sufficiently large, it can happen that observed fluctuations are not levelled and the observed r-value can be quite distinct from the expected value. This is illustrated by the example in the top plots of Fig. 4.22, which shows the results for a wino-bino point (P3). In this case, the two most sensitive analyses (CMS-SUS-21-002 and ATLAS-SUSY-2019-08) saw an excess, which reduced their observed exclusion. Although the other two analyses entering the combination could mitigate the excess, they are not sufficiently sensitive to have a significant impact on the final r-value. As a result, although the combination is expected to exclude the point, $r_{\text{obs}}^{\text{comb}} < 1$.

Finally, there are cases where the combination decreases the *observed* r-value when compared to the most sensitive analysis. This can take place when the latter observes an under-fluctuation, thus increasing r_{obs}, while the combination reduces this effect, resulting in a smaller value for r_{obs}. This is illustrated by the bottom plots in Fig. 4.22, which shows the results for a mixed scenario (P4). The most sensitive analysis in this case is the ATLAS-SUSY-2018-41 search, which recorded an under-fluctuation, resulting in an exclusion: $r_{\text{obs}} = 1.11 > 1 > r_{\text{exp}}$. Once the CMS-SUS-21-002 search, which saw an excess, is included, the combined r_{obs} is reduced below one and the point is no longer excluded. This result is strengthened once the remaining third analysis entering the combination is included. It is expected, however, that once additional sensitive analyses enter the combination, such effects will be reduced (if they did not record an excess), and the combined observed exclusion will be quite similar to its expected value.

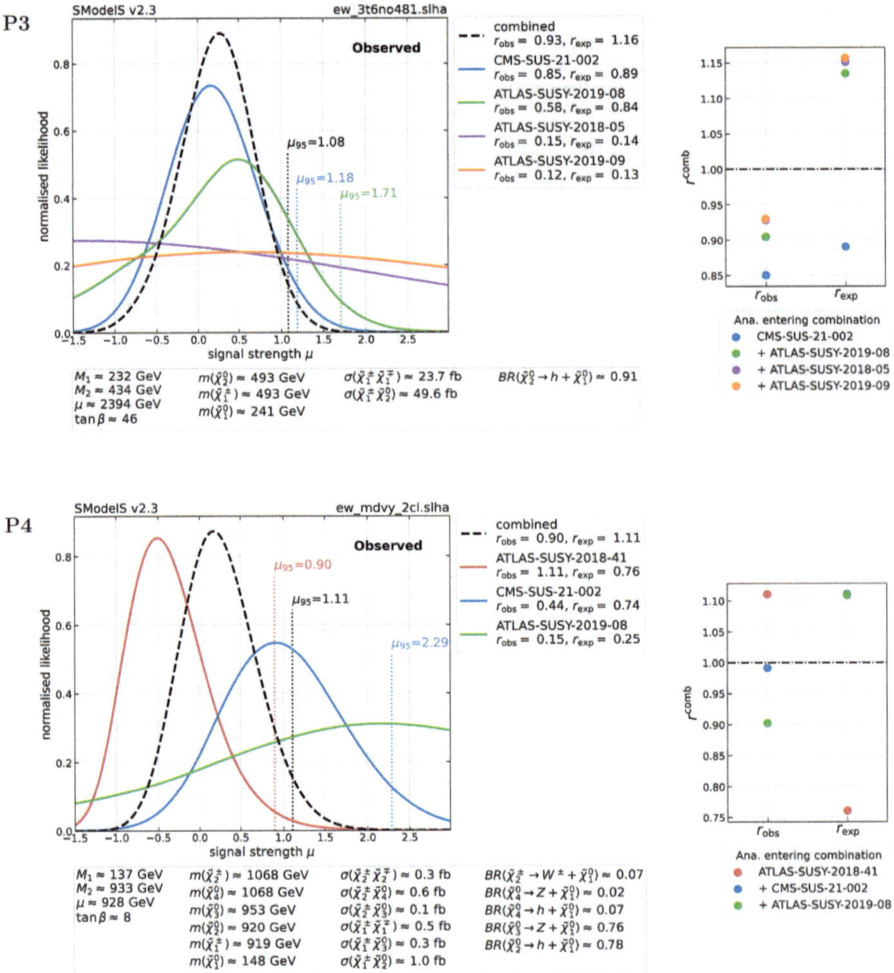

Fig. 4.22 As Fig. 4.21 but for two sample points, P3 (top) and P4 (bottom), not excluded by the combination

A global overview of the impact of analysis combination is shown in Fig. 4.23 for excluded points with a bino-like LSP. Red (blue) points correspond to a gain (loss) in exclusion power when compared to the most sensitive analysis. The numbered points highlighted in the plot correspond to the points in Figs. 4.21 and 4.22. The majority of points display an increase in exclusion power, as expected. However, for 700 GeV $\lesssim m_{\tilde{\chi}_1^\pm} \lesssim$ 900 GeV, there is a region where the observed r-value for the combination is smaller than the one obtained using only the most sensitive analysis. This is due to the most sensitive analysis (ATLAS-SUSY-2018-41) recording an under-fluctuation, which increased its observed exclusion power. When combined with the other analyses, which recorded an excess, the combined

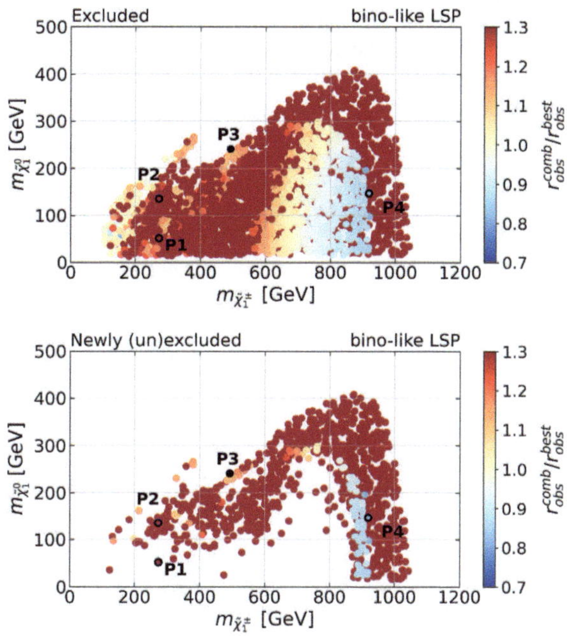

Fig. 4.23 Impact of the combination on the exclusion power with respect to the most sensitive analysis. All the points featuring a bino-like LSP that are excluded by the combination or by the most sensitive analysis are shown on the top panel. The subset of points whose exclusion status changed with the combination is shown on the bottom panel. The circled points denote points P1, P2 and P4, for which the likelihoods are plotted in Figs. 4.21 or 4.22. Point P3 (black dot) is also indicated for completeness, although it is not excluded

r_{obs} is reduced, as illustrated by the example shown in Fig. 4.22 (bottom plot). Finally, the bright yellow regions display points for which the impact of combining analysis is negligible. For the points around $m_{\tilde{\chi}_1^\pm} \simeq 700$ GeV this happens because the analyses entering the combination have observed under and over-fluctuations, thus pulling the combined likelihood in opposite directions. The combined result in this case does not significantly increase the constraining power when compared to the most sensitive analysis. The yellow region at lower masses, $m_{\tilde{\chi}_1^\pm} \lesssim 200$ GeV, are so highly constrained by the $3\ell + \not{E}_T$ search ATLAS-SUSY-2019-09, that adding new analyses to the combination almost does not change the result.

The bottom plot of Fig. 4.23 shows only the points which were not excluded by the most sensitive analysis but are now excluded by the combination (orange and red) and the points which were excluded by the most sensitive analysis but are now not excluded by the combination (blue). We can see that the blue points are concentrated in a small region around $m_{\tilde{\chi}_1^\pm} \simeq 900$ GeV, where red points are also found. All these blue points feature a higgsino-like NLSP, while the red points in the same region feature a wino-like NLSP. This behaviour can be explained because the ATLAS hadronic search (ATLAS-SUSY-2018-41) is the analysis the most sensitive to the higgsino-bino scenario in this region and, as already pointed out above, the under-fluctuations recorded by this analysis result in an increase in its constraining power with respect to the combination. On the other hand, in this region, the most sensitive analysis to a wino-bino scenario is the CMS hadronic search (CMS-SUS-21-002), which recorded an excess. A large number of points hence become excluded when it is combined with the ATLAS hadronic search.

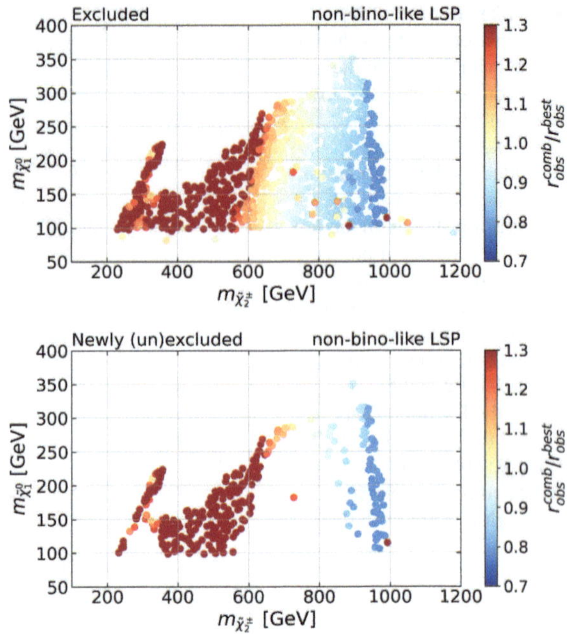

Fig. 4.24 As Fig. 4.23 but in the $m_{\tilde{\chi}_1^0}$ vs. $m_{\tilde{\chi}_2^\pm}$ plane, for points featuring a non-bino-like LSP

Figure 4.24 shows the same plots but for non-bino-like LSP points in the plane of $m_{\tilde{\chi}_1^0}$ vs. $m_{\tilde{\chi}_2^\pm}$. Most of these feature a higgsino-like LSP and correspond to the wino-higgsino or mixed scenarios. The three most sensitive analyses in this case are the $3\ell + \not{E}_T$ search ATLAS-SUSY-2019-09 for 200 GeV $\lesssim m_{\tilde{\chi}_2^\pm} \lesssim$ 400 GeV, the CMS hadronic search for 400 GeV $\lesssim m_{\tilde{\chi}_2^\pm} \lesssim$ 600 GeV, and the ATLAS hadronic search (ATLAS-SUSY-2018-41) for $m_{\tilde{\chi}_2^\pm}$ between 600 GeV and 1000 GeV. As seen for the bino-like LSP points, the first two mass regions display an increase in constraining power due to analysis combination. Once again, for the high mass region ($m_{\tilde{\chi}_2^\pm} \gtrsim$ 700 GeV), the under-fluctuations recorded by the hadronic ATLAS search result in an increase of constraining power, thus explaining the blue points in Fig. 4.24.

The bottom plot shows only the set of points which have their exclusion status modified by the combination. The low-mass region shows the gain in exclusion power (red points), while the high-mass region displays points which become unexcluded by the combination (blue points). Unlike the bino-like LSP scenario, the blue points are approximately clustered into two disjoint regions. This behaviour is caused by the `minmassgap` value chosen and the mass compression procedure (see discussion in Sect. 4.4.3), which compresses topologies with $m_{\tilde{\chi}_2^0} - m_{\tilde{\chi}_1^0} < 10$ GeV. Only for $m_{\tilde{\chi}_2^\pm} \gtrsim$ 900 GeV this mass difference falls below 10 GeV and some complex topologies generated by two-step cascade decays are compressed to single-step topologies, thus increasing the sensitivity of the analyses. As a result, even though the analysis combination reduces r_{obs}, its values are still sufficiently large to not alter the exclusion status of the points in this region, resulting in a lack of

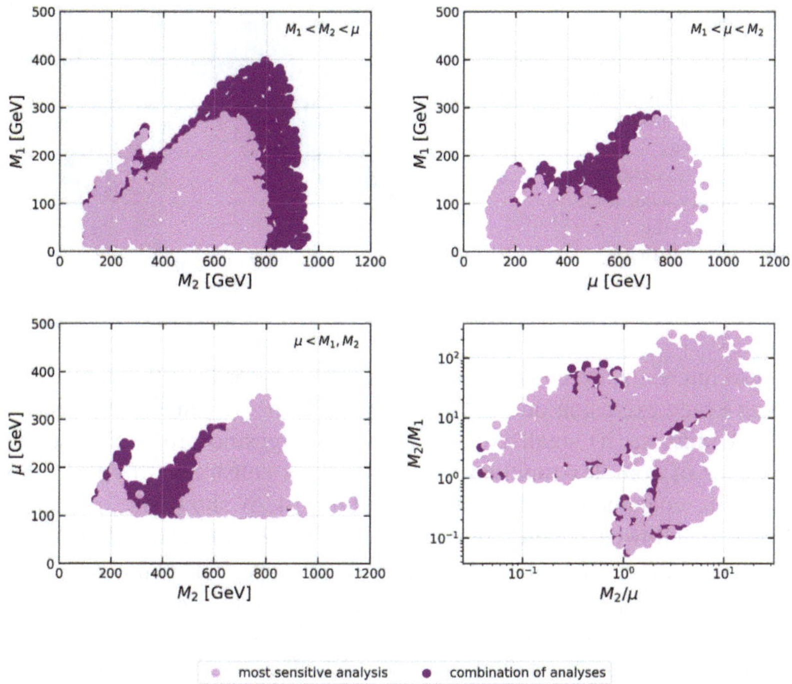

Fig. 4.25 Observed exclusion from the most sensitive analysis and from the combination, in terms of M_1, M_2, and μ. Besides mixed scenarios, the top left plot corresponds to the wino-bino scenario, the top right plot to the higgsino-bino scenario, and the bottom left plot to the wino-higgsino scenario. The bottom right plot covers all cases

points around 900 GeV. For larger $m_{\tilde{\chi}_2^\pm}$ the signal is suppressed and even with mass compression the combined result can no longer exclude points, resulting in the second strip of blue points seen around $m_{\tilde{\chi}_2^\pm} \simeq 1$ TeV. In the end, (148) 1226 points are newly (un)excluded by the combination, regardless of the LSP nature.

The impact of analysis combination on the EW-ino parameters is shown in Fig. 4.25. Dark purple points show points excluded by the combination, while light pink points show the exclusion considering only the most sensitive analysis. The top left and top right plots correspond to points mostly in the wino-bino and higgsino-bino scenarios, respectively. The bottom left plot mostly depicts points with a higgsino-like LSP, also showing up in Fig. 4.24. Finally, the bottom right plot contains the three other plots as well as the few points with $M_2 \lesssim M_1, \mu$. We can clearly see that the gain in exclusion power at high masses only takes place in the wino-bino case, as discussed above. In the other scenarios, the gain is situated at intermediate masses. Projected onto the M_2/M_1 vs. M_2/μ plane, the footprint of excluded points does not increase. Note, however, that the number of excluded points increases by 35% with the combination.

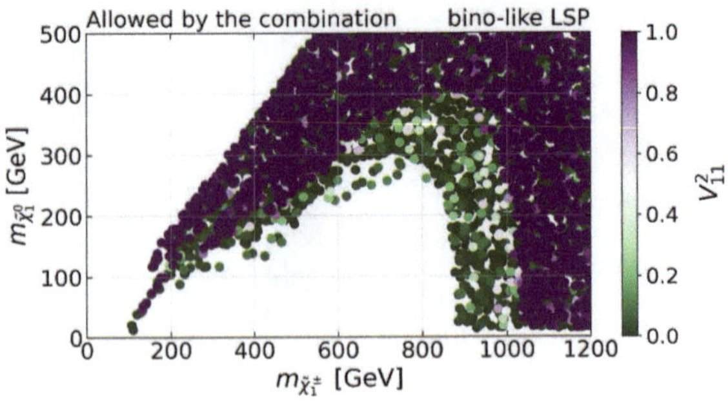

Fig. 4.26 Points with a bino-like LSP not excluded by the combination, identified by the wino admixture of the lightest chargino. Purple points correspond to scenarios where the lightest chargino is mainly wino-like, and green points to the scenarios where it is mainly higgsino-like

It is also interesting to consider which points or scenarios still escape exclusion. Points can evade exclusion due to three major reasons: (i) small cross sections for large BSM masses, (ii) small cross sections for higgsino production, and (iii) a large number of complex topologies which are not constrained by the simplified model results contained in the database. Figure 4.26 shows the points with a bino-like LSP allowed by the combination ($r_{\text{obs}}^{\text{comb}} < 1$). The colour represents the $\tilde{\chi}_1^+$ wino content, quantified through V_{11}, the upper left component of the V matrix defined in Eq. (1.88). Purple points are therefore points with a wino-like NLSP, while green points feature a higgsino-like NLSP. As expected, the allowed parameter space for points with a wino-like NLSP is smaller, due to their larger production cross sections. We also see that several points avoid exclusion at low masses if $m_{\tilde{\chi}_1^0} \lesssim m_{\tilde{\chi}_1^\pm}$. These points display mixed scenarios, where the number of complex topologies is large, thus diluting the signal going into the simple 1-step decay topologies constrained by the database. The most important observation from Fig. 4.26 is, however, that there is a sizeable region which is definitely excluded. This region extends up to $m_{\tilde{\chi}_1^\pm} \approx 900$ GeV for higgsino-like NLSP and up to $m_{\tilde{\chi}_1^\pm} \approx 1$ TeV for wino-like NLSP. Such a region does not exist for non-bino-like LSP points.

Last but not least, it is interesting to see which fluctuations exist in the data with respect to the expected background. To this end, Fig. 4.27 shows the distribution of $r_{\text{obs}}/r_{\text{exp}}$ across the EW-ino dataset. Values below one indicate excesses in the data, while values above one signal under-fluctuations. With enough data, if there is no BSM signal, one expects $r_{\text{obs}}/r_{\text{exp}}$ to be normal distributed around unity. In Fig. 4.27, the results for the most sensitive individual analysis show a large spread in $r_{\text{obs}}/r_{\text{exp}}$, with only a subdominant number of points having $r_{\text{obs}} \approx r_{\text{exp}}$. Overall, the over-fluctuations outweigh the under-fluctuations with 54% of points having $r_{\text{obs}}/r_{\text{exp}} < 1$. In the combination, on the other hand, large fluctuations are almost always suppressed, and the distribution of $r_{\text{obs}}/r_{\text{exp}}$ is more centred towards one. Nonetheless, the tendency for excesses still remains: $r_{\text{obs}} < r_{\text{exp}}$ occurs for 72% of the points. This is even more true when considering only the points not excluded by the combination: here, $r_{\text{obs}} < r_{\text{exp}}$ occurs for 75% of the points. One should note, however, that in this case the combination often includes only a small number of analyses, which is not necessarily sufficient to mitigate fluctuations from individual

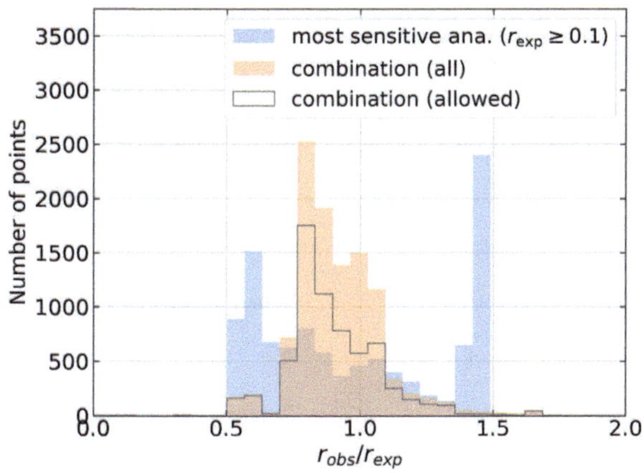

Fig. 4.27 Ratio of r_{obs}/r_{exp} from the most sensitive analysis (blue histogram), and from the combination of analyses (orange histogram) considering all scan points. Also shown is r_{obs}/r_{exp} for the points allowed by the combination (grey steps). Only the points with $r_{exp} \geq 0.1$ fill the "most sensitive ana." histogram in order to have a fair comparison with the combination. If $r_{obs}/r_{exp} > 1$, it means the model point is influenced by an under-fluctuation in the data, otherwise, by an excess

searches. It will be exciting to see how this tendency will evolve with Run 3 of the LHC.

4.4.3 Impact of Mass Compression

This section explains the chosen value for the `minmassgap` parameter, which controls mass compression in SModelS, and its influence on the results of this study.

The decay of an intermediate state to a nearly degenerate one typically results in the generation of soft final states that are beyond experimental detection capabilities. Thus, the soft states can usually be ignored and the topology can be simplified (compressed). Given that the simplified-model results of the experimental analyses that are considered focus exclusively on direct production with a single decay in each branch, the adoption of more simplified topologies translates to an increasing sensitivity, thereby enhancing the exclusion power. A detailed description of this so-called mass compression is given in [68] and in the SModelSonline documentation.

In the present analysis, the non-bino-like LSP points are sensitive to the mass compression since, in such scenarios, the masses of $\tilde{\chi}_1^0$, $\tilde{\chi}_1^{\pm}$ and $\tilde{\chi}_2^0$ can be close to each other, as illustrated by the wino-higgsino example in Fig. 2.9. The distributions of those points with respect to the difference between the masses of $\tilde{\chi}_1^{\pm}$ and $\tilde{\chi}_1^0$ (coral histogram), and the masses of $\tilde{\chi}_2^0$ and $\tilde{\chi}_1^0$ (blue histogram) are shown in Fig. 4.28. In order to concentrate on the region where LHC results are

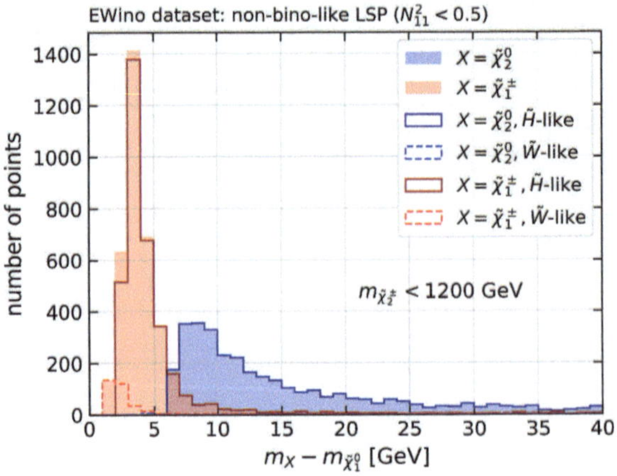

Fig. 4.28 Number of points as a function of the mass difference between $\tilde{\chi}_1^\pm$ and $\tilde{\chi}_1^0$ (coral) and between $\tilde{\chi}_2^0$ and $\tilde{\chi}_1^0$ (blue). The plot is restricted to non-bino-like LSP points with $m_{\tilde{\chi}_2^\pm} < 1200$ GeV. The portion of higgsino-like points among the blue points is indicated by a solid dark blue line, while the points that are higgsino-like among the coral points are contoured by a solid dark red line. Additionally, the portion of the coral (blue) points that are wino-like is highlighted with a dashed red (blue) line

potentially sensitive (cf. Fig. 4.20), a cutoff of 1200 GeV on the $\tilde{\chi}_2^\pm$ mass is imposed. The portion of the $m_{\tilde{\chi}_1^\pm} - m_{\tilde{\chi}_1^0}$ histogram under the dashed red line represents wino-like LSP points. This portion is small because in wino-LSP scenarios the $\tilde{\chi}_1^\pm$ is typically long-lived and, as mentioned above, points with a total decay width smaller than 10^{-11} GeV were removed. The region under the dark red line represents the higgsino-like LSP points. The blue histogram showing the $\tilde{\chi}_2^0$–$\tilde{\chi}_1^0$ mass difference comes almost entirely from higgsino-like LSP points (recall that higgsinos correspond to a triplet of 2 neutralinos and 1 chargino). Here, the small number of wino-like points spread across the whole range of the distribution, and can barely be seen in the figure. While the $\tilde{\chi}_1^\pm$–$\tilde{\chi}_1^0$ mass difference in Fig. 4.28 peaks below 5 GeV, many points have $m_{\tilde{\chi}_1^\pm} - m_{\tilde{\chi}_1^0} \in [5, 10]$ GeV. Moreover, the bulk of the $\tilde{\chi}_2^0$–$\tilde{\chi}_1^0$ mass difference lies between 6 and 15 GeV. It shows that, with the default choice of `minmassgap` = 5 GeV in SModelS, many non-bino-like LSP points would not be mass-compressed, which could lead to overly conservative results.

The question to consider before adjusting the `minmassgap` parameter is from which mass difference onward the decay products of an additional step in the decay chain will be hard enough to have a significant impact on the cut acceptances so that the EMs in the database are no longer valid. Note that the efficiencies can either increase or decrease once the soft particles pass a given analysis event selection. For instance, in a leptonic search that requires at least n leptons and is agnostic to additional jets, the efficiencies tend to increase if the soft particles contain leptons. On the other hand, if a veto is applied on the number of leptons and/or jets, more events will be discarded and the resulting efficiency will decrease.

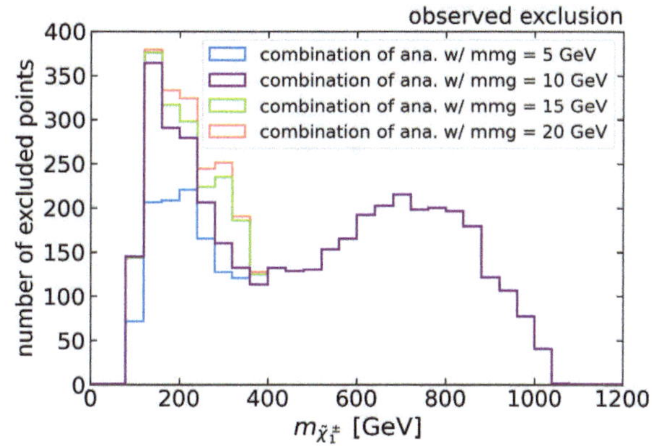

Fig. 4.29 Observed exclusions based on the best combination of analyses. The number of excluded points is shown as a function of the mass of $\tilde{\chi}_1^\pm$ for different choices of the `minmassgap` parameter (abbreviated as "mmg" in the legend): 5 GeV, 10 GeV, 15 GeV, and 20 GeV

The impact of the mass difference was studied for a particular leptonic ATLAS search [5] (ATLAS-SUSY-2019-08), which sets limits on $W^\pm(\to \ell^\pm \nu)\, h(\to b\bar{b}) + \not{E}_T$ from chargino-neutralino production, by means of the MadAnalysis 5 recast code [69]. It was found that for mass differences between 5 and 10 GeV, the efficiencies are not significantly affected by the presence of additional soft states from compressed decays. However, once the mass difference was increased above 10 GeV, significant changes in the efficiencies appeared, leading to an impact on the excluded parameter space. Thus, a `minmassgap` of 10 GeV was chosen in the presented analysis. This ensures that more points are compressed, making them sensitive to the study, cf. Fig. 4.28. At the same time, the associated decay products are soft enough, preserving the validity of the corresponding EMs.

The impact of changing the `minmassgap` parameter on the observed exclusions using the best combination of analyses is shown in Fig. 4.29. The plot shows the number of excluded points with respect to the mass of the $\tilde{\chi}_1^\pm$. The SModelS results obtained with a `minmassgap` of 5, 10, 15 and 20 GeV are shown in blue, purple, green and salmon, respectively.[10] As anticipated, the number of excluded points increases with increasing `minmassgap`, going from 3665 excluded points at 5 GeV to 4124 at 10 GeV and to 4332 (4427) at 15 (20) GeV. Moreover, the results differ only in the region of light charginos, as the mass compression only affects points for which the mass difference between the LSP and at least one other EW-ino is below the `minmassgap` and $m_{\tilde{\chi}_1^0} < 500$ GeV in the dataset.

For completeness, Fig. 4.30 shows how the observed exclusion in terms of M_1, M_2, and μ depends on the `minmassgap` parameter. Shown are the regions excluded by the combination of analyses for `minmassgap` $= 5$, 10, and 15 GeV, in beige, purple and light blue, respectively, in comparison with Fig. 4.25. Since changing `minmassgap` only affects the results for a higgsino-like LSP, only shown here are the case where $\mu < M_1, M_2$ in the plane of μ vs. M_2 (left panel) and the ratio plot of M_2/M_1 vs. M_2/μ (right panel). One can see that, compared to `minmassgap` $=$

[10] The results for `minmassgap` $= 15$ and 20 GeV are shown for illustration only.

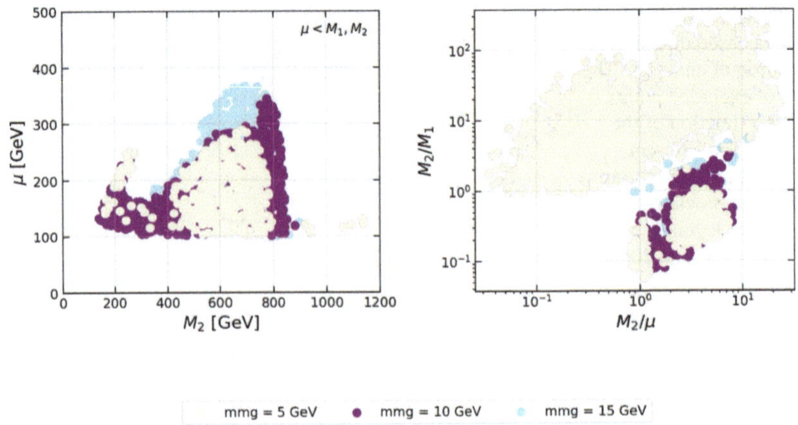

Fig. 4.30 Observed exclusion using the best combination of analyses for various choices of the mass compression parameter (abbreviated as "mmg" in the legend), in terms of M_1, M_2, and μ

5 GeV, a value of `minmassgap` = 10 GeV gives a much better coverage of scenarios with higgsino-like LSPs, in both the low- and the high-mass regions.

4.5 Conclusions and Outlooks

The EW-ino sector of the MSSM is difficult to constrain at the LHC in a generic manner because mixing effects lead to large variations in production cross sections and decay branching ratios. As a consequence, limits on charginos and neutralinos established in the context of simplified models do not hold in general. A reinterpretation in realistic theoretical scenarios, combining the wealth of experimental results from searches in different final states, is necessary. The version 2.3 of the SModelS software package provides a framework to do this, with the possibility to combine approximately uncorrelated analyses and a significant database containing the latest available experimental results for full Run 2 luminosity.

The presented study used the set of EW-ino analyses in the SModelS database v2.3 to constrain the EW-ino sector of the MSSM. To this end, 18K points with promptly decaying EW-inos from a large scan over M_1, M_2, μ, and $\tan\beta$ were confronted against the experimental data (points with long-lived charginos, as typical for wino-LSP scenarios, were not considered, as they are constrained by disappearing-track searches).

First, a comparative study was performed, without combining analyses, between the SModelS database v2.3 and v2.1 [7] to assess the impact of the newly implemented EW-inos searches. For promptly decaying EW-inos, the reach in the lightest chargino mass increases from about 650 GeV in v2.1 to about 1 TeV in v2.3. The total number of excluded points increases by more than a factor of four. Then, the impact of the combination of analyses was studied. In a simple approach to begin with, combining only the two ATLAS and CMS hadronic searches. It showed a first example of how the combination of approximately independent analyses helps average out statistical fluctuations and thus leads to more robust limits. Then, in a global likelihood analysis dynamically finding the most sensitive combination of approximately uncorrelated analyses among 16 EW-ino searches. The dynamic determination of the most sensitive combination is necessary because not all searches considered in this study are approximately uncorrelated, and therefore different sets of combinations are possible. Consequently, since the sensitivity of each individual analysis changes from point to point in the scan, so does the sensitivity of any possible combination. From the 16 searches considered, 13 are from Run 2, and 9 (7 ATLAS and 2 CMS) for full Run 2 luminosity. Besides the "conventional" EW-ino searches in final states with leptons, this also comprises the new ATLAS and CMS searches for EW-inos in fully hadronic final states.

This approach permitted to highlight the various most sensitive combinations and how they populate the parameter space, the effect of individual analyses on the global likelihood, and the combination's impact on the exclusion power compared to an analysis-by-analysis approach. It also showed how the combination in the high-mass region is dominated by the two ATLAS and CMS hadronic searches, which recorded opposite fluctuations, leading to a fair augmentation of excluded points, while unexcluding only a few. For lighter masses, the combination's behaviour is less intuitive, especially because of the increased sensitivity to other analyses, resulting in an augmented number of analyses entering the combination. All in all, the combination of analyses increases the number of points (expected to be) excluded by (48%) 35% compared to the most sensitive analysis. More importantly, it mitigates the sensitivity to fluctuations in the data, therefore leading to more robust constraints.

One aspect that could not be covered in the study is the small excess seen in searches for compressed EW-inos by both ATLAS [6, 41] and CMS [56, 70], see Sect. 2.3.4. The reason why this case could not comprehensively be treated was that only one of the relevant analyses, ATLAS-SUSY-2019-09 [41], was implemented in the SModelS database at the time of the study. The other ATLAS search, ATLAS-SUSY-2018-16 [6], does provide extensive material on HEPData, but was only implemented with database v3.0 [71]. The two CMS searches, CMS-SUS-18-004 [56] and CMS-SUS-19-012 [70], on the other hand, do not provide any auxiliary material for reinterpretation. Efforts to recast [56, 70] and produce EMs for SModelS are ongoing.

References

1. Araz JY et al (2023) Strength in numbers: optimal and scalable combination of LHC new-physics searches. SciPost Phys 14(4):077. https://doi.org/10.21468/SciPostPhys.14.4.077. arXiv:2209.00025 [hep-ph]
2. Feike A et al (2024) Combination and reinterpretation of LHC SUSY searches. arXiv:2403.11715 [hep-ph]
3. Athron P et al (2019) Combined collider constraints on neutralinos and charginos. Eur Phys J C 79(5):395. https://doi.org/10.1140/epjc/s10052-019-6837-x. arXiv:1809.02097[hep-ph]
4. Ananyev V et al (2023) Collider constraints on electroweakinos in the presence of a light gravitino. Eur Phys J C 83(6):493. https://doi.org/10.1140/epjc/s10052-023-11574-z. arXiv:2303.09082 [hep-ph]
5. Aad G et al (2020) Search for direct production of electroweakinos in final states with one lepton, missing transverse momentum and a Higgs boson decaying into two b-jets in pp collisions at $\sqrt{s} = 13$ TeV with the ATLAS detector. Eur Phys J C 80(8):691. https://doi.org/10.1140/epjc/s10052-020-8050-3. arXiv:1909.09226 [hep-ex]
6. Aad G et al (2020) Searches for electroweak production of supersymmetric particles with compressed mass spectra in $\sqrt{s} = 13$ TeV pp collisions with the ATLAS detector. Phys Rev D 101(5):052005. https://doi.org/10.1103/PhysRevD.101.052005. arXiv:1911.12606[hep-ex]
7. Alguero G et al (2022) Constraining new physics with SModelS version 2. J High Energy Phys 08:068. https://doi.org/10.1007/JHEP08(2022)068. arXiv:2112.00769 [hep-ph]
8. Altakach MM et al (2023) SModelS v2.3: enabling global likelihood analyses. SciPost Phys 15(5):185. https://doi.org/10.21468/SciPostPhys.15.5.185. arXiv:2306.17676 [hep-ph]
9. Altakach MM et al (2024) Global LHC constraints on electroweak-inos with SModelS v2.3. SciPost Phys 16:101. https://doi.org/10.21468/SciPostPhys.16.4.101. arXiv:2312.16635 [hep-ph]
10. Altakach M et al (2023) EW-ino scan points from "SModelS v2.3: enabling global likelihood analyses" paper. Zenodo. https://doi.org/10.5281/zenodo.8275263
11. Altakach MM et al (2024) Datasets used in "Global LHC constraints on electroweak- inos with SModelS v2.3". Zenodo. https://doi.org/10.5281/zenodo.10471553
12. Heinrich L et al (2021) pyhf: pure-Python implementation of HistFactory statistical models. J. Open Source Softw. 6(58):2823. https://doi.org/10.21105/joss.02823
13. ATLAS Collaboration (2021) Implementation of simplified likelihoods in HistFactory for searches for supersymmetry. Tech. Rep. ATL-PHYS-PUB-2021-038
14. CMS Collaboration (2017) Simplified likelihood for the re-interpretation of public CMS results. Tech. Rep. CMS-NOTE-2017-001. https://cds.cern.ch/record/2242860. CERN, Geneva
15. Aad G et al (2014) Search for direct production of charginos, neutralinos and sleptons in final states with two leptons and missing transverse momentum in pp collisions at $\sqrt{s} = 8$ TeV with the ATLAS detector. J High Energy Phys 05:071. https://doi.org/10.1007/JHEP05(2014)071. arXiv:1403.5294 [hep-ex]
16. Dumont B (2021) Implementation of a search for electroweakinos in the di-lepton + missing energy channel (20.3 fb^{-1}; 8 TeV; ATLAS-SUSY-2013-11). https://doi.org/10.14428/DVN/MPMPNN. Version V1
17. Aad G et al (2014) Search for direct production of charginos and neutralinos in events with three leptons and missing transverse momentum in $\sqrt{s} = 8$ TeV pp collisions with the ATLAS detector. J High Energy Phys 1404:169. https://doi.org/10.1007/JHEP04(2014)169. arXiv:1402.7029[hep-ex]
18. ATLAS Collaboration (2014) Search for direct production of charginos and neutralinos in events with three leptons and missing transverse momentum in $\sqrt{s} = 8$ TeV pp collisions with the ATLAS detector. HEPData (Collection). https://doi.org/10.17182/hepdata.63114
19. Aaboud M et al (2018) Search for electroweak production of supersymmetric particles in final states with two or three leptons at $\sqrt{s} = 13$ TeV with the ATLAS detector. Eur Phys J C78(12):995. https://doi.org/10.1140/epjc/s10052-018-6423-7. arXiv:1803.02762 [hep-ex]

20. ATLAS Collaboration (2018) Search for electroweak production of supersymmetric particles in final states with two or three leptons at $\sqrt{s} = 13$ TeV with the ATLAS detector (Version 1). HEPData (Collection). https://doi.org/10.17182/hepdata.81996.v1
21. Aaboud M et al (2018) Search for chargino-neutralino production using recursive jigsaw reconstruction in final states with two or three charged leptons in proton-proton collisions at $\sqrt{s} = 13$ TeV with the ATLAS detector. Phys Rev D98(9):092012. https://doi.org/10.1103/PhysRevD.98.092012. arXiv:1806.02293 [hep-ex]
22. ATLAS Collaboration (2018) Search for chargino-neutralino production using recursive jigsaw reconstruction in final states with two or three charged leptons in proton-proton collisions at $\sqrt{s}$ =13 TeV with the ATLAS detector (Version 1) HEPData (Collection). https://doi.org/10.17182/hepdata.83419.v1
23. Aad G et al (2023) Searches for new phenomena in events with two leptons, jets, and missing transverse momentum in 139 fb^{-1} of $\sqrt{s} = 13$ TeV *pp* collisions with the ATLAS detector. Eur Phys J C 83(6):515. https://doi.org/10.1140/epjc/s10052-023-11434-w. arXiv:2204.13072 [hep-ex]
24. ATLAS Collaboration (2022) Searches for new phenomena in events with two leptons, jets, and missing transverse momentum in 139 fb^{-1} of $\sqrt{s} = 13$ TeV pp collisions with the ATLAS detector (Version 1). HEPData (Collection). https://doi.org/10.17182/hepdata.116034.v1
25. ATLAS Collaboration (2022) "llh jsons.tar" of "Searches for new phenomena in events with two leptons, jets, and missing transverse momentum in 139 fb^{-1} of $\sqrt{s} = 13$ TeV pp collisions with the ATLAS detector" (Version 1). HEPData (Other). https://doi.org/10.17182/hepdata.116034.v1/r34
26. Aad G et al (2020) Search for chargino-neutralino production with mass splittings near the electroweak scale in three-lepton final states in $\sqrt{s} = 13$ TeV *pp* collisions with the ATLAS detector. Phys Rev D101(7):072001. https://doi.org/10.1103/PhysRevD.101.072001. arXiv:1912.08479 [hep-ex]
27. Jackson P, Rogan C, Santoni M (2017) Sparticles in motion: analyzing compressed SUSY scenarios with a new method of event reconstruction. Phys Rev D95(3):035031. https://doi.org/10.1103/PhysRevD.95.035031. arXiv:1607.08307[hep-ph]
28. Jackson P, Rogan C (2017) Recursive jigsaw reconstruction: HEP event analysis in the presence of kinematic and combinatoric ambiguities. Phys Rev D 96(11):112007. https://doi.org/10.1103/PhysRevD.96.112007. arXiv:1705.10733 [hep-ph]
29. ATLAS Collaboration (2020) Search for chargino-neutralino production with mass splittings near the electroweak scale in three-lepton final states in $\sqrt{s} = 13$ TeV pp collisions with the ATLAS detector (Version 2). HEPData (Collection). https://doi.org/10.17182/hepdata.91127.v2
30. ATLAS Collaboration (2020) "likelihoods ANA-SUSY-2018-06 3L-RJ-mimic.tar.gz" of "Search for chargino-neutralino production with mass splittings near the electroweak scale in three-lepton final states in $\sqrt{s} = 13$ TeV pp collisions with the ATLAS detector" (Version 2). HEPData (Other). https://doi.org/10.17182/hepdata.91127.v2/r3
31. Aad G et al (2020) Search for electroweak production of charginos and sleptons decaying into final states with two leptons and missing transverse momentum in $\sqrt{s} = 13$ TeV *pp* collisions using the ATLAS detector. Eur Phys J C80(2):123. https://doi.org/10.1140/epjc/s10052-019-7594-6. arXiv:1908.08215 [hep-ex]
32. ATLAS Collaboration (2022) Search for electroweak production of charginos and sleptons decaying into final states with two leptons and missing transverse momentum in $\sqrt{s} = 13$ TeV pp collisions using the ATLAS detector (Version 4). HEPData (Collection). https://doi.org/10.17182/hepdata.89413.v4
33. ATLAS Collaboration (2022) "likelihoods.tar.gz" of "Search for electroweak production of charginos and sleptons decaying into final states with two leptons and missing transverse momentum in $\sqrt{s} = 13$ TeV pp collisions using the ATLAS detector" (Version 4). HEPData (Other). https://doi.org/10.17182/hepdata.89413.v4/r5

34. Alguero G, Kraml S, Waltenberger W (2021) A SModelS interface for pyhf likelihoods. Comput Phys Commun 264:107909. https://doi.org/10.1016/j.cpc.2021.107909. arXiv:2009.01809 [hep-ph]
35. Aad G et al (2021) Search for charginos and neutralinos in final states with two boosted hadronically decaying bosons and missing transverse momentum in *pp* collisions at $\sqrt{s}$ = 13 TeV with the ATLAS detector. Phys Rev D 104(11):112010. https://doi.org/10.1103/PhysRevD.104.112010. arXiv:2108.07586 [hep-ex]
36. ATLAS Collaboration (2021) Search for charginos and neutralinos in final states with two boosted hadronically decaying bosons and missing transverse momentum in pp collisions at $\sqrt{s}$ = 13 TeV with the ATLAS detector (Version 1). HEPData (Collection). https://doi.org/10.17182/hepdata.104458.v1
37. Aad G et al (2023) Search for direct pair production of sleptons and charginos decaying to two leptons and neutralinos with mass splittings near the W-boson mass in $\sqrt{s}$ = 13 TeV *pp* collisions with the ATLAS detector. J High Energy Phys 06:031. https://doi.org/10.1007/JHEP06(2023)031. arXiv:2209.13935 [hep-ex]
38. ATLAS Collaboration (2022) Search for direct pair production of sleptons and charginos decaying to two leptons and neutralinos with mass splittings near the W-boson mass in $\sqrt{s}$ = 13 TeV pp collisions with the ATLAS detector (Version 1). HEPData (Collection). https://doi.org/10.17182/hepdata.134068.v1
39. Aad G et al (2020) Search for direct production of electroweakinos in final states with one lepton, missing transverse momentum and a Higgs boson decaying into two *b*-jets in *pp* collisions at $\sqrt{s}$ = 13 TeV with the ATLAS detector. Eur Phys J C80(8):691. https://doi.org/10.1140/epjc/s10052-020-8050-3. arXiv:1909.09226 [hep-ex]
40. ATLAS Collaboration (2020) Search for direct production of electroweakinos in final states with one lepton, missing transverse momentum and a Higgs boson decaying into two b-jets in (pp) collisions at $\sqrt{s}$ = 13 TeV with the ATLAS detector (Version 2). HEPData (Collection). https://doi.org/10.17182/hepdata.90607.v2
41. Aad G et al (2021) Search for chargino-neutralino pair production in final states with three leptons and missing transverse momentum in $\sqrt{s}$ = 13 TeV *pp* collisions with the ATLAS detector. Eur Phys J C 81(12):1118. https://doi.org/10.1140/epjc/s10052-021-09749-7. arXiv:2106.01676 [hep-ex]
42. ATLAS Collaboration (2022) Search for chargino-neutralino pair production in final states with three leptons and missing transverse momentum in $\sqrt{s}$ = 13 TeV pp collisions with the ATLAS detector (Version 2). HEPData (Collection). https://doi.org/10.17182/hepdata.95751.v2
43. ATLAS Collaboration (2022) "statistical models.tar.gz" of "Search for chargino-neutralino pair production in final states with three leptons and missing transverse momentum in $\sqrt{s}$ = 13 TeV pp collisions with the ATLAS detector" (Version 2). HEPData (Other). https://doi.org/10.17182/hepdata.95751.v2/r3
44. Khachatryan V et al (2015) Constraints on the pMSSM, AMSB model and on other models from the search for long-lived charged particles in proton-proton collisions at $\sqrt{s}$ = 8 TeV. Eur Phys J C75(7):325. https://doi.org/10.1140/epjc/s10052-015-3533-3. arXiv:1502.02522 [hep-ex]
45. Chatrchyan S et al (2014) Search for new physics in the multijet and missing transverse momentum final state in proton-proton collisions at $\sqrt{s}$ = 8 TeV. J High Energy Phys 06:055. https://doi.org/10.1007/JHEP06(2014)055. arXiv:1402.4770 [hep-ex]
46. Bein S, Sengupta D (2021) Implementation of a search for gluinos and squarks in the multijet + missing energy channel (19.5 fb^{-1}; 8 TeV; CMS-SUS-13-012). https://doi.org/10.14428/DVN/UDR1SA. Version V1
47. Sirunyan AM et al (2018) Search for electroweak production of charginos and neutralinos in multilepton final states in proton-proton collisions at $\sqrt{s}$ = 13 TeV. J High Energy Phys 03:166. https://doi.org/10.1007/JHEP03(2018)166. arXiv:1709.05406 [hep-ex]

48. Fuks B, Mondal S (2021) Implementation of a search for electroweakinos in the multi-lepton + missing energy channel (35.9 fb-1; 13 TeV; CMS-SUS-16-039). https://doi.org/10.14428/DVN/E8WN3G. Version V3
49. Sirunyan AM et al (2018) Search for new physics in events with two soft oppositely charged leptons and missing transverse momentum in proton-proton collisions at $\sqrt{s} = 13$ TeV. Phys Lett B 782:440–467. https://doi.org/10.1016/j.physletb.2018.05.062. arXiv:1801.01846 [hep-ex]
50. Fuks B (2020) Re-implementation of the soft lepton + missing energy analysis of CMS (35.9 fb^{-1}; CMS-SUS-16-048). https://doi.org/10.14428/DVN/YA8E9V. Version V2
51. Sirunyan AM et al (2018) Combined search for electroweak production of charginos and neutralinos in proton-proton collisions at $\sqrt{s} = 13$ TeV. J High Energy Phys 03:160. https://doi.org/10.1007/JHEP03(2018)160. arXiv:1801.03957 [hep-ex]
52. Sirunyan AM et al (2018) Search for new phenomena in final states with two opposite-charge, same-flavor leptons, jets, and missing transverse momentum in *pp* collisions at $\sqrt{s} = 13$ TeV. J High Energy Phys 03:076. https://doi.org/10.1007/s13130-018-7845-2. arXiv:1709.08908 [hep-ex]
53. Sirunyan AM et al (2017) Search for electroweak production of charginos and neutralinos in WH events in proton-proton collisions at $\sqrt{s} = 13$ TeV. J High Energy Phys 11:029. https://doi.org/10.1007/JHEP11(2017)029. arXiv:1706.09933 [hep-ex]
54. Sirunyan AM et al (2018) Search for Higgsino pair production in *pp* collisions at $\sqrt{s} = 13$ TeV in final states with large missing transverse momentum and two Higgs bosons decaying via $H \rightarrow b\bar{b}$. Phys Rev D 97(3):032007. https://doi.org/10.1103/PhysRevD.97.032007. arXiv:1709.04896 [hep-ex]
55. Sirunyan AM et al (2018) Search for supersymmetry with Higgs boson to diphoton decays using the razor variables at $\sqrt{s} = 13$ TeV. Phys Lett B779:166–190. https://doi.org/10.1016/j.physletb.2017.12.069. arXiv:1709.00384 [hep-ex]
56. Tumasyan A et al (2022) Search for supersymmetry in final states with two or three soft leptons and missing transverse momentum in proton-proton collisions at $\sqrt{s} = 13$ TeV. J High Energy Phys 04:091. https://doi.org/10.1007/JHEP04(2022)091. arXiv:2111.06296 [hep-ex]
57. Tumasyan A et al (2022) Search for higgsinos decaying to two Higgs bosons and missing transverse momentum in proton-proton collisions at $\sqrt{s} = 13$ TeV. J High Energy Phys 05:014. https://doi.org/10.1007/JHEP05(2022)014. arXiv:2201.04206 [hep-ex]
58. CMS Collaboration (2022) Search for higgsinos decaying to two Higgs bosons and missing transverse momentum in proton-proton collisions at $\sqrt{s} = 13$ *TeV (Version 2)*. HEPData (Collection). https://doi.org/10.17182/hepdata.114414.v2
59. Buckley A et al (2019) The simplified likelihood framework. J High Energy Phys 04:064. https://doi.org/10.1007/JHEP04(2019)064. arXiv:1809.05548 [hep-ph]
60. Ford W (2022) Reproducing a CMS higgsino search from public data. Talk at the 7th Workshop on '(Re)interpretation of the LHC results for new physics', 12–15 Dec 2022, CERN. https://indico.cern.ch/event/1197680/contributions/5164689/
61. Tumasyan A et al (2023) Search for electroweak production of charginos and neutralinos at s =13TeV in final states containing hadronic decays of WW, WZ, or WH and missing transverse momentum. Phys Lett B 842:137460. https://doi.org/10.1016/j.physletb.2022.137460. arXiv:2205.09597 [hep-ex]
62. CMS Collaboration (2024) Search for electroweak production of charginos and neutralinos at $\sqrt{s}$ =13 TeV in final states containing hadronic decays of WW, WZ, or WH and missing transverse momentum. HEPData (Collection). https://doi.org/10.17182/hepdata.127766
63. Skands PZ et al (2004) SUSY Les Houches accord: interfacing SUSY spectrum calculators, decay packages, and event generators. J High Energy Phys 0407:036. https://doi.org/10.1088/1126-6708/2004/07/036. arXiv:hep-ph/0311123 [hep-ph]
64. Ambrogi F et al (2018) SModelS v1.1 user manual: improving simplified model constraints with efficiency maps. Comput Phys Commun 227:72–98. https://doi.org/10.1016/j.cpc.2018.02.007. arXiv:1701.06586 [hep-ph]

65. Waltenberger W, Lessa A, Kraml S (2021) Artificial proto-modelling: building precursors of a next standard model from simplified model results. J High Energy Phys 03:207. https://doi.org/10.1007/JHEP03(2021)207. arXiv:2012.12246 [hep-ph]
66. Fuks B et al (2012) Gaugino production in proton-proton collisions at a center-of-mass energy of 8 TeV. J High Energy Phys 10:081. https://doi.org/10.1007/JHEP10(2012)081. arXiv:1207.2159 [hep-ph]
67. Fuks B et al (2013) Precision predictions for electroweak superpartner production at hadron colliders with Resummino. Eur Phys J C 73:2480. https://doi.org/10.1140/epjc/s10052-013-2480-0. arXiv:1304.0790 [hep-ph]
68. Kraml S et al (2014) SModelS: a tool for interpreting simplified-model results from the LHC and its application to supersymmetry. Eur Phys J C 74:2868. https://doi.org/10.1140/epjc/s10052-014-2868-5. arXiv:1312.4175 [hep-ph]
69. Goodsell M (2020) Re-implementation of the H (into b bbar) + 1 lepton + missing transverse momentum analysis (139 fb^{-1}; ATLAS-SUSY-2019-08). https://doi.org/10.14428/DVN/BUN2UX. Version V6
70. Tumasyan A et al (2022) Search for electroweak production of charginos and neutralinos in proton-proton collisions at $\sqrt{s} = 13$ TeV. J High Energy Phys 04:147. https://doi.org/10.1007/JHEP04(2022)147. arXiv:2106.14246 [hep-ex]
71. SModelS Collaboration (2024) SModelS database v3.0.0. Zenodo. https://doi.org/10.5281/zenodo.13354582

Chapter 5
Characterising LHC Dispersed Signals

Abstract Given the absence of clear hints for new physics in the multiple channel-by-channel searches at the Large Hadron Collider, it is now more important than ever to adopt a different approach and try a more comprehensive strategy to see where beyond the Standard Model physics may hide. Instead of a typical top-down method, a more global, more model-independent search for new physics can be envisaged through an anomaly detection procedure based on a bottom-up approach. This chapter will focus on the "proto-modelling" project, whose specificity is to use simplified models to look for potential dispersed signals that may have been missed in the common analysis-by-analysis interpretations of the data. It uses SModelS and its database to find small excesses in the experimental results and to construct collections of simplified models maximally violating the Standard Model hypothesis while maintaining consistency with the ATLAS and CMS constraints. This chapter discusses improvements and novelties made to the "proto-modelling" machine (initially published as a proof-of-concept) and presents a first test run of this machine.

Despite experimental and theoretical motivations for BSM physics, no experimental evidence has yet been recorded that could shed light on new physics. Moreover, should signals for new physics be detected, at the LHC or elsewhere, this would require mapping the signals into the parameter space of the underlying theory. This inverse problem will not be trivial for two reasons. First, the number of SM extensions is large and still growing (see e.g. [1, 2] for a short overview), many of which can have a significant number of free parameters and several, non-minimal variations. Second, too little information (i.e. observables) will likely be available to directly infer the theoretical model from the data. These issues were discussed and studied in, e.g., [3–9].

Therefore, given the absence of clear hints for new physics in the multiple channel-by-channel searches at the LHC, it is now more important than ever to adopt a different approach and try a more comprehensive strategy to see where BSM physics may hide. Instead of a typical top-down method, a more global, more model-independent search for new physics can be envisaged through a procedure

T. Pascal, *Beyond the Standard Model at the Large Hadron Collider: From Simplified Model Constraints to Artificial Proto-Modelling for Dispersed Signals*, Springer Theses, https://doi.org/10.1007/978-3-032-09391-2_5

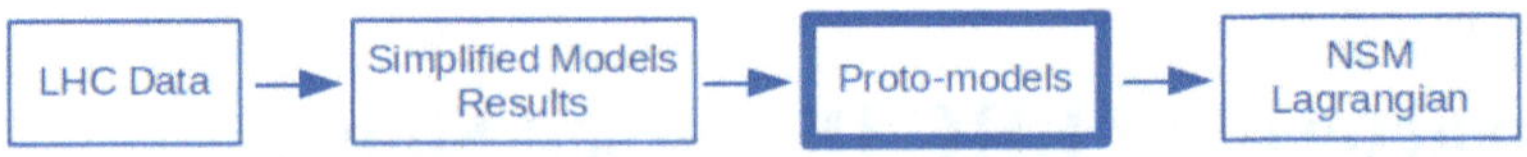

Fig. 5.1 Schematic representation of how proto-models could guide the search for BSM physics. The experimental data from the LHC is used to constrain simplified models, which in turn are used to build minimal proto-models maximally violating the SM hypothesis while evading current LHC constraints. If successful, this may pave the way to the NSM Lagrangian. Taken from [26]

that resembles anomaly detection, based on a bottom-up approach. For examples of anomaly detection procedures, see, e.g., [5, 10–25]. This chapter will focus on the "proto-modelling" project [26], whose specificity is to use simplified models to look for potential dispersed signals that may have been missed in the common analysis-by-analysis interpretations of the data. It uses SModelS and its database to find small excesses in the experimental results and to construct "proto-models" maximally violating the SM hypothesis while maintaining consistency with the ATLAS and CMS constraints.

More precisely, proto-models are sets of simplified models obtained by varying the number of BSM particles, the mass of the particles, their production cross sections and their decay branching ratios. It is important to stress that they are not required to satisfy a SUSY model, and are not intended to be ultraviolet-complete models or a low-energy effective field theory. Their purpose is to guide the experimental and theoretical search for new physics, and hopefully be an intermediate step in the construction of the Next Standard Model (NSM) Lagrangian. The overall idea is represented in Fig. 5.1.

A proof-of-concept has been presented in [26]. The following discussion is a continuation of this work, improving the initial concept with new features and improved statistical treatments. This chapter is organised as follows. Section 5.1 describes the working principle as initially presented in [26]. Section 5.2 introduces the novelties and improvements with respect to [26]. Section 5.3 presents the first tests and preliminary results. Finally, Sect. 5.4 provides a summary and conclusions.

My contribution concerns almost all the presented points of Sect. 5.2. For instance, I have worked on the improved combinability matrix, a better initialisation of the production cross sections and the overall structure of the algorithm. Moreover, I implemented the new critics, and I carried out the test runs and analysed the results as described in Sect. 5.3. The work is still ongoing but is expected to lead to a journal publication within a few months.

5.1 Initial Concept

As originally described in [26], the proto-modelling algorithm is made of three building blocks: the builder, the critic, and the combiner. This section will present their working principle.

5.1.1 The Builder

The builder's task is to generate proto-models. As mentioned above, proto-models are defined by their BSM particle content, the particle masses, production cross sections and decay branching ratios. They are not meant to be fully consistent theoretical models and are not bounded by any higher-level theoretical assumptions. They are built by randomly changing the number of BSM particles or their simplified model parameters. The parameters are varied using a Markov Chain Monte Carlo (MCMC)-type walk in the parameter space (of varying dimensionality!).

The proto-models are required to follow some constraints:

- all BSM particles are odd under a $\mathbb{Z}_2$ symmetry: they can only be pair-produced and can only decay into an odd number of BSM particles;
- the lightest BSM particle (LBP) is stable and electrically and color neutral (and so, a dark matter candidate);
- all BSM particles decay promptly, except for the LBP;
- BSM particles with a mass above the LHC reach are not considered (here, only particles with a mass below 2.4 TeV are considered).

They come from the fact that SModels v1.2.4, used in [26], could only constrain models with a $\mathbb{Z}_2$-like symmetry and leading to collider signatures with missing transverse energy. The SModels database v1.2.4 comprised Run 1 and Run 2 results from 40 ATLAS and 46 CMS experimental searches for a variety of final states with missing transverse energy.

5.1.1.1 Particle Content

The particle degrees of freedom (e.g. the spin and the multiplicity of states) are mostly relevant for the production cross sections. Since the latter are free parameters here, the spin and the multiplicity of the BSM particles are disregarded. Twenty SUSY-inspired particles are considered for building proto-models:

- four light quark partners X_q ($q \in \{u, d, s, c\}$). Only a single partner is considered for each light quark. Should there be other (mass-degenerate) partners (e.g. another chirality), this would be accounted for in the rescaling of the production cross sections;
- four heavy quark partners X_b^i, X_t^i ($i \in \{1, 2\}$). Two partners are considered for the heavy quarks in order to have enough degrees of freedom to accommodate the numerous searches targeting their signatures;
- one gluon partner X_g. A new color-octet particle, similar to the gluino in SUSY;
- five electroweak partners X_W^i, X_Z^j ($i \in \{1, 2\}$, $j \in \{1, 2, 3\}$). The X_W^i are electrically charged, while the X_Z^j are electrically neutral. The lightest neutral state (X_Z^1) is assumed to be the LBP;

- three electrically charged lepton partners X_ℓ ($\ell \in \{e, \mu, \tau\}$). As for light quarks, each charged lepton can only have up to one partner;
- three neutrino partners X_{ν_ℓ} ($\nu_\ell \in \{\nu_e, \nu_\mu, \nu_\tau\}$). Again, only a single partner can be introduced for each neutrino flavor.

States with multiple partners are mass ordered, e.g., $m(X_t^2) > m(X_t^1)$. Moreover, to avoid "picking-up" phase-space regions only because they are poorly covered, the masses of the colored particles are further constrained. Concretely, most of the simplified model results in the SModelS database v1.2.4 do not constrain new colored states with a mass below 310 GeV. Therefore, X_q and X_g are required to be heavier than this value. In addition, the region 200 GeV $> m(X_t) - m(X_Z^1) >$ 150 GeV is forbidden when X_t is below 280 GeV. Indeed, events in this "corridor" region become too similar to $t\bar{t}$ events, so many CMS analyses do not make any statement for it, see [27] for example.

5.1.1.2 Production Cross Sections

Each production cross section can vary freely. Nonetheless, to have sensible starting values, they are initialised using MSSM-like ones and are then freely rescaled using a signal strength multiplier κ. Thus, the free parameters are not the cross sections but the signal strength multipliers (SSMs). For instance, the cross section for the pair-production of X_g is

$$\sigma(pp \to X_g X_g) = \kappa_{X_g X_g} \times \sigma(pp \to \tilde{g}\tilde{g}), \tag{5.1}$$

where the gluino cross section is evaluated at the X_g mass. The cross sections were evaluated during runtime using Pythia 8.2 [28] and NLLFast 3.1 [29–35].

5.1.1.3 Decay Branching Ratios

Except for the LBP, which is stable, each BSM particle should decay according to its quantum numbers. The considered decays are listed in Table 5.1. Not all possible decays are allowed here. In particular, only on-shell decays were considered, and X_W and X_Z were not allowed to decay into $X_\ell + \nu/\ell$, $X_\nu + \ell/\nu$ or $X_{W/Z} + \gamma$. Of course, a given decay is turned off if it is kinematically forbidden (or if the daughter particle is not in the proto-model). The branching ratios are free parameters but are always rescaled to add up to unity.

Table 5.1 The BSM particles considered for building proto-models in [26] and their decay channels

Particle	Decay channels	Particle	Decay channels
X_q	$qX_Z^j,\ q'X_W^i,\ qX_g$	X_W^1	WX_Z^j
X_t^1	$tX_Z^j,\ bX_W^i,\ WX_b^1,\ tX_g$	X_W^2	$WX_Z^j,\ ZX_W^1,\ hX_W^1$
X_b^1	$bX_Z^j,\ tX_W^i,\ WX_t^1,\ bX_g$	$X_Z^{j\neq 1}$	$WX_W^i,\ ZX_Z^k,\ hX_Z^k$
X_t^2	$tX_Z^j,\ bX_W^i,\ ZX_t^1,\ WX_b^1,\ tX_g$	X_ℓ	$\ell X_Z^j,\ \nu_\ell X_W^i$
X_b^2	$bX_Z^j,\ tX_W^i,\ ZX_b^1,\ WX_t^1,\ bX_g$	X_{ν_ℓ}	$\nu_\ell X_Z^j,\ \ell X_W^i$
X_g	$q\bar{q}X_Z^i,\ q\bar{q}'X_W^i,\ b\bar{b}X_Z^i,\ t\bar{t}X_Z^j,\ btX_W^i,\ qX_q,\ bX_b,\ tX_t$		

5.1.2 Test Statistic

Once a new proto-model is built, the goal is to know if it performs better than the previous one, while evading LHC constraints. To this end, the following test statistic is computed:

$$K := \max_{c \in C}\left[2\ln\left(\frac{L^c_{\text{BSM}}(\hat{\mu}) \cdot \pi_{\text{BSM}}}{L^c_{\text{SM}} \cdot \pi_{\text{SM}}}\right)\right]. \tag{5.2}$$

Here, c represents a combination of analyses among the set C of all the allowed combinations (see the combiner, below); $L^c_{\text{BSM}}(\hat{\mu})$ is the new proto-model maximal likelihood for a combination c; L^c_{SM} is the corresponding SM likelihood: $L^c_{\text{SM}} = L^c_{\text{BSM}}(\mu = 0)$; and π_{BSM} and π_{SM} are the priors for the new proto-model and the SM, respectively. Since π_{SM} is common to all proto-models and does not impact the comparison between them, then

$$\pi_{\text{SM}} := 1\,. \tag{5.3}$$

Regarding the proto-model prior, it is used to prevent the builder from constructing proto-models over-fitting the data. It penalises the test statistic for newly introduced particles, branching ratios and signal strength multipliers:

$$\pi_{\text{BSM}} := \exp\left[-\left(\frac{n_{\text{particles}}}{a_1} + \frac{n_{\text{BRs}}}{a_2} + \frac{n_{\text{SSMs}}}{a_3}\right)\right], \tag{5.4}$$

with $n_{\text{particles}}$ the number of BSM particles in the new proto-model, n_{BRs} the number of branching ratios not equal to unity, and n_{SSMs} the number of SSMs (the particle-particle and particle-antiparticle SSMs can be different). The parameters a_1, a_2 and a_3 are equal to 2, 4 and 8, respectively. This way, one free parameter in the Akaike information criterion [36] corresponds to one particle with one non-trivial decay and two production modes. Finally, note that the prior is not normalized in the space of all proto-models.

The test statistic K_{new} of the new proto-model is compared to the one of the previous proto-model K_{old}. If $K_{\text{new}} > K_{\text{old}}$, the new proto-model is kept and another one is built from it. Otherwise, the new proto-model is kept with the probability

$$\exp\left[\frac{1}{2}(K_{\text{new}} - K_{\text{old}})\right] . \tag{5.5}$$

If the new proto-model is not kept, the step is reverted, i.e. all the changes with respect to the previous proto-model are undone. In the end, the proto-model violating the SM the most while being consistent with the data is the one with the highest K.

5.1.3 The Critic

The critic checks the viability of the proto-model against the results implemented in the SModelS database. The following quantity was computed:

$$\mu_{\max} := 1.3\frac{\sigma_{95}^{\text{obs}}}{\sigma} , \tag{5.6}$$

where σ_{95}^{obs} is the UL on the signal cross section obtained from the most sensitive SR (see Sect. 3.2), σ is the relevant signal cross section, and the factor 1.3 is here to allow for a 30% violation of the 95% CL observed UL. This violation is accepted because when checking for constraints from a large set of results, a few are statistically allowed to be violated. Despite being approximate, this method has the advantage of taking into account the constraints from analyses for which a likelihood can be built and for which it cannot. The combined likelihood $L^{c}_{\text{BSM}}(\hat{\mu})$ of Eq. (5.2) was then truncated above $\mu_{\max}$; that is to say, for a given combination c,

$$\hat{\mu} = \min\left(\hat{\mu}_c, \mu_{\max}\right) , \tag{5.7}$$

with $\hat{\mu}_c$ the value maximising L^{c}_{BSM}.

5.1.4 The Combiner

The combiner is called to find the set of all the permitted analysis combinations C and to compute the maximum of the combined likelihood $L^{c}_{\text{BSM}}(\hat{\mu})$ for each one of them. As explained in Sect. 3.3.1, the global likelihood is simply the product of likelihoods from individual analyses/SRs. Therefore, only analyses/SRs which are approximately uncorrelated can be combined, and all the analyses within the same combination must be approximately uncorrelated with all the other analyses of the combination. Two results are considered approximately uncorrelated if they belong to different runs, different experiments or if they target clearly different final states in their SRs (e.g. fully hadronic vs. fully leptonic final states). Moreover, all the permitted analysis combinations also have to be found dynamically, and all the results allowed to be added to the combination are added.

All the results for which a likelihood can be built are used. Following [37], an approximated likelihood was built from a truncated Gaussian for UL-type results with UL maps on the observed and expected cross sections. For EM-type results, the SRs were not combined, due to a lack of correlation information, and only the SR with the highest $\sigma_{95}^{\text{obs}}/\sigma_{95}^{\text{exp}}$ was entering the combination.

5.2 Novelties and Improvements

An important aim of this thesis was to bring the proof-of-concept presented in [26] to a higher level of maturity. A number of extensions and improvements were thus carried out. The most relevant ones are as follows:

- **SModelS database v3.0:** the next iteration will benefit from the newly released version 3.0 of the SModelS database, containing a total of 125 analyses: 45 ATLAS and 44 CMS Run 2 results (of which 42 use full Run 2 luminosity), and 16 ATLAS and 20 CMS Run 1 results [38, 39] (see Appendix B for a detailed list). Among them, 21 (11 using full Run 2 luminosity) are searching for LLPs, resonances or monojet signatures and are therefore insensitive to the proto-model signals. Moreover, the statistical treatment is also improved and benefits from the developments discussed in Sect. 3.3.
- **Combination of SRs:** the SModelS database v3.0 comprises a total of 9 analyses for which a statistical model is provided, 10 analyses for which a multivariate Gaussian can be built, and 1 analysis for which a skewed multivariate Gaussian can be built. Due to the significant number of analyses for which the correlations among the SRs are known, the combination of SRs is used when available. This gain in accuracy represents an important step forward.
- **No more likelihood from truncated Gaussians:** they give a too crude approximation and are not required anymore since the number of analyses for which a likelihood can be built using EMs has drastically increased,
- **Combinability matrix:** it has been updated with the new database entries, and the analyses for which no likelihood could be built were removed from it. More attention has also been given to analysis combinability. For analyses belonging to the same run and experiment, the overlap of the SRs has been carefully scrutinised through the SR selection cuts. The new combinability matrices for ATLAS and CMS Run 2 results are shown in Fig. 5.2.
- **New decays:** off-shell decays for the X_W^1 and X_Z^2 are introduced, allowing for the algorithm to test proto-models with compressed mass spectra in the EW sector. In addition to the decays of Table 5.1, the following decays are now accessible:

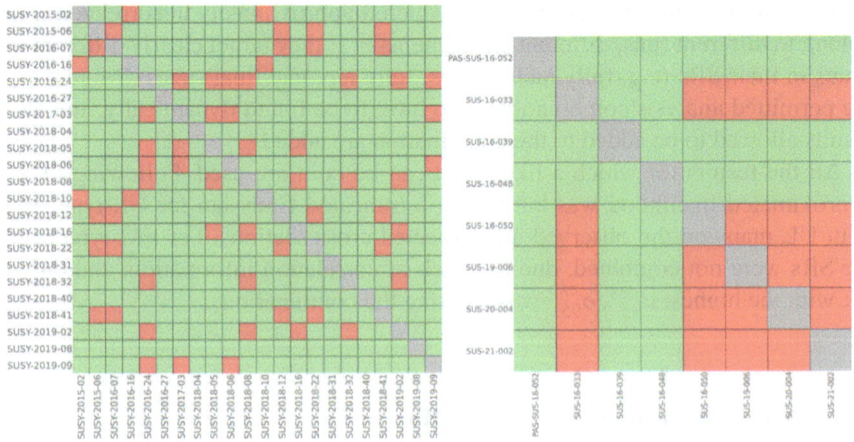

Fig. 5.2 Combinability matrices for ATLAS (left) and CMS (right) Run 2 results. The combinability of the ATLAS EW-ino searches is slightly different from the one of Figure 4.17 because the study of Sect. 4.4 contained EW-ino signatures only

$$X_W^1 \to q\bar{q}' X_Z^1\,,\ \ell\nu_\ell X_Z^1\,; \tag{5.8}$$

$$X_Z^2 \to q\bar{q} X_Z^1\,,\ b\bar{b} X_Z^1\,,\ \ell\ell X_Z^1\,, \tag{5.9}$$

with $q, q' \in \{u, d, s, c\}$, and $\ell \in \{e, \mu, \tau\}$. The $X_Z^2 \to \nu_\ell \nu_\ell X_Z^1$ decays lead to invisible products and are not considered. For the moment, the leptonic decays are forced to be flavor-democratic; for light quarks, the flavor is ignored. In the future, the flavor universality may be alleviated. A penalty term for undemocratic flavor decays should then be added. The X_W^1 and X_Z^2 off-shell decays are turned on if the mass difference between these particles and the LBP is lower than $m_{W^\pm}$ and m_Z, respectively. In that case, the corresponding on-shell decays are turned off. When a X_W^1 or a X_Z^2 is added to the proto-model or when their masses are changed, there are 10% chances for them to get to the off-shell region. Moreover, when one of their masses is changed to the off-shell region, the other EW-ino has 20% chance to be changed to this mass too (and to be added to the proto-model if it was absent). This is to constrain the builder to probe off-shell regions of the parameter space. Otherwise, the off-shell region would be less probed due to volume effect, especially the region where both the X_W^1 and the X_Z^2 are off-shell.

- **Cross section initialisation:** the initialisation is now done using the reference cross sections from [40]. If the required cross section is not available, a similar one is used instead. For example, the 8 TeV cross sections $\sigma(pp \to \tilde{\chi}^0\tilde{\chi}^0)$ are not given by [40], so the $\sigma(pp \to X_Z X_Z)$ cross sections are initialised with $\sigma(pp \to \tilde{\chi}_1^\pm \tilde{\chi}_1^\mp)$; for consistency, the same is done for 13 TeV cross sections.
- **Code structure:** the overall structure has been adapted to the discussed changes. A simplified flowchart is shown in Fig. 5.3. After an initialisation phase, a new proto-model is built from the builder, which is then passed to the combiner to

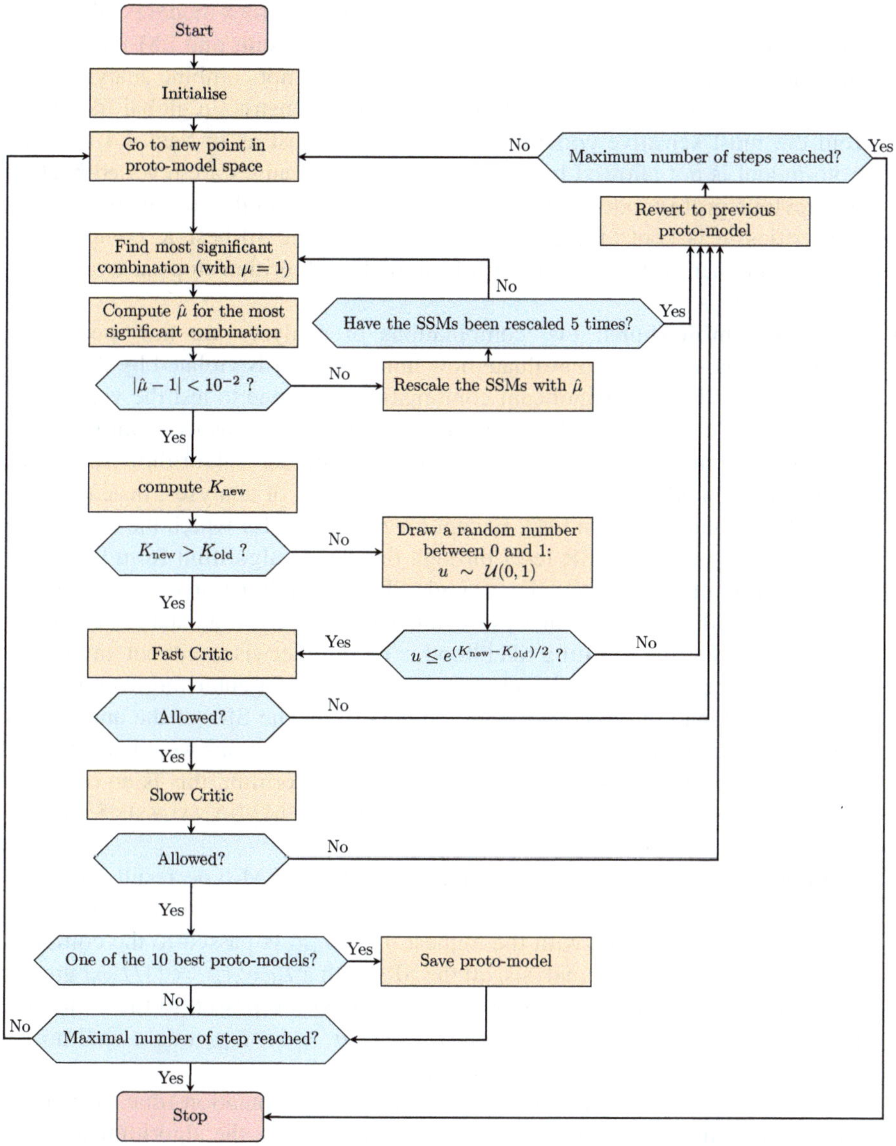

Fig. 5.3 Simplified flowchart of the new code architecture. See the text for details

find the most significant combination of analyses, i.e. the one maximising K. This new K is compared to the one of the previous proto-model, just as in the initial concept. If the new proto-model is not kept, all the changes are reverted back to the previous proto-model and another proto-model is built from it. If it is kept, it is then passed to the critic, which checks that it is not excluded by the results implemented in the SMODELS database. This is actually done through two critics now, each giving a binary output: either the proto-model is allowed or it is not. The first critic is a fast one: it uses UL-type results and EM-type results for which no combination of SRs is possible; it does not combine analyses. The second critic is slower: it uses EM-type results to construct a global likelihood from the most sensitive combination of analyses, just as in Sect. 4.4.1. If the proto-model is not allowed by both critics, all the changes are reverted back to the previous proto-model and another proto-model is built from it. If it passes both critics, it is kept (and saved if it has one of the 10 best K computed so far), and another proto-model is built from it. The algorithm stops after a certain number of steps.

- **Analysis combinations:** two combinations of analyses are now performed during a given step: one to evaluate how much the SM is violated by the proto-model (this is the most significant combination), and one to test the viability of the proto-model against the LHC results (using the most sensitive combination).

  Both combinations are found using the "pathfinder" algorithm from [41], which has been adapted to find the best combination of analyses, instead of the optimal SR combination in [41]. Except for 8 analyses for which the correlation has been obtained at the SR level through the TACO algorithm from [41],[1] the combinability of two datasets (i.e. a single SR or a combination of SRs within an analysis) is addressed at the analysis level. That is to say, if a database entry can combine its SRs, the resulting likelihood is used, otherwise, without any means to combine the SRs, only a single SR per analysis can enter the combination, with its combinability assessed from the cuts of all the SRs of the analysis. A more refined combination could be possible if the correlations were known at the SR level for each analysis. Thanks to the TACO algorithm, this is an objective that should be achievable for the analyses for which a MADANALYSIS 5 recast is available.
- **Finding the most significant combination:** all the EM-type results sensitive to the proto-model signals are used. Here, when the combination of SRs is not possible, not only the SR with the highest $\sigma_{95}^{\text{obs}}/\sigma_{95}^{\text{exp}}$ is passed to the combiner anymore. Now, within an analysis, all the SRs with a $L_{\text{BSM}}(\mu = 1)/L_{\text{SM}}$ greater or equal to 5% of the highest $L_{\text{BSM}}(\mu = 1)/L_{\text{SM}}$ are passed to the combiner. This is because the SR with the highest $L_{\text{BSM}}(\mu = 1)/L_{\text{SM}}$ might well not be the one increasing the most the global $L^c_{\text{BSM}}(\hat{\mu})/L^c_{\text{SM}}$.

  Algorithm-wise, a trick is made to reduce the computation time. Instead of finding the combination that maximises $L^c_{\text{BSM}}(\hat{\mu})/L^c_{\text{SM}}$, the algorithm searches for the combination maximising $L^c_{\text{BSM}}(\mu = 1)/L^c_{\text{SM}}$, computes $\hat{\mu}$ for that combination, and rescales the SSMs with the obtained $\hat{\mu}$. Since the sensitivity of

[1] They are ATLAS-SUSY-2013-02, ATLAS-SUSY-2013-04, ATLAS-SUSY-2013-11, ATLAS-SUSY-2015-06, ATLAS-SUSY-2016-07, CMS-SUS-13-011, CMS-SUS-13-012, CMS-SUS-16-033. This is also true for ATLAS-SUSY-2018-04, ATLAS-SUSY-2018-31, ATLAS-SUSY-2018-32, CMS-SUS-16-039 and CMS-SUS-19-006; however, the full or simplified likelihoods (combining SRs) built for these analyses are always used (hence, their inter-analysis correlations are never considered at the SR level).

the analyses to the proto-model signals depends on the SSMs, it is possible that the most significant combination of analyses changed with the SSMs rescaling. Thus, the procedure consisting of finding the most significant combination is done until it finds a combination with a $\hat{\mu}$ close to 1. If found, it means it is the combination that maximises K, since $L^c_{\text{BSM}}(\mu = 1) = L^c_{\text{BSM}}(\mu = \hat{\mu})$.

Furthermore, new penalty terms are introduced into the BSM prior π_{BSM}. The idea is to penalise proto-models that have constraints coming from only one source (e.g. only ATLAS or CMS results), or that have abnormally small or large SSMs. They are very naive terms for the moment and will very likely change in the future, therefore, they will not be described in detail here.

- **Fast critic:** it is used to quickly discard proto-models violating the LHC constraints implemented in the SMODELS database. The exclusion is only based on UL-type results. However, if an analysis has no UL-type result but has an EM-type result implemented, the latter is also used. For each EM-type result, only the most sensitive SR is used. No likelihood is built; thus, only an approximated result is computed, but it has the advantage of being fast.

  In [26], a constraint was put on $\hat{\mu}$ using only the highest observed r-value, where an arbitrarily chosen 30% violation of the UL was allowed. The critic now gives a binary output: either the proto-model is excluded or it is not. This decision should take into account the total number of analyses providing a sensitive constraint and allow for small violations of the ULs. Therefore, the maximal number of ULs that are allowed by the critic to be violated has to be computed. The number of violated results is modelled as a binomial distribution with a p-value of 0.05 (since the limits are given as 95% CL ULs) and an n corresponding to the number of sensitive results constraining the proto-model. Here, the sensitive results are the ones with an $r_{\text{exp}} \geq 0.7$. For UL-type results without an UL map on the expected cross section, the expected r-value is taken equal to the observed one (this works because no abnormally large deviation from the SM has been observed).

  The maximal number of results allowed to exclude the proto-model corresponds to the lowest number of analyses for which the cumulative distribution function of the binomial distribution is above 66%. If the actual number of results excluding the proto-model is below this value, the proto-model is not excluded; if it is above, the proto-model is excluded. Moreover, to find the actual number of results excluding the proto-model, a 38% violation of the UL is still allowed. That is to say, an UL- or EM-type result excludes the tested proto-model if it gives $r_{\text{obs}} > 1.38$. This value of 38% has been obtained by creating fake EM-type result databases sampled from the SM distribution and computing the 1 sigma deviation of the $r_{\text{obs}}/r_{\text{exp}} \equiv \sigma_{95}^{\text{exp}}/\sigma_{95}^{\text{obs}}$ distribution (computed at the SR level).
- **Slow critic:** if the proto-model survives the first critic, it is then passed to a second critic that tests it against the limit from the most sensitive combination of analyses. Here, the same procedure as the one described in Sect. 4.4.1 is used. Although much more robust than the fast critic, it may be much more time-consuming. This is why it is done at the very end of the step.

Similarly to the method used to find the most significant combination, all the EM-type results sensitive to the proto-model signals are used. However, if the combination of SRs is not possible for an analysis, only the most sensitive SR is used. This is because the combination is done at the analysis level and the most sensitive SR is always the one increasing the most the sensitivity of the global combination. Moreover, only the analyses sensitive enough, i.e. with a $\mu_{95}^{\text{exp}} \leq 10$, are used.

- **Others:** other improvements include, for instance, an initialisation and a generation of new proto-models allowing for the coverage of the most phase-space regions as possible; or a way to decrease the computation time, e.g., by using the simplified models instead of the full ones in a first approximation, or by collapsing the full statistical models into covariance matrices. These are currently work in progress.

5.3 Test Runs

5.3.1 Phase-Space Region of Interest

As already mentioned, although no important deviation from the SM has been reported by experimental collaborations, small excesses have been recorded in several ATLAS and CMS searches, notably in the compressed EW-ino mass region as discussed in Sect. 2.3.4. On the ATLAS side, the search for events with 2 soft OS SF $\ell + \not{E}_T$ (ATLAS-SUSY-2018-16) [42] observed a small excess in this region (see the top panel of Fig. 2.22, where its results are combined with an ATLAS search for 3 $\ell + \not{E}_T$ (ATLAS-SUSY-2019-09) [43]). On the CMS side, two other analyses observed a small excess in the same region: one searching for 2 or 3 low p_T $\ell + \not{E}_T$ (CMS-SUS-18-004) [44], and one searching for 3 or 4 $\ell + \not{E}_T$, with up to 2 hadronically decaying τ, or 2 same-sign light ℓ (CMS-SUS-19-012) [45] (see Fig. 2.23). All these analyses use the full luminosity of Run 2. These excesses can be interpreted, e.g., as the production of higgsino-like EW-inos with chargino-LSP mass splittings of roughly 5–15 GeV.

Even if this kind of excess is expected to appear in 5% of the searches, the fact that it appears in different experiments and search channels, and lies at the right spot to provide a viable dark matter candidate as well as an explanation for the measured value of the magnetic dipole moment of the muon [46] makes it worth to investigate. It would thus be interesting to see what proto-model could be constructed to fit these excesses. However, only the ATLAS analysis published enough reinterpretation material for SModelS to construct a likelihood. Indeed, EMs are available for this analysis, as well as a full statistical model [47], while only UL maps are available for the CMS 2-3 $\ell + \not{E}_T$ search, and neither UL maps nor EMs are available for the CMS 3-4 $\ell + \not{E}_T$ search.

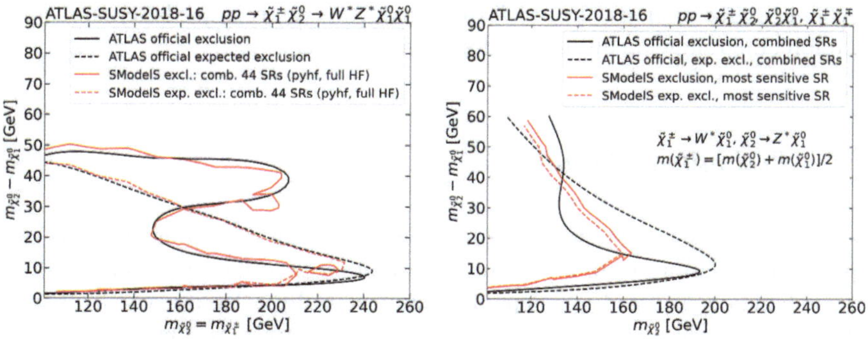

Fig. 5.4 SModelS validation plots for ATLAS-SUSY-2018-16, a 2 soft OS SF $\ell + \not{E}_T$ search using Run 2 full luminosity. In the left panel, a wino-bino scenario is considered, where $m_{\tilde{\chi}_2^0} = m_{\tilde{\chi}_1^\pm}$; while in the right panel, a higgsino-LSP scenario is considered, where $m_{\tilde{\chi}_1^\pm} = [m_{\tilde{\chi}_2^0} + m_{\tilde{\chi}_1^0}]/2$. In both cases, the $\tilde{\chi}_1^\pm$ and $\tilde{\chi}_2^0$ are decaying into $W^*\tilde{\chi}_1^0$ and $Z^*\tilde{\chi}_1^0$, respectively, with $\tilde{\chi}_1^0$ being the LSP. The ATLAS limits can be correctly reproduced in the wino-bino scenario thanks to the EMs for the exclusive SRs and the full statistical model. However, the ATLAS limits cannot be correctly reproduced in the higgsino-LSP scenario because only EMs for inclusive SRs are available for this case

Nonetheless, the SModelS ability to reproduce the experimental limits of the ATLAS search differs depending on the scenario considered. EMs for the exclusive SRs (extracted from the HEPData patchset) and a full statistical model are available on HEPData for scenarios where the $\tilde{\chi}_2^0$ and $\tilde{\chi}_1^\pm$ are (almost) mass-degenerate. As shown in the left panel of Fig. 5.4, it allows SModelS to correctly reproduce the experimental limits. However, for scenarios where the produced particles ($\tilde{\chi}_2^0$, $\tilde{\chi}_1^\pm$ and $\tilde{\chi}_1^0$) are not mass-degenerate, only EMs for inclusive SRs are available, without any means to combine them. The SModelS exclusion is then based on the most sensitive SR only. As shown in the right panel of Fig. 5.4, in this case SModelS cannot correctly reproduce the experimental limit. What is more, SModelS also cannot reproduce the excess. Therefore, it is to be expected for the algorithm to favour mass-degenerate EW-ino over non-mass-degenerate ones in this region. This is a pity because naturalness argues that these deviations from the SM should preferably be studied in natural SUSY frameworks, where the lightest EW-inos are dominantly higgsino-like, see e.g. [48–50].

Moreover, the study [51], based on MadAnalysis 5, analysed the consistency of these small excesses with monojet searches (ATLAS-EXOT-2018-06 [52] and CMS-EXO-20-004 [53]) in the context of light compressed higgsinos. While the best-fit points from the monojet searches were shown to be compatible with the excesses of the soft-lepton searches, no global likelihood could be built due to missing information from the experiments. Indeed, only UL maps on the observed cross section are available for the ATLAS search. The monojet results in the current SModelS (v3.0) database are for new resonance only; the production of "homegrown" EMs for monojets from EW-ino production, as relevant here, is ongoing. Therefore, the following discussion do not include them.

5.3.2 Preliminary Results

As a first physics test of the revised machinery, a run consisting of 1000 steps is performed, starting from a proto-model point close to the remaining excess of the combination of the 2 soft leptons search with the 3 lepton search. Because SModelS reproduces the 2 soft lepton excess better for mass-degenerate $\tilde{\chi}_2^0$ and $\tilde{\chi}_1^{\pm}$ (compared to non-mass-degenerate), the initial masses of X_W^1 and X_Z^2 are taken to be mass-degenerate. More precisely, the masses of the X_W^1, X_Z^2 and X_Z^1 are taken to be 250, 250, and 235 GeV, respectively. This roughly corresponds to the remaining excess of the aforementioned combination when interpreted in terms of simplified models featuring a wino-bino scenario (see Figure 16b of [43]). All the other BSM particles are absent of this initial proto-model (but can be added during the walk). The five decay channels of Eqs. (5.8) and (5.9) are allowed (and only them). The production cross sections and the branching ratios are initialised to correspond to that of a pure wino-bino scenario (with the preservation of lepton universality for the branching ratios).

The resulting best proto-model, i.e. the one with the highest K, is described in Table 5.2 and is labelled Best Point 1. It features a X_t^1, a X_Z^2 and a X_Z^1 of around 1124, 427 and 83 GeV, respectively. The relevant production cross sections have a signal strength multiplier not very far from 1, and the branching ratios are reasonably balanced. However, Best Point 1 does not feature off-shell particles. It might be because more interesting excesses were found elsewhere, or because the limited number of runs (only 1) and steps (only a 1000). Among the 1000 proto-models constructed, only 12 contained simultaneously an off-shell X_W^1 and an off-shell X_Z^2 (42 contained an off-shell X_W^1 but no off-shell X_Z^2, and 77 contained an off-shell X_Z^2 but no off-shell X_W^1).

The analyses entering the most significant combination are the $hh + \not{E}_T$ search CMS-SUS-20-004 [54], the $t\bar{t} + 1\ell + \not{E}_T$ search ATLAS-SUSY-2016-16 [55], and

Table 5.2 Description of best point 1. Only relevant cross sections are shown

Best point 1, $\boldsymbol{K \approx 4.14, Z \approx 3.0}$			
Particle	Mass [GeV]	Cross section (13 TeV)	Branching ratio
X_t^1	1124	$\sigma(pp \to X_t^1 X_t^1) \approx 6$ fb	$\mathrm{BR}(X_t^1 \to tX_Z^2) \approx 0.59$
		$\kappa_{X_t^1 X_t^1} \approx 2$	$\mathrm{BR}(X_t^1 \to tX_Z^1) \approx 0.41$
X_Z^2	427	$\sigma(pp \to X_Z^2 X_Z^2) \approx 125$ fb	$\mathrm{BR}(X_Z^2 \to ZX_Z^1) \approx 0.70$
		$\kappa_{X_Z^2 X_Z^2} \approx 3$	$\mathrm{BR}(X_Z^2 \to hX_Z^1) \approx 0.30$
X_Z^1	83	–	–

Table 5.3 Description of best point 2. Only relevant cross sections are shown. The hadronic decay of the off-shell X_W^1 are counted as a whole, while the leptonic decays are counted individually

Best point 2, $\boldsymbol{K \approx 6.31}$, $\boldsymbol{Z \approx 3.9}$			
Particle	Mass [GeV]	Cross section (13 TeV)	Branching ratio
X_b^1	1292	$\sigma(pp \to X_b^1 X_b^1) \approx 1$ fb $\kappa_{X_b^1 X_b^1} \approx 1$	$\mathrm{BR}(X_b^1 \to t X_W^1) = 1$
X_W^2	568	$\sigma(pp \to X_W^2 X_W^2) \approx 18$ fb $\kappa_{X_W^2 X_W^2} \approx 1$	$\mathrm{BR}(X_W^2 \to h X_W^1) \approx 0.50$ $\mathrm{BR}(X_W^2 X_W^2 \to W X_Z^1) \approx 0.50$
X_W^1	244	$\sigma(pp \to X_W^1 X_W^1) \approx 18$ pb $\kappa_{X_W^1 X_W^1} \approx 41$	$\mathrm{BR}(X_W^1 \to q\bar{q}' X_Z^1) \approx 0.67$ $\mathrm{BR}(X_W^1 \to \ell \nu_\ell X_Z^1) \approx 0.11$
X_Z^1	241	–	–

the all-hadronic $t\bar{t} + \not{E}_T$ search ATLAS-SUSY-2018-12 [56]. They are all from Run 2. Only the CMS analysis has the material allowing for SR combination. This combination leads to $K \approx 4.14$ and a significance of $Z \approx 3.0$. The latter is obtained by using the asymptotic formulae $Z = \sqrt{2 \ln \left(L^c_{\mathrm{BSM}}(\hat{\mu}) / L^c_{\mathrm{SM}} \right)}$ [57].

To refine this proto-model, a second run of 1000 steps is performed, starting from Best Point 1. The resulting best proto-model, labelled Best Point 2 is described in Table 5.3. Interestingly, Best Point 2 neither features a X_t^1 nor a X_Z^2. Instead, it contains a X_b^1, a X_W^2, a X_W^1, and a X_Z^1 of around 1292, 568, 244, and 241 GeV, respectively. The $X_b^1 X_b^1$ and $X_W^2 X_W^2$ production cross sections have a signal strength multiplier close to unity. However, the $X_W^1 X_W^1$ production cross section features a large signal strength multiplier of 41. This is unusual, particularly because the cross sections are initialised using wino-like reference cross sections (which are significantly heavier than the higgsino-like cross sections). Regarding the branching ratios, the X_b^1 is only allowed to decay into $t X_W^1$, the X_W^2 can decay into $h X_W^1$ or $W X_Z^1$ with almost the same probability, and the X_W^1 can decay hadronically 67% of the time, and leptonically in 33% of the time (with preservation of lepton universality).

One important point is that due to the small $\Delta m(X_W^1, X_Z^1)$, the X_W^1 decay products can be too soft to be detected. Therefore, SModelS can compress the topology (see the discussion in the third paragraph of Sect. 4.2) and remove intermediate X_W^1 decays. For instance, the $X_W^2 \to h X_W^1 \to h q\bar{q}' X_Z^1$ decay chain can be compressed into the simpler 1-step decay $X_W^2 \to h X_Z^1$. Nonetheless, if an analysis is sensitive to a non-compressed topology, it will also be used to constrain the model.

The five analyses entering the most significant combination of Best Point 2 are listed in Table 5.4. They are all from Run 2. Two of them were already in the most significant combination of the starting point, Best Point 1. The are the all-hadronic $+ \not{E}_T$ search ATLAS-SUSY-2016-07 [58], the $hh + \not{E}_T$ search CMS-SUS-20-004, the $h + 1\ell + \not{E}_T$ search ATLAS-SUSY-2019-08 [59], the $t\bar{t} + 1\ell + \not{E}_T$

Table 5.4 Analyses entering the most significant combination of best point 2, by decreasing sensitivity. They are all from Run 2. The column "Lumi." gives the integrated luminosity in fb^{-1}. The "Sensitive Channels" column shows the compressed and non-compressed topologies produced by the proto-model to which the corresponding analysis is sensitive. The EMs for the non-compressed topology of ATLAS-SUSY-2019-08 were obtained through the MADANALYSIS 5 recast [60]. The column "Comb." specifies whether and how SRs are combined: "pyhf" means a full or SIMPLIFY'ed HISTFACTORY model is used through interface with PYHF; "SLv1" ("SLv2") means that a covariance matrix is used in the Simplified Likelihood scheme of [61] and [62]; and a "–" means that only the SR maximising the combination's significance is used

ID	Short description ($+\not{E}_T$)	Lumi.	Sensitive channels	Comb.
ATLAS-SUSY-2016-07 [58]	0ℓ + jets	36.1	$X^1_W X^1_W \to q\bar{q}'q\bar{q}'X^1_Z X^1_Z$	–
CMS-SUS-20-004 [54]	$2\,h(bb)$, EW	137.0	$X^2_W X^2_W \to hhX^1_Z X^1_Z$	SLv2
ATLAS-SUSY-2019-08 [59]	$1\ell + h(bb)$, EW	139.0	$X^2_W X^2_W \to WhX^1_Z X^1_Z$ $X^2_W X^2_W \to WhX^1_W X^1_Z \to Whq\bar{q}'X^1_Z X^1_Z$	PYHF
ATLAS-SUSY-2016-16 [55]	1ℓ stop	36.1	$X^1_b X^1_b \to t\bar{t}X^1_Z X^1_Z$	–
ATLAS-SUSY-2018-16 [42]	2 soft ℓ + jets, EW (hino)	139.0	$X^1_W X^1_W \to \ell\nu_\ell\ell\nu_\ell X^1_Z X^1_Z$	–

search ATLAS-SUSY-2016-16, and the 2 OS SF soft $\ell + \not{E}_T$ search which recorded a small excess, ATLAS-SUSY-2018-16. However, regarding the latter, the result entering the combination is the one assuming non-mass-degenerate light EW-inos. Therefore, no combination of SRs is possible. Moreover, the proto-model does not really fit the recorded excess of this analysis since the generated signal does not lie at the excess spot. The all-hadronic $+ \not{E}_T$ search ATLAS-SUSY-2016-07 and the 2 OS SF $\ell + \not{E}_T$ search ATLAS-SUSY-2018-16 are the only one to probe off-shell signatures of the proto-model (through the $X^1_W X^1_W \rightarrow q\bar{q}'q\bar{q}'$ and the $X^1_W X^1_W \rightarrow \ell\nu_\ell\ell\nu_\ell$ channels, respectively). Nevertheless, compared to the first run, many more steps probed the off-shell region this time. Among the 1000 steps, 456 contained an off-shell X^1_W (without an off-shell X^2_Z), only 1 contained an off-shell X^2_Z (without an off-shell X^1_W), and only 6 contained both off-shell X^1_W and X^2_Z.

The likelihoods of these five analyses as well as the combined likelihood of the most significant combination are shown in the top plot of Fig. 5.5. The quantity of interest here is the ratio of likelihoods $L_{\text{BSM}}(\mu = 1)/L_{\text{SM}}$. The most significant combination is mostly driven by the three most sensitive analyses: ATLAS-SUSY-2016-17 (light blue), CMS-SUS-20-004 (gold) and ATLAS-SUSY-2019-08 (green). As shown in Table 5.4, they are searching for events with $\not{E}_T$ and jets, hh or Wh (with or without additional jets) in their final states. Regarding ATLAS-SUSY-2016-16 (brown), the analysis is sensitive to final states with $t\bar{t} + \not{E}_T$, but the proto-model signal lies at the sensitivity edge of the search. Finally, as far as ATLAS-SUSY-2018-16 (teal) is concerned, it is sensitive to final states with 2 OS SF $\ell + \not{E}_T$, but the proto-model signal also lies at the sensitivity edge of the search. Despite having observed a bigger excess (characterised by a likelihood maximised at higher μ), the selected SR of this search is less contributing to the likelihood ratio $L_{\text{BSM}}(\mu = 1)/L_{\text{SM}}$ due to a lesser fit to the proto-model. In the end, the most significant combination leads to $K \approx 6.31$ and $Z \approx 3.9$.

Regarding the fast critic, four UL-type results are sensitive to the proto-model signals (with $r_{\text{exp}} \geq 0.7$). Although two of them give $r_{\text{obs}} > 1$, none significantly exclude it (i.e. with $r_{\text{obs}} > 1.38$). Regarding the slow critic, the analyses entering the most sensitive combination are the same as the ones entering the most significant combination. Their likelihoods and the likelihood of the most sensitive combination are shown in the bottom plot of Fig. 5.5. The individual likelihoods of the top and bottom plots are identical if the analyses have a way to combine their SRs. Otherwise, the SR entering the most significant combination is not necessary the one entering the most sensitive combination. However, only the ATLAS-SUSY-2018-16 search (teal) is in that case. For the two other analyses without the possibility

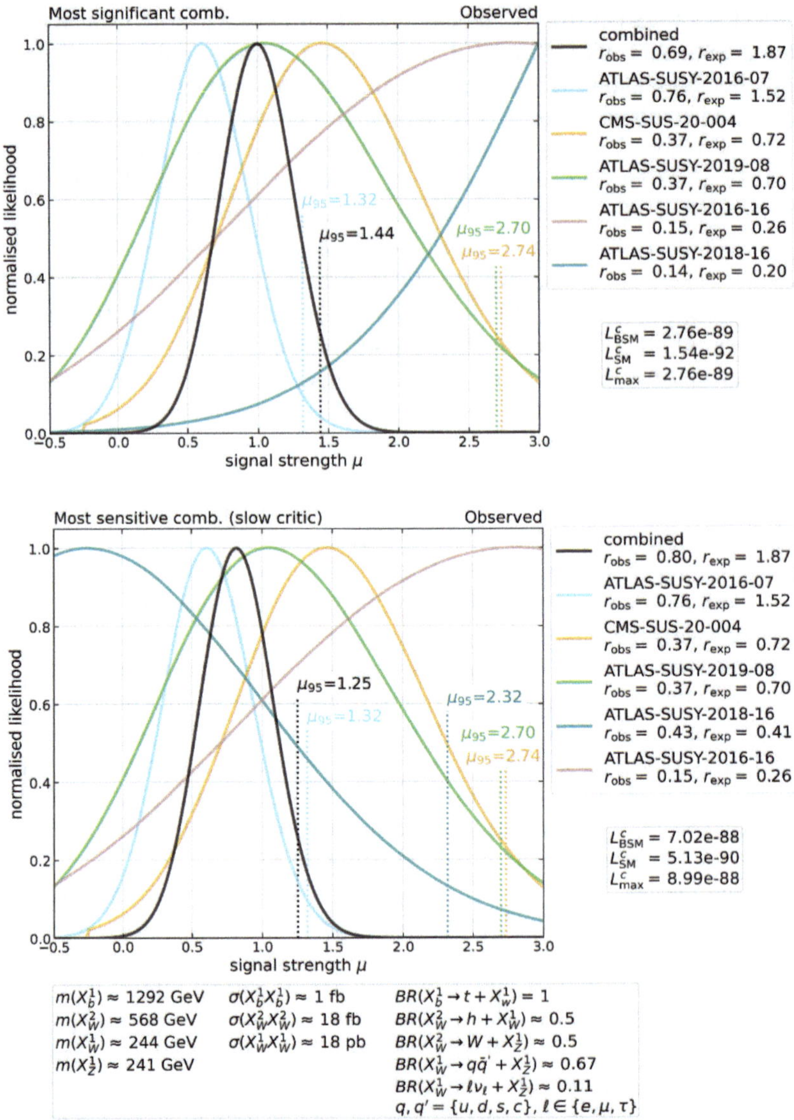

Fig. 5.5 Visualisation of the likelihoods entering the most significant (top) and most sensitive (bottom) combinations, as well as the combined likelihoods, of Best Point 2

to combine their SRs, their most significant SR is also their most sensitive SR. It results that the r_{exp} of the most significant and of the most sensitive combinations are almost identical, but the most sensitive combination has a slightly higher r_{obs}. Since the r_{obs} of the most sensitive combination is below 1, Best Point 2 is not excluded.

5.4 Conclusions and Outlooks

Searches for new physics typically adopt a top-bottom approach, where theoretically motivated models are confronted to experimental data one after the other. In the absence of a clear BSM signal, and with the ever-increasing number of models, it becomes relevant to adopt a more model-independent, data-driven approach. The approach discussed here relies on SModelS features and its large database of simplified model results to search for feeble signals dispersed across several search channels, and to explain them by means of proto-models (a collection of simplified models not tied to any higher-level theoretical assumption). The presented work consisted in the improvement of this proto-modelling machine.

These improvements consisted in the inclusion of likelihoods with combined SRs, of off-shell decays for light EW-inos, of two new boolean critics (of which one uses the most sensitive combination to constrain the proto-model), and of an interface to a new algorithm used to find the most significant and the most sensitive combination of analyses. In addition, the combinability treatment was refined, with a careful scrutiny to SR selection cuts when needed. Moreover, to implement these modifications, the code structure had to be reworked. Nevertheless, the presented version is not the final one and needs further work and adjustments. For instance, the fast critic is currently only based on the number of analyses sensitive to the tested proto-model, and the number of analyses excluding it. It could be relevant to also take the amount of violation into account (e.g. exclude the proto-model if any analysis gives a very large r_{obs}).

As a first test, a run of 1000 steps was performed, starting from the off-shell region, close to the small excess recorded by the 2 OS SF soft $\ell + \not\!\!E_T$ search ATLAS-SUSY-2018-16. The resulting best proto-model, with a $K \approx 4.14$ and a $Z \approx 3.0$ did not fit this excess, and contained on-shell particles only. This best fitting point served as a starting point for a second run of 1000 steps. The resulting best proto-model of this second run, with a $K \approx 6.31$ and a $Z \approx 3.9$, contained a X_b^1, a X_W^2, a X_W^1, and a X_Z^1 of around 1292, 568, 244, and 241 GeV, respectively. Interestingly, even if starting from the on-shell region, the algorithm found a best-fit model containing both on- and off-shell particles. However, the 2 OS SF soft $\ell + \not\!\!E_T$ excess was still not properly fitted. The signal of this proto-model indeed falls into the analysis sensitivity reach, but not at the excess spot. This proto-model mainly fitted small excesses recorded in the $hh + \not\!\!E_T$, $Wh + \not\!\!E_T$, and multijet channels. It has to be kept in mind that this proto-model featured a large $X_W^1 X_W^1$ production cross section. Future runs should include a penalty term to prevent such behaviours.

It has also to be noted that 2 runs of only 1000 steps each are far from being enough to perform a throughout walk over the parameter space. More walkers, with longer walks, would be required to have a more robust result. Moreover, a procedure to ensure that each relevant corner of the parameter space have been visited by the walkers should be implemented. Future runs should also include the possibility to construct proto-models without a $\mathbb{Z}_2$-like symmetry and to probe additional signatures, such as monojet signatures.

Finally, it is worth noting that the biggest limitation of this procedure is the number of topologies constrained by the experimental searches. Indeed, if each analysis is sensitive to a single topology, the machine will only find the same excesses as the ones obtained by the experimental collaborations. Full potential can thus only be achieved if each experimental analysis constrains multiple topologies. More importantly, the presented proto-modelling machine heavily relies on likelihoods, and therefore on the availability of EMs and ways to combine SRs. The possibility to build additional EMs from recasting tools utilising analysis cuts would therefore be highly valuable. An internal tool for this purpose is in development.

References

1. Lykken JD (2010) Beyond the standard model. arXiv:1005.1676 [hep-ph]
2. Dobrescu BA (2014) Beyond the standard model. In: 2011 European School of High-Energy Physics. CERN, Geneva, pp 119–149. https://doi.org/10.5170/CERN-2014-003.119
3. Binetruy P et al (2004) Relating incomplete data and incomplete theory. Phys Rev D 70:095006. https://doi.org/10.1103/PhysRevD.70.095006. arXiv:hep-ph/0312248
4. Arkani-Hamed N et al (2006) Supersymmetry and the LHC inverse problem. J High Energy Phys 08:070. https://doi.org/10.1088/1126-6708/2006/08/070. arXiv:hep-ph/0512190
5. Knuteson B, Mrenna S (2006) BARD: interpreting new frontier energy collider physics. arXiv:hep-ph/0602101
6. Berger CF et al (2009) General features of supersymmetric signals at the ILC: solving the LHC inverse problem. Phys Rev D 80:095018. https://doi.org/10.1103/PhysRevD.80.095018. arXiv:0712.2965 [hep-ph]
7. Altunkaynak B, Holmes M, Nelson BD (2008) Solving the LHC inverse problem with dark matter observations. J High Energy Phys 10:013. https://doi.org/10.1088/1126-6708/2008/10/013. arXiv:0804.2899 [hep-ph]
8. Bornhauser N, Drees M (2012) Mitigation of the LHC inverse problem. Phys Rev D 86:015025. https://doi.org/10.1103/PhysRevD.86.015025. arXiv:1205.6080 [hep-ph]
9. Gainer JS, Matchev KT, Park M (2015) The hierarchy solution to the LHC inverse problem. J High Energy Phys 06:014. https://doi.org/10.1007/JHEP06(2015)014. arXiv:1504.03689 [hep-ph]
10. Abbott B et al (2000) Search for new physics in $e\mu X$ data at DØ using SLEUTH: a quasi-model-independent search strategy for new physics. Phys Rev D 62:092004. https://doi.org/10.1103/PhysRevD.62.092004. arXiv:hep-ex/0006011
11. Arkani-Hamed N et al (2007) MARMOSET: the path from LHC data to the new standard model via on-shell effective theories. arXiv:hep-ph/0703088
12. Aaltonen T et al (2009) Global search for new physics with 2.0 fb^{-1} at CDF. Phys Rev D 79:011101. https://doi.org/10.1103/PhysRevD.79.011101. arXiv:0809.3781[hep-ex]
13. Aaron FD et al (2009) A general search for new phenomena at HERA. Phys Lett B 674:257–268. https://doi.org/10.1016/j.physletb.2009.03.034. arXiv:0901.0507 [hep-ex]
14. Collins JH, Howe K, Nachman B (2018) Anomaly detection for resonant new physics with machine learning. Phys Rev Lett 121(24):241803. https://doi.org/10.1103/PhysRevLett.121.241803. arXiv:1805.02664 [hep-ph]
15. Collins JH, Howe K, Nachman B (2019) Extending the search for new res- onances with machine learning. Phys Rev D 99(1):014038. https://doi.org/10.1103/PhysRevD.99.014038. arXiv:1902.02634 [hep-ph]
16. Aaboud M et al (2019) A strategy for a general search for new phenomena using data-derived signal regions and its application within the ATLAS experiment. Eur Phys J C 79(2):120. https://doi.org/10.1140/epjc/s10052-019-6540-y. arXiv:1807.07447 [hep-ex]

17. Aad G et al (2024) Search for new phenomena in two-body invariant mass distributions using unsupervised machine learning for anomaly detection at s =13 TeV with the ATLAS detector. Phys Rev Lett 132(8):081801. https://doi.org/10.1103/PhysRevLett.132.081801. arXiv:2307.01612 [hep-ex]
18. CMS Collaboration (2021) MUSiC, a model unspecific search for new physics, in *pp* collisions at $\sqrt{s} = 13$ TeV. Eur Phys J C 81(7):629. https://doi.org/10.1140/epjc/s10052-021-09236-z. arXiv:2010.02984[hep-ex]
19. Benkendorfer K, Pottier LL, Nachman B (2021) Simulation-assisted decorrelation for resonant anomaly detection. Phys Rev D 104(3):035003. https://doi.org/10.1103/PhysRevD.104.035003. arXiv:2009.02205 [hep-ph]
20. Nachman B, Shih D (2020) Anomaly detection with density estimation. Phys Rev D 101:075042. https://doi.org/10.1103/PhysRevD.101.075042. arXiv:2001.04990 [hep-ph]
21. Das R, Kasieczka G, Shih D (2023) Residual ANODE. arXiv:2312.11629 [hep-ph]
22. Hallin A et al (2022) Classifying anomalies through outer density estimation. Phys Rev D 106(5):055006. https://doi.org/10.1103/PhysRevD.106.055006. arXiv:2109.00546 [hep-ph]
23. Birman M et al (2022) Data-directed search for new physics based on symmetries of the SM. Eur Phys J C 82(6):508. https://doi.org/10.1140/epjc/s10052-022-10454-2. arXiv:2203.07529 [hep-ph]
24. Kasieczka G et al (2024) Complete optimal non-resonant anomaly detection. arXiv:2404.07258 [hep-ph]
25. Diaz MA et al (2024) Bayesian active search on parameter space: a 95 GeV Spin-0 resonance in the $(B - L)$ SSM. arXiv:2404.18653 [hep-ph]
26. Waltenberger W, Lessa A, Kraml S (2021) Artificial proto-modelling: building precursors of a next standard model from simplified model results. J High Energy Phys 03:207. https://doi.org/10.1007/JHEP03(2021)207. arXiv:2012.12246 [hep-ph]
27. The CMS Collaboration (2019) Search for supersymmetry in proton-proton collisions at 13 TeV in final states with jets and missing transverse momentum. J High Energy Phys 10:244. https://doi.org/10.1007/JHEP10(2019)244. arXiv:1908.04722 [hep-ex]
28. Sjöstrand T et al (2015) An introduction to PYTHIA 8.2. Comput Phys Commun 191:159–177. https://doi.org/10.1016/j.cpc.2015.01.024. arXiv:1410.3012 [hep-ph]
29. Beenakker W et al (1997) Squark and gluino production at hadron colliders. Nucl Phys B 492:51–103. https://doi.org/10.1016/S0550-3213(97)80027-2. arXiv:hep-ph/9610490
30. Beenakker W et al (1998) Stop production at hadron colliders. Nucl Phys B 515:3–14. https://doi.org/10.1016/S0550-3213(98)00014-5. arXiv:hep-ph/9710451
31. Kulesza A, Motyka L (2009) Threshold resummation for squark-antisquark and gluino-pair production at the LHC. Phys Rev Lett 102:111802. https://doi.org/10.1103/PhysRevLett.102.111802. arXiv:0807.2405 [hep-ph]
32. Kulesza A, Motyka L (2009) Soft gluon resummation for the production of gluino-gluino and squark-antisquark pairs at the LHC. Phys Rev D 80:095004. https://doi.org/10.1103/PhysRevD.80.095004. arXiv:0905.4749 [hep-ph]
33. Beenakker W et al (2009) Soft-gluon resummation for squark and gluino hadroproduction. J High Energy Phys 12:041. https://doi.org/10.1088/1126-6708/2009/12/041. arXiv:0909.4418 [hep-ph]
34. Beenakker W et al (2010) Supersymmetric top and bottom squark production at hadron colliders. J High Energy Phys 08:098. https://doi.org/10.1007/JHEP08(2010)098. arXiv:1006.4771 [hep-ph]
35. Beenakker W et al (2001) Squark and Gluino Hadroproduction. Int J Mod Phys A 26:2637–2664. https://doi.org/10.1142/S0217751X11053560. arXiv:1105.1110 [hep-ph]
36. Akaike H (1974) A new look at the statistical model identification. IEEE Trans Autom Control 19(6):716–723. https://doi.org/10.1109/TAC.1974.1100705
37. Azatov A, Contino R, Galloway J (2012) Model-independent bounds on a light higgs. J High Energy Phys 04:127. [Erratum: JHEP 04, 140 (2013)]. https://doi.org/10.1007/JHEP04(2012)127. arXiv:1202.3415 [hep-ph]

38. SModelS Collaboration (2024) SModelS database v3.0.0 Zenodo. https://doi.org/10.5281/zenodo.13354582
39. A detailed list of the analyses implemented in the current SModelS database can be found at https://smodels.github.io/docs/ListOfAnalyses
40. Borschensky C et al (2023) LHC SUSY cross section working group. https://twiki.cern.ch/twiki/bin/view/LHCPhysics/SUSYCrossSections. Accessed 13 Dec 2023
41. Araz JY et al (2023) Strength in numbers: optimal and scalable combination of LHC new-physics searches. SciPost Phys 14(4):077. https://doi.org/10.21468/SciPostPhys.14.4.077. arXiv:2209.00025 [hep-ph]
42. Aad G et al (2020) Searches for electroweak production of supersymmetric particles with compressed mass spectra in $\sqrt{s} = 13$ TeV *pp* collisions with the ATLAS detector. Phys Rev D 101(5):052005. https://doi.org/10.1103/PhysRevD.101.052005. arXiv:1911.12606[hep-ex]
43. Aad G et al (2021) Search for chargino-neutralino pair production in final states with three leptons and missing transverse momentum in $\sqrt{s} = 13$ TeV *pp* collisions with the ATLAS detector. Eur Phys J C 81(12):1118. https://doi.org/10.1140/epjc/s10052-021-09749-7. arXiv:2106.01676 [hep-ex]
44. Tumasyan A et al (2022) Search for supersymmetry in final states with two or three soft leptons and missing transverse momentum in proton-proton collisions at $\sqrt{s} = 13$ TeV. J High Energy Phys 04:091. https://doi.org/10.1007/JHEP04(2022)091. arXiv:2111.06296 [hep-ex]
45. Tumasyan A et al (2022) Search for electroweak production of charginos and neutralinos in proton-proton collisions at $\sqrt{s} = 13$ TeV. J High Energy Phys 04:147. https://doi.org/10.1007/JHEP04(2022)147. arXiv:2106.14246 [hep-ex]
46. Baum S et al (2023) Lighting up the LHC with dark matter. J High Energy Phys 11:037. https://doi.org/10.1007/JHEP11(2023)037. arXiv:2303.01523 [hep-ph]
47. ATLAS Collaboration (2022) Searches for electroweak production of supersymmetric particles with compressed mass spectra in $\sqrt{s} = 13$ TeV pp collisions with the ATLAS detector (Version 5). HEPData (Collection). https://doi.org/10.17182/hepdata.91374.v5
48. Casas JA et al (2017) What is a natural SUSY scenario? J High Energy Phys 06:070. https://doi.org/10.1007/JHEP06(2015)070. arXiv:1407.6966 [hep-ph]
49. Baer H et al (2020) Status of weak scale supersymmetry after LHC run 2 and ton-scale noble liquid WIMP searches. Eur Phys J ST 229(21):3085–3141. https://doi.org/10.1140/epjst/e2020-000020-x. arXiv:2002.03013 [hep-ph]
50. Baer H et al (2024) Winos from natural SUSY at the high luminosity LHC. Phys Rev D 109(1):015027. https://doi.org/10.1103/PhysRevD.109.015027. arXiv:2310.10829 [hep-ph]
51. Agin D et al (2024) Monojets reveal overlapping excesses for light compressed higgsinos. Phys Lett B 853:138597. https://doi.org/10.1016/j.physletb.2024.138597. arXiv:2311.17149[hep-ph]
52. Aad G et al (2021) Search for new phenomena in events with an energetic jet and missing transverse momentum in *pp* collisions at $\sqrt{s} = 13$ TeV with the ATLAS detector. Phys Rev D 103(11):112006. https://doi.org/10.1103/PhysRevD.103.112006. arXiv:2102.10874[hep-ex]
53. Tumasyan A et al (2021) Search for new particles in events with energetic jets and large missing transverse momentum in proton-proton collisions at $\sqrt{s} = 13$ TeV. J High Energy Phys 11:153. https://doi.org/10.1007/JHEP11(2021)153. arXiv:2107.13021 [hep-ex]
54. Tumasyan A et al (2022) Search for higgsinos decaying to two Higgs bosons and missing transverse momentum in proton-proton collisions at $\sqrt{s} = 13$ TeV. J High Energy Phys 05:014. https://doi.org/10.1007/JHEP05(2022)014. arXiv:2201.04206 [hep-ex]
55. Aaboud M et al (2018) Search for top-squark pair production in final states with one lepton, jets, and missing transverse momentum using 36 fb^{-1} of $\sqrt{s} = 13$ TeV *pp* collision data with the ATLAS detector. J High Energy Phys 06:108. https://doi.org/10.1007/JHEP06(2018)108. arXiv:1711.11520 [hep-ex]
56. Aad G et al. (2020) Search for a scalar partner of the top quark in the all-hadronic $t\bar{t}$ plus missing transverse momentum final state at $\sqrt{s} = 13$ TeV with the ATLAS detector. Eur Phys J C 80(8):737. https://doi.org/10.1140/epjc/s10052-020-8102-8. arXiv:2004.14060[hep-ex]

57. Cowan G et al (2011) Asymptotic formulae for likelihood-based tests of new physics. Eur Phys J C71:1554. https://doi.org/10.1140/epjc/s10052-011-1554-0. arXiv:1007.1727 [physics.data-an]. [Erratum: Eur. Phys. J.C73,2501(2013)]. https://doi.org/10.1140/epjc/s10052-013- 2501-z
58. Aaboud M et al (2018) Search for squarks and gluinos in final states with jets and missing transverse momentum using 36 fb^{-1} of $\sqrt{s} = 13$ TeV pp collision data with the ATLAS detector. Phys Rev D97(11):112001. https://doi.org/10.1103/PhysRevD.97.112001. arXiv:1712.02332 [hep-ex]
59. Aad G et al (2020) Search for direct production of electroweakinos in final states with one lepton, missing transverse momentum and a Higgs boson decaying into two b-jets in pp collisions at $\sqrt{s} = 13$ TeV with the ATLAS detector. Eur Phys J C80(8):691. https://doi.org/10.1140/epjc/s10052-020-8050-3. arXiv:1909.09226 [hep-ex]
60. Goodsell MD (2021) Implementation of the ATLAS-SUSY-2019-08 analysis in the MadAnalysis 5 framework (electroweakinos with a Higgs decay into a $b\bar{b}$ pair, one lepton and missing transverse energy; 139 fb^{-1}). Mod Phys Lett A 36(01):2141006. https://doi.org/10.1142/S0217732321410066
61. CMS Collaboration (2017) Simplified likelihood for the re-interpretation of public CMS results. Tech. Rep. CMS-NOTE-2017-001. CERN, Geneva. https://cds.cern.ch/record/2242860
62. Buckley A et al (2019) The simplified likelihood framework. J High Energy Phys 04:064. https://doi.org/10.1007/JHEP04(2019)064. arXiv:1809.05548 [hep-ph]

Chapter 6
Conclusions and Outlooks

Abstract This chapter summarises the main results of this thesis, with an emphasis on the global constraints on the electroweak-ino sector of the Minimal Supersymmetric Standard Model obtained by dynamically combining different ATLAS and CMS results, as well as the improvements and prospects for the "proto-modelling" machine.

The Standard Model of particle physics is one of the most accomplished theories of modern physics. Nonetheless, strong evidence points to BSM physics, both on the experimental and the theoretical side. Many BSM theories try to address these issues, but no compelling signal from any of them has been observed to date, and this despite the intensive search program carried out by several experiments.

In particular, the LHC experiments are strenuously searching for signatures of SUSY particles, and in absence of any discovery, the masses of SUSY particles have been stringently constrained. However, these searches are pursued in an channel-by-channel approach and the derived limits generally hold only in the context of simplified models. In the simplified model framework, only a handful of SUSY particles are assumed to be kinematically accessible and they most of time have trivial mixings and branching ratios. Therefore, their limits are not straightforwardly applicable to full theories. It is then important to develop techniques to confront full theories to experimental results.

Several reinterpretation tools exist to this extent and can be classified into two categories. The tools of the first one reproduce the experimental analysis through a Monte Carlo simulation. They require a lightweight version of the data treatment performed in the experimental analysis (e.g. signal and background selections). The tools of the second one aim at reusing the simplified model results and they require some simplified form of analysis selection, such as the efficiency and acceptance maps. The work of this thesis heavily relies on a tool belonging to the second category: the SModelS software package. The SModelS tool is a public tool whose purpose is to constrain theories by decomposing them into a collection of simplified models that are then confronted to the simplified model

T. Pascal, *Beyond the Standard Model at the Large Hadron Collider: From Simplified Model Constraints to Artificial Proto-Modelling for Dispersed Signals*, Springer Theses, https://doi.org/10.1007/978-3-032-09391-2_6

results implemented in its database. Since SModelS does not depend on any Monte Carlo simulation, it is fast and hence suitable for large scans.

The work of this thesis led to the extension of the SModelS framework in several ways, on both the code and the database side; they are described in Sect. 3.3 and Appendix B, respectively. Code-wise, the combination of approximately uncorrelated analyses is now possible, enabling the confrontation of models against different experimental results in a consistent way through global likelihoods. Moreover, the statistical treatment was improved for likelihoods constructed using full statistical models. They can now emulate signal leaking into the control regions and include an uncertainty on the signal. Regarding the database, many new analyses were implemented. The SModelS database v3.0 now contains a total of 125 results: 45 ATLAS and 44 CMS Run 2 results (of which 42 use full Run 2 luminosity), and 16 ATLAS and 20 CMS Run 1 results.

In Chap. 4, the impact of the improved statistical treatment and of the newly implemented analyses were first studied by comparing v2.1 and v2.3 constraints on the prompt EW-ino sector of the MSSM without analysis combination. More precisely, a random scan over the EW-ino parameters of the Lagrangian was performed, giving around 18K points to which the searches for prompt EW-ino signatures implemented in the SModelS database were sensitive. The reach in the lightest chargino mass increased from about 650 GeV in v2.1 to about 1 TeV in v2.3, with, on overall, around four times more points excluded. The sensitivity and constraining power were mostly driven by the ATLAS and CMS Run 2 hadronic searches. The most sensitive one was the CMS search, but the most constraining one was the ATLAS one. This is due to the over- and under-fluctuations recorded by the CMS and ATLAS searches, respectively.

The same set of models points were then constrained by the combination of these two analyses. It resulted in a significant increase in the sensitivity in the high EW-ino masses region. It also helped averaging out statistical fluctuations, leading to a substantial increase in the exclusion power with respect to the most sensitive analysis. Nevertheless, combining analyses did not increase the exclusion power of every model point. Indeed, several points were unexcluded by the combination. This happened when the point was excluded by the ATLAS under-fluctuations but driven away from exclusion by the CMS over-fluctuation when combined. In any case, the combination always results in more robust, realistic constraints.

A more complete phenomenological study was subsequently presented, where the same set of points was used to set global LHC constraints on the EW-ino sector of the MSSM through the dynamical combination of 16 ATLAS and CMS Run 1 and Run 2 searches for prompt EW-ino signatures. This time, the combination was not tied to two fixed analyses, but the most sensitive combination of analyses was dynamically found for each model point. Two analyses were assumed to be approximately uncorrelated, and hence combinable, if they were from different runs, different experiments, or if they the selection cuts of their SRs did not overlap. At high chargino masses, the sensitivity and the exclusion were mostly driven by the ATLAS and CMS hadronic searches once again. For models with compressed mass spectra, the sensitivity and the exclusion power were mostly driven by the ATLAS three leptons plus missing transverse energy search from Run 2.

The results were presented according to the nature of the LSP. For models with a bino-like LSP, $m_{\tilde{\chi}_1^\pm}$ were excluded up to around 1 TeV when $m_{\tilde{\chi}_1^0} \approx 0$; but no points with $m_{\tilde{\chi}_1^0} \gtrsim 400$ GeV were excluded. The combination permitted to have for the first time a clear excluded region, whose range depends on the nature of the NLSP. Indeed, the production cross section of the higgsino-like particles being smaller than for wino-like particles, the exclusion limits of higgsino-bino models are around 100 GeV weaker than for wino-bino models. For models with a non-bino-like LSP, $m_{\tilde{\chi}_2^\pm}$ were excluded up to around 950 GeV when $m_{\tilde{\chi}_1^0} \approx 100$ GeV (the respect of the LEP constraints on light charginos forbid almost all non-bino-like LSP points with $m_{\tilde{\chi}_1^0} \lesssim 100$ GeV); but no points with $m_{\tilde{\chi}_1^0} \gtrsim 350$ GeV were excluded. The non-bino-like LSP points are slightly less constrained compared to the bino-like LSP points. This is because the wino-higgsino scenarios lead to longer decay chains, thus diluting the signal going into the simple 1-step decay topologies constrained by the database. Moreover, because of this, no clear excluded region could be drawn for models featuring a non-bino-like LSP.

Additionally, the exclusion power of the combination was compared to the one of the most sensitive analysis. For both bino-like and non-bino-like LSP models, some points that were not excluded by the most sensitive analysis were excluded by the combination, and on the contrary, some points that were excluded by the most sensitive analysis were not excluded by the combination. The increase of exclusion power was more important for models featuring a bino-like LSP because, for theses models, the ATLAS hadronic search was the most sensitive analysis at high chargino masses. So when combining it with the other analyses, the exclusion power mostly decreased but was still enough to exclude the corresponding model points. All in all, the number of points (expected to be) excluded by the most sensitive analysis increased by (48%) 35% when the combination of analyses was considered instead.

In Chap. 5, a method to go beyond the top-bottom approach typically adopted in BSM searches was presented. This method was intended to be data-driven and was designed to construct simple BSM models based on the small over-fluctuations observed in the simplified model results. More precisely, it searched for signs of new physics by constructing so-called proto-models, which are collections of simplified models not bounded by any higher-level theoretical assumptions, and by confronting them against the simplified model results in the SModelS database. Proto-models were constructed to find mutually consistent over-fluctuations from different LHC search channels, through the dynamical combination of analyses, while being consistent with the current limits. They were meant to be as model-independent as possible and were thus an interesting tool for bringing out underlying theories, should they be hidden in different search channels. Since it uses SModelS functionalities and its database, the particle content of the proto-models was SUSY-inspired. This chapter described the concept and how the statistical treatment and database of SModelS were exploited to improve this method.

A major upgrade was about the quantity of analyses in the database, and more specifically the number of analyses for which a likelihood could be constructed (often with the possibility to combine SRs within an analysis). Another important upgrade concerned the attention given to analysis combinability; instead of only considering analyses with trivially orthogonal search channels to be combinable, the SR overlap was carefully scrutinised through the SR selection cuts when necessary. Moreover, proto-models were given the possibility to account for off-shell decays for the lightest EW-ino-like particles. Furthermore, even if the structure of the proto-modelling machine remained the same (with a builder building proto-models, a combiner combining the analyses, and a critic setting constraints on the proto-models), the code architecture was considerably improved. The main modifications had to do with the inclusion of the "pathfinder" algorithm, in charge of finding the best combination of analyses, and with the critic. In the improved version, now two critics are used: a "fast" one, mainly relying on UL-type results, and a "slow" one, acting as an adversarial machine to the builder.

Following these novelties and improvements, a first application to the LHC searches for the production of EW-inos with compressed mass spectra in the SModelS database was studied. It consisted in a first run of 1000 steps, starting from a proto-model featuring only off-shell $\tilde{\chi}_2^0$-like and $\tilde{\chi}_1^\pm$-like particles decaying into a $\tilde{\chi}_1^0$-like particle. This starting proto-model was chosen as so to be close to the small excess recorded by the soft dilepton search ATLAS-SUSY-2018-16. The resulting best-fitting point then served as a starting point for a second run of 1000 steps. The resulting best-fitting proto-model of this second run was then analysed. It contained a $\tilde{b}_1$-like, a $\tilde{\chi}_2^\pm$-like, a $\tilde{\chi}_1^\pm$-like, and a $\tilde{\chi}_1^0$-like of around 1292, 568, 244, and 241 GeV, respectively. Interestingly, it contained an off-shell $\tilde{\chi}_1^\pm$-like particle but did not fit the excess observed in the off-shell EW-ino region. Instead, the proto-model fitted small excesses recorded in analyses searching for events with $\not{E}_T$ and jets, hh or Wh (with or without additional jets) in their final states. The resulting best-performing proto-model had an approximate significance of 3.9. It has to be kept in mind, however, that these runs were preliminary results originating from test runs. Further numerical studies will be needed, for example to be sure the full space of proto-models is well covered.

As a concluding remark, the absence of clear BSM signals and the increasing diversity of theoretically-motivated scenarios motivate a change of perspective in the quest for new physics. The proto-modelling machine goes in this direction, proposing a more model-independent, data-driven approach. Appropriate reinterpretation materials play a crucial role in this endeavour, as likelihoods can only be accurately reconstructed when EMs and statistical models are available.

Appendix A
LHC Constraints on Long-Lived Sparticles and RPV Couplings After Run 2

This appendix adds additional information to the thesis. It contains an overview of ATLAS and CMS limits on the masses of supersymmetric particles for models featuring long-lived particles and/or R-parity violating couplings after Large Hadron Collider Run 2.

The analyses presented in Sect. 2.3 assumed the prompt decay of the produced sparticles within RPC models. This appendix also proposes an overview of the ATLAS and CMS constraints on simplified models but focusing on searches targeting long-lived sparticles and/or non-vanishing RPV couplings.

A.1 Searches for Long-Lived Sparticles within RPC Models

Long-lived sparticles are predicted by many SUSY models.[1] Here, LLPs denote the particles that decay significantly far from the LHC collision regions. They produce unconventional signatures and may escape the detector without depositing all their energies. A schematic representation of the possible signatures is pictured in Fig. A.1.

[1] Three factors can give a particle a large mean lifetime (or decay length): 1- a weak coupling to its decay products, 2- a small mass difference between the particle and its decay products (e.g. almost mass-degenerate higgsinos with a higgsino-like LSP), 3- the only allowed decay(s) is (are) via heavier mediator(s).

T. Pascal, *Beyond the Standard Model at the Large Hadron Collider: From Simplified Model Constraints to Artificial Proto-Modelling for Dispersed Signals*, Springer Theses, https://doi.org/10.1007/978-3-032-09391-2

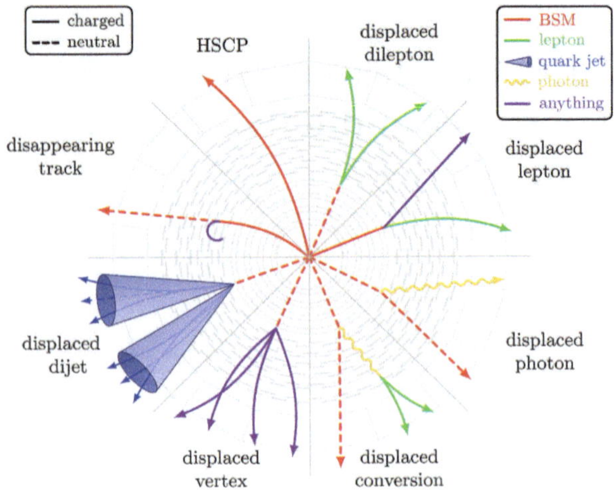

Fig. A.1 Schematic representation of the various unconventional signatures a long-lived particle can leave in LHC detectors, with the CMS detector as background. "HSCP" stands for heavy stable charged particle. Taken from [1]

A.1.1 Gluino Production

Starting with gluinos, long-lived gluinos are motivated for example by split SUSY scenarios [2–4], where the mass of the squarks they decay into are much higher than theirs. ATLAS limits on models featuring a long-lived $\tilde{g}$ are presented in Fig. A.2. Here, the $\tilde{g}$ is assumed to hadronise into a R-hadron, and to decay into $q\bar{q}\tilde{\chi}_1^0$, with $m_{\tilde{\chi}_1^0} = 100$ GeV. All the models are excluded for $m_{\tilde{g}} \lesssim 2$ TeV. Masses are excluded up to ~ 2.3 TeV on average for $c\tau_0(\tilde{g}) \gtrsim 10^{-2}$ m.

On the CMS side, similarly to the displaced vertices result of ATLAS (blue line in Fig. A.2), a search for displaced jets using 35.9 fb^{-1} of LHC Run 2 data [5] excludes $m_{\tilde{g}}$ up to ~ 2.3 TeV for a $c\tau_0(\tilde{g})$ between 10^{-2} and 10^{-1} m, and up to ~ 1.75 TeV for $c\tau_0(\tilde{g}) = 10^{-3}$ m. Additionally, a preliminary CMS search for heavy long-lived charged particles with large ionization energy loss excludes the mass of stable[2] $\tilde{g}$ up to ~ 2.1 TeV under the assumption that 10% of them form R-hadrons [6].

Moreover, many CMS constraints on gluinos were derived assuming a long-lived wino-like $\tilde{\chi}_1^{\pm}$. Some of them, coming from an analysis searching for hadronic final states with and without disappearing tracks [7], are shown in Fig. A.3. Here, BR$(q\bar{q}\tilde{\chi}_1^0) = 1/3$, and BR$(q\bar{q}\tilde{\chi}_1^{\pm}) = 2/3$. The chargino further decays into $\pi^{\pm}\tilde{\chi}_1^0$. Two results are presented: when $c\tau_0(\tilde{\chi}_1^{\pm}) = 10$ cm (left panel) and when $c\tau_0(\tilde{\chi}_1^{\pm}) = 2$ m (right panel). The first one excludes $m_{\tilde{g}}$ up to ~ 2.45 TeV for light $\tilde{\chi}_1^0$, and up to ~ 1.6 TeV for compressed spectra. The second one reduces the exclusion to ~ 2 TeV for light $\tilde{\chi}_1^0$ but increases it up to ~ 2.3 TeV for

[2] Here, and in the following, the stability of a particle is assessed with respect to the detector scale, i.e. a particle is stable if its mean proper decay length is above the detector dimensions.

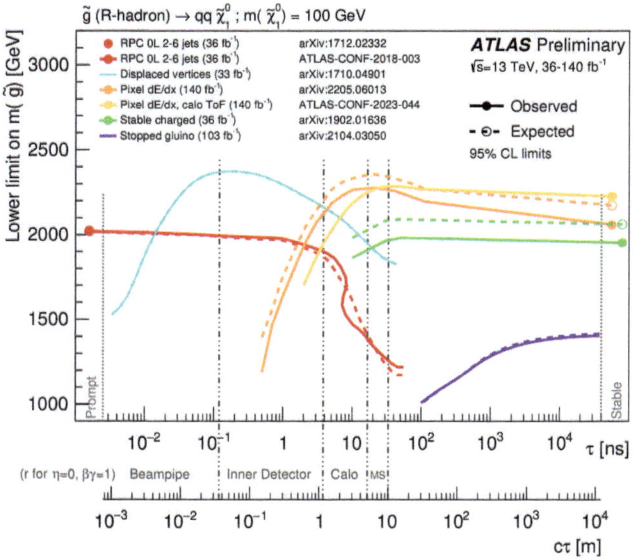

Fig. A.2 Summary of ATLAS exclusion limits on direct pair production of long-lived gluinos, in the $m_{\tilde{g}}$ vs. $\tau_0(\tilde{g})$, after LHC Run 2. A conversion between τ_0 and $c\tau_0$ is shown on the x-axis. Limits are obtained using split SUSY models with the $\tilde{g}$ forming R-hadrons and decaying into $q\bar{q}\tilde{\chi}_1^0$, with $m_{\tilde{\chi}_1^0} = 100$ GeV. The coloured dots and circles represent limits obtained on prompt and stable $\tilde{g}$. The yellow, orange and green limits stop at 90, 300 and 15 m, respectively, and are interpolated to their stable results. The labels between the two x-axis refer to parts of the ATLAS detector. "Calo" and "MS" refer to the calorimeters and the muon spectrometer, respectively. Taken from [9]

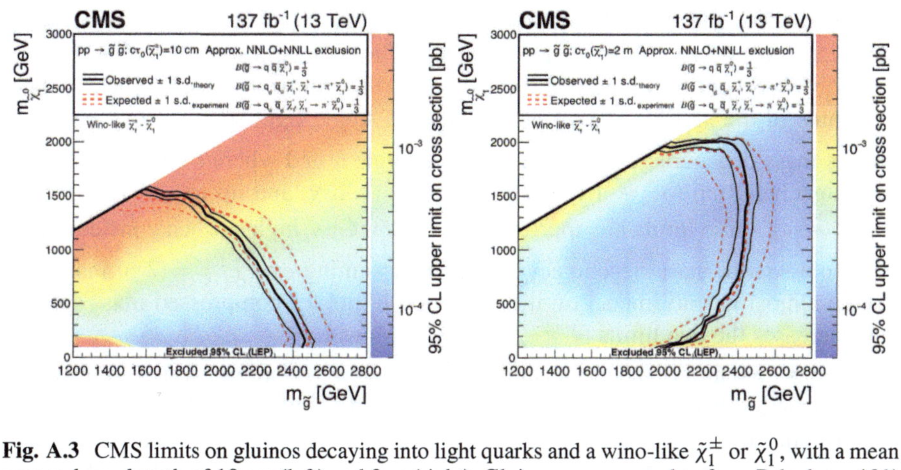

Fig. A.3 CMS limits on gluinos decaying into light quarks and a wino-like $\tilde{\chi}_1^{\pm}$ or $\tilde{\chi}_1^0$, with a mean proper decay length of 10 cm (left) and 2 m (right). Gluinos are assumed to form R-hadrons 10% of the time. Taken from [7]

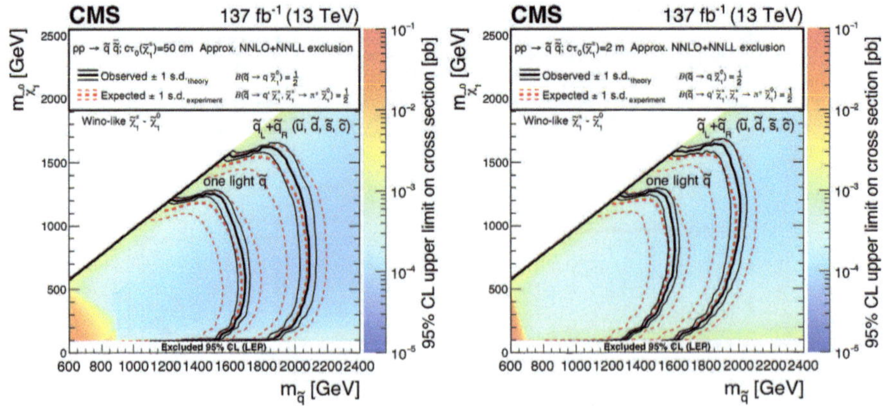

Fig. A.4 CMS constraints on light-flavor squarks decaying into light quarks and a purely wino $\tilde{\chi}_1^\pm$ or $\tilde{\chi}_1^0$, with a mean proper decay length of 50 cm (left) and 2 m (right). Constraints are derived assuming a single light squark or an eightfold squark mass degeneracy. Taken from [7]

more compressed spectra ($\sim$ 2 TeV for nearly mass degenerate $\tilde{g}$ and $\tilde{\chi}_1^0$). ATLAS constraints on these models are similar to the ones of CMS [8].

A.1.2 Squark Production

Analysis [7] also constrained light-flavor squark pair production, with an equal probability for the $\tilde{q}$ to decay into $q\tilde{\chi}_1^0$ and $q'\tilde{\chi}_1^\pm$. Again, the $\tilde{\chi}_1^\pm$ is assumed to decay into $\pi^\pm\tilde{\chi}_1^0$. Some of the exclusion limits are presented in Fig. A.4: for $c\tau_0(\tilde{\chi}_1^\pm) = 50$ cm (left panel) and for $c\tau_0(\tilde{\chi}_1^\pm) = 2$ m (right panel). Each of them set limits in the $m_{\tilde{\chi}_1^0}$ vs. $m_{\tilde{q}}$ plane assuming a single light squark or an eightfold squark mass degeneracy. For models with a single $\tilde{q}$ and where $c\tau_0(\tilde{\chi}_1^\pm) = 50$ cm, $m_{\tilde{q}}$ is excluded up to $\sim$ 1.6 TeV for $m_{\tilde{\chi}_1^0} \lesssim 1$ TeV, and up to $\sim$ 1.2 TeV for $m_{\tilde{q}} \approx m_{\tilde{\chi}_1^0}$. These limits are approximatively 350 GeV higher when considering an eightfold squark mass degeneracy scenario. Assuming $c\tau_0(\tilde{\chi}_1^\pm) = 2$ m instead of 50 cm almost has no impact on the exclusion limits for compressed mass spectra, but decreases the two limits of $\sim$ 350 GeV for low $m_{\tilde{\chi}_1^0}$.

Lets us now turn on sbottoms and stops. Long-lived $\tilde{t}$ are motivated for example by electroweak baryogenesis [10, 11], where the mass difference between the $\tilde{t}$ (the NLSP) and the $\tilde{\chi}_1^0$ (the LSP) is small. Using 36.1 fb^{-1} of LHC Run 2, ATLAS excludes $\tilde{b}$ and $\tilde{t}$ R-hadrons up to 1.25 TeV and 1.34 TeV, respectively [12]. On the CMS side, the preliminary search for heavy long-lived charged particles with large ionization energy loss excludes $m_{\tilde{t}}$ up to $\sim$ 1.45 TeV when assuming 10% of them to form R-hadrons [6].

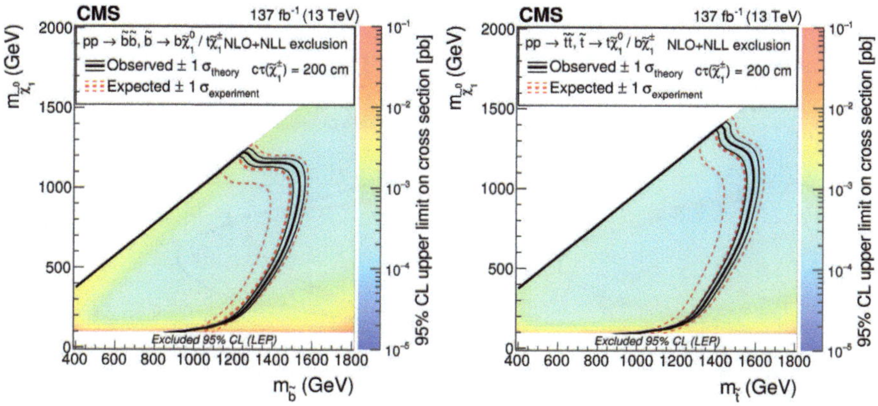

Fig. A.5 Left: CMS constraints in the $m_{\tilde{\chi}_1^0}$ vs. $m_{\tilde{b}}$ plane on the pair production of $\tilde{b}$, each decaying into $b\tilde{\chi}_1^0$ or $\tilde{t}\tilde{\chi}_1^{\pm}$ with the same probability and with $\tilde{\chi}_1^{\pm} \to \pi^{\pm}\tilde{\chi}_1^0$ after 2 m on average. The $\tilde{\chi}_1^{\pm}$ and $\tilde{\chi}_1^0$ are wino-like and bino-like, respectively. Right: CMS constraints in the $m_{\tilde{\chi}_1^0}$ vs. $m_{\tilde{t}}$ plane on the pair production of $\tilde{t}$, each decaying into $t\tilde{\chi}_1^0$ or $\tilde{b}\tilde{\chi}_1^{\pm}$ with the same probability and with $\tilde{\chi}_1^{\pm} \to \pi^{\pm}\tilde{\chi}_1^0$ after 2 m on average. The $\tilde{\chi}_1^{\pm}$ and $\tilde{\chi}_1^0$ are wino-like and bino-like, respectively. Taken from [13]

Selected CMS limits on the pair production of $\tilde{b}$ and $\tilde{t}$ with a long-lived, purely wino $\tilde{\chi}_1^{\pm}$ and a purely bino $\tilde{\chi}_1^0$ are presented in Fig. A.5. The presented constraints are derived for models where the $\tilde{b}$ ($\tilde{t}$) has the same probability to decay into $b\tilde{\chi}_1^0$ and $t\tilde{\chi}_1^{\pm}$ ($t\tilde{\chi}_1^0$ and $b\tilde{\chi}_1^{\pm}$). The $\tilde{\chi}_1^{\pm}$ is assumed to further decay into $\pi^{\pm}\tilde{\chi}_1^0$ with $c\tau_0 = 2$ m. Sbottoms with a mass below $\sim$ 850 GeV are excluded if $m_{\tilde{\chi}_1^0} = 100$ GeV. The exclusion limit then rapidly increases with $m_{\tilde{\chi}_1^0}$, reaching $\sim$ 1.5 TeV for $m_{\tilde{\chi}_1^0} \approx$ 1.1 TeV, and $\sim$ 1.25 TeV for nearly mass degenerate $\tilde{b}$ and $\tilde{\chi}_1^0$. The exclusion limit on $m_{\tilde{t}}$ is similar for low $m_{\tilde{\chi}_1^0}$ but is $\sim$ 200 GeV higher for compressed mass spectra.

A.1.3 Slepton Production

Regarding sleptons, GMSB models favour long-lived NLSP sleptons ($\tilde{\tau}$ or mass degenerate $\tilde{e}$, $\tilde{\mu}$ and $\tilde{\tau}$) because of their small couplings to the nearly massless LSP gravitino. Figure A.6 shows selected ATLAS and CMS limits on long-lived sleptons, either when considered individually or when the three generations are mass degenerate (co-NLSP scenario). For both ATLAS and CMS, the exclusion limit obtained when considering only $\tilde{e}$ is similar to the one obtained when considering only $\tilde{\mu}$. The limit on $\tilde{\tau}$ alone is reduced compared to them. The co-NLSP scenario comprises all the excluded regions and extends them where the limits are the higher in mass. ATLAS exclusion limits go a bit higher in slepton mass compared to CMS, but CMS limits reach lower mean proper decay lengths.

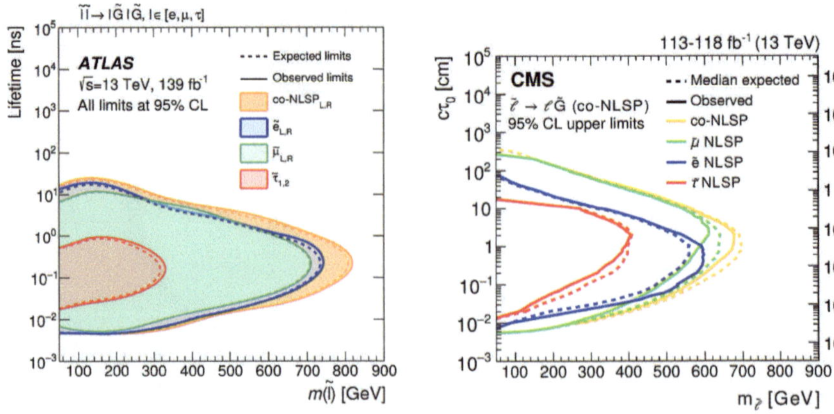

Fig. A.6 ATLAS (left) and CMS (right) limits on long-lived sleptons in the τ_0 vs. $m_{\tilde{\ell}}$ plane. The conversion from τ_0 to $c\tau_0$ is shown in the y-axis of right plot. Sleptons are either considered individually (blue, green and red contours) or all together with the same mass (orange contours). In every scenario they decay into their SM counterparts and a $\tilde{G}$. The different slepton chiral and mixed states are assumed to be mass degenerate. Taken from [14] (left) and [15] (right)

Moreover, an ATLAS analysis [16] excludes $m_{\tilde{\mu}} = 100$ GeV for a $c\tau_0(\tilde{\mu})$ between $6 \times 1.8 \times 10^{-4}$ and 3×10^{-1} m. The exclusion contour peaks at ~ 520 GeV for $c\tau_0(\tilde{\mu}) = 3 \times 10^{-3}$ m. In addition, another ATLAS analysis [17] excludes $m_{\tilde{\tau}}$ between ~ 200 and ~ 360 GeV for a $c\tau_0(\tilde{\tau})$ between 6×10^{-1} and 6 m. This search has been extended in a more recent, preliminary, analysis [18] excluding $m_{\tilde{\tau}}$ between ~ 260 and ~ 480 GeV for a $c\tau_0(\tilde{\tau})$ between 3 and 60 m. Finally, a CMS preliminary search [6] excludes directly pair-produced stable $\tilde{\tau}$ up to around 670 GeV.

A.1.4 Electroweak-ino Production

On the EW-ino side, long-lived charginos can arise in AMSB models, where the purely wino $\tilde{\chi}_1^{\pm}$ and $\tilde{\chi}_1^0$ are the NLSP and the LSP, respectively. In that case, the $\tilde{\chi}_1^{\pm}$ always decays into $\pi^{\pm}\tilde{\chi}_1^0$. Figure A.7 shows ATLAS limits on these models. The exclusion limit on $m_{\tilde{\chi}_1^0}$ increases with $c\tau_0(\tilde{\chi}_1^0)$. At most, the disappearing track search (blue lines) excludes $m_{\tilde{\chi}_1^{\pm}}$ up to ~ 900 GeV for $c\tau_0(\tilde{\chi}_1^{\pm}) = 3 \times 10^{-1}$ m.[3] For $c\tau_0(\tilde{\chi}_1^{\pm}) > 3$ m, the search for large ionisation energy loss (orange lines) excludes $m_{\tilde{\chi}_1^{\pm}}$ up to ~ 1.05 TeV. It also excludes a stable $\tilde{\chi}_1^{\pm}$ up to the same value, similarly

[3] Assuming purely higgsino $\tilde{\chi}_1^{\pm}$ and $\tilde{\chi}_1^0$ (instead of pure wino) decreases the limits by around 150 GeV.

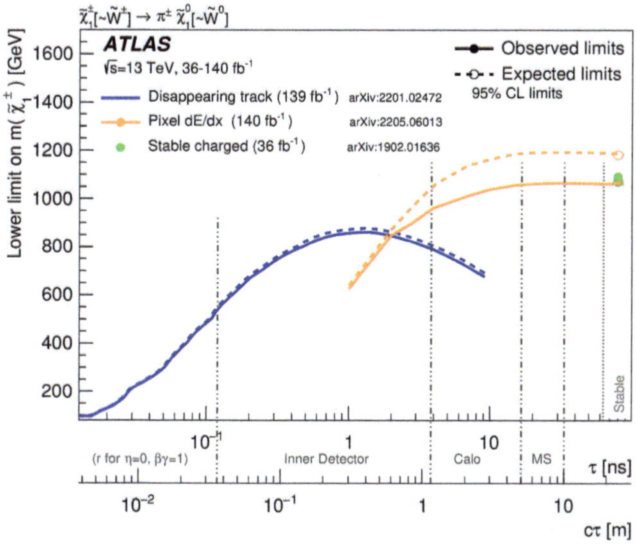

Fig. A.7 Summary of ATLAS limits on long-lived $\tilde{\chi}_1^{\pm}$, after LHC Run 2. Constraints are presented in the $m_{\tilde{\chi}_1^{\pm}}$ vs. $\tau_0(\tilde{\chi}_1^{\pm})$ plane assuming an AMSB model with $\mu > 0$ and $\tan\beta = 5$. The $\tilde{\chi}_1^{\pm}$ and the $\tilde{\chi}_1^0$ are pure winos. The $\tilde{\chi}_1^{\pm}$ is pair-produced and decays into $\pi^{\pm}\tilde{\chi}_1^0$. The coloured dots and circles represent limits obtained on stable $\tilde{\chi}_1^{\pm}$. The labels between the two x-axis refer to parts of the ATLAS detector. "Calo" and "MS" refer to the calorimeters and the muon spectrometer, respectively. Taken from [9]

to the search for stable charged particles using 36 fb^{-1} of LHC Run 2 data (green dots). The search for large ionisation energy loss recorded a small excess, leading to an observed limit approximately 150 GeV lower than the expected one. This excess was scrutinised in a more recent search considering measurements of time-of-flight to the calorimeter as well as large ionisation energy loss in the pixel detector and found compatibility with the SM [18].

As already mentioned in the prompt search section, the ATLAS search for disappearing tracks [8] constrains purely higgsino $\tilde{\chi}_2^{\pm}$, $\tilde{\chi}_1^{\pm}$ and $\tilde{\chi}_1^0$ with very compressed mass spectra. It excludes a mass difference between the $\tilde{\chi}_1^{\pm}$ and the $\tilde{\chi}_1^0$ up to around 0.37 GeV for $m_{\tilde{\chi}_1^{\pm}} \approx 100$ GeV, and up to 0.15 GeV for $m_{\tilde{\chi}_1^{\pm}} \approx$ 700 GeV.

Figure A.8 shows selected CMS constraints on purely wino $\tilde{\chi}_1^{\pm}$ mean proper lifetime and mass. The left panel represents limits assuming the direct production of a $\tilde{\chi}_1^{\pm}$ and a $\tilde{\chi}_1^0$ in an AMSB model similar to the one used to derive limits shown in Fig. A.7. A $m_{\tilde{\chi}_1^{\pm}}$ of 100 GeV is excluded for a $c\tau_0(\tilde{\chi}_1^{\pm})$ between 6×10^{-3} to 90 m. The exclusion area narrows when $m_{\tilde{\chi}_1^{\pm}}$ increases and peaks at $(m_{\tilde{\chi}_1^{\pm}}, c\tau_0(\tilde{\chi}_1^{\pm})) \approx$ (875 GeV, 1 m).

The right panel shows limits assuming the pair production of $\tilde{g}$, with $\mathrm{BR}(q\bar{q}\tilde{\chi}_1^0) = 1/3$, and $\mathrm{BR}(q\bar{q}\tilde{\chi}_1^{\pm}) = 2/3$. The $\tilde{\chi}_1^{\pm}$ and $\tilde{\chi}_1^0$ are pure winos. The

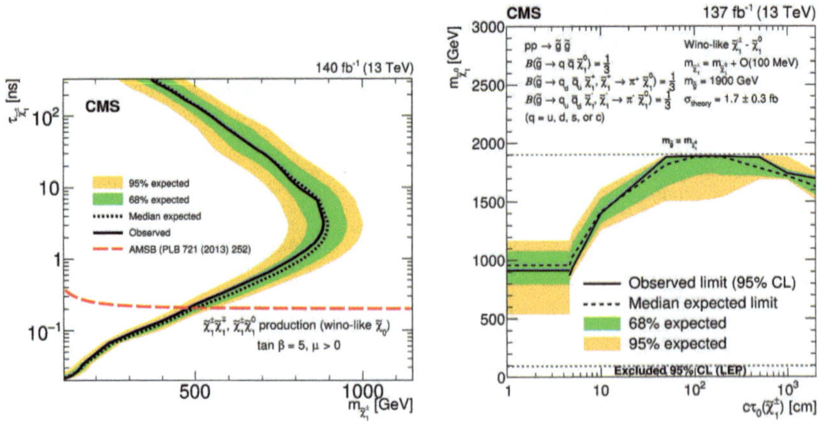

Fig. A.8 Left: CMS constraints in the $\tau_0(\tilde{\chi}_1^{\pm})$ vs. $m_{\tilde{\chi}_1^{\pm}}$ plane. As for the limits shown in Fig. A.7, an AMSB model is assumed with purely wino $\tilde{\chi}_1^{\pm}$ and $\tilde{\chi}_1^0$, $\tilde{\chi}_1^{\pm}$ decaying into $\pi^{\pm}\tilde{\chi}_1^0$, $\mu > 0$ and $\tan\beta = 5$. However, here the $\tilde{\chi}_1^{\pm}\tilde{\chi}_1^0$ production is also considered. The red dashed line indicates the predicted $\tau_0(\tilde{\chi}_1^{\pm})$ for the model considered [19]. Taken from [20]. Right: CMS constraints in the $m_{\tilde{\chi}_1^0}$ vs. $c\tau_0(\tilde{\chi}_1^{\pm})$ plane. The considered process is the pair production of $\tilde{g}$ (not $\tilde{\chi}_1^{\pm}$), with $\mathrm{BR}(q\bar{q}\tilde{\chi}_1^0) = 1/3$, and $\mathrm{BR}(q\bar{q}\tilde{\chi}_1^{\pm}) = 2/3$. The $\tilde{\chi}_1^{\pm}$ always decays into $\pi^{\pm}\tilde{\chi}_1^{\pm}$. The squarks are the ones from the first to generations. The $\tilde{\chi}_1^{\pm}$ and $\tilde{\chi}_1^0$ are wino-like and nearly mass degenerate. The gluino mass is fixed at 1.9 TeV. The horizontal line at $m_{\tilde{\chi}_1^0} = 1900$ GeV bound the mass range in which the $\tilde{g}$ decays are kinematically allowed. Taken from [7] (right)

$\tilde{\chi}_1^{\pm}$ is therefore long-lived and decays into $\pi^{\pm}\tilde{\chi}_1^0$. The limits are derived assuming $m_{\tilde{g}} = 1.9$ TeV. The $\tilde{\chi}_1^{\pm}$ and the $\tilde{\chi}_1^0$ are excluded up to ~ 900 GeV for a $c\tau_0(\tilde{\chi}_1^{\pm})$ between 10^{-2} and 20 m. All the kinematically available phase-space is excluded for a $c\tau_0(\tilde{\chi}_1^{\pm})$ between around 5×10^{-1} and 6 m.

Similarly to ATLAS, the CMS search for disappearing tracks [13] constrains purely wino $\tilde{\chi}_1^{\pm}$ and $\tilde{\chi}_1^0$, and purely higgsino $\tilde{\chi}_2^{\pm}$,$\tilde{\chi}_1^{\pm}$ and $\tilde{\chi}_1^0$ with very compressed mass spectra. When considering pure winos, $\Delta(\tilde{\chi}_1^{\pm}, \tilde{\chi}_1^0)$ is excluded up to around 0.23 GeV for $m_{\tilde{\chi}_1^{\pm}} \approx 100$ GeV, and up to around 0.14 GeV for $m_{\tilde{\chi}_1^{\pm}} \approx 1$ TeV. When considering pure higgsinos, $\Delta(\tilde{\chi}_1^{\pm}, \tilde{\chi}_1^0)$ is excluded up to around 0.39 GeV, and up to around 0.14 GeV for $m_{\tilde{\chi}_1^{\pm}} \approx 950$ GeV.

Similarly to long-lived $\tilde{\ell}$, a long-lived $\tilde{\chi}_1^0$ is realised in GMSB models where the $\tilde{\chi}_1^0$ is the NLSP and the $\tilde{G}$ is the LSP. Since the $\tilde{\chi}_1^0$ leaves no track in the detectors, ATLAS and CMS analyses typically search for high energetic jets and/or displaced decay vertices. Some of their results are shown in Fig. A.9.

The left panel shows results of an ATLAS search for diphoton and dielectron final states with displaced vertices and large $\not{E}_T$, assuming the direct pair production of a purely higgsino $\tilde{\chi}_1^0$, and different values for $\mathrm{BR}(\tilde{\chi}_1^0 \to h\tilde{G})$. When $\mathrm{BR}(\tilde{\chi}_1^0 \to h\tilde{G}) = 1$, $m_{\tilde{\chi}_1^0}$ is excluded up to ~ 100 GeV for a $c\tau_0(\tilde{\chi}_1^0)$ between 6×10^{-2} and

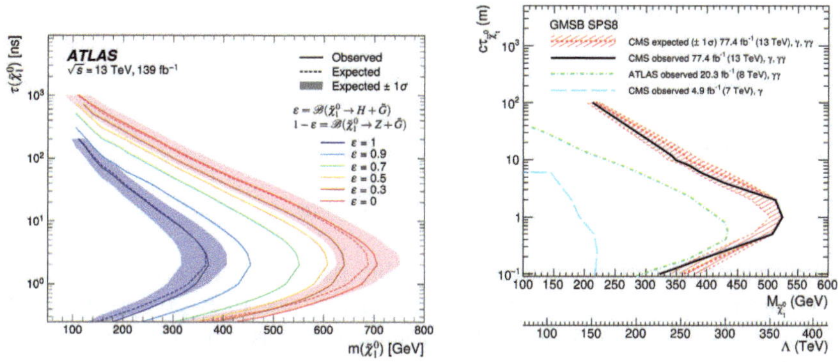

Fig. A.9 Left: ATLAS limits in the $\tau_0(\tilde{\chi}_1^0)$ vs. $m_{\tilde{\chi}_1^0}$ plane, assuming a GMSB scenario where pair-produced higgsinos decay into soft fermions and long-lived $\tilde{\chi}_1^0$. The $\tilde{\chi}_1^0$ can further decay into $Z\tilde{G}$ or $h\tilde{G}$. Limits are derived assuming different branching ratios. Taken from [21]. Right: CMS limits, using 77.4 fb^{-1} of LHC Run 2 data, in the $c\tau_0(\tilde{\chi}_1^0)$ vs. $m_{\tilde{\chi}_1^0}$ plane, assuming the "Snowmass Points and Slopes 8" (SPS8) benchmark model [22]. In this model, the $\tilde{\chi}_1^0$ mass is linearly related to the SUSY breaking scale Λ and the two quantities appear on the x axis. The green and blue lines show limits set by a previous ATLAS 8 TeV [23] and a previous CMS 7 TeV [24] searches, respectively. Taken from [25]

60 m, and up to $\sim$ 360 GeV for $c\tau_0(\tilde{\chi}_1^0) \approx 6 \times 10^{-1}$ m. The sensitivity increases with BR($\tilde{\chi}_1^0 \to Z + \tilde{G}$) and so is the exclusion limit. When BR($\tilde{\chi}_1^0 \to h\tilde{G}) = 0$, $m_{\tilde{\chi}_1^0}$ is excluded up to $\sim$ 100 GeV for a $c\tau_0(\tilde{\chi}_1^0)$ between 6×10^{-2} and 300 m, and up to $\sim$ 700 GeV for $c\tau_0(\tilde{\chi}_1^0) \approx 6 \times 10^{-1}$ m.

The right panel shows CMS results based on the "Snowmass Points and Slopes 8" (SPS8) benchmark model [22]. It assumes a GMSB scenario with the pair production of $\tilde{g}$ or light-flavor $\tilde{q}$, with $\tilde{g} \to q\tilde{q}$ and $\tilde{q} \to q'\tilde{\chi}_1^{\pm}$, $q\tilde{\chi}_1^0$. The $\tilde{\chi}_1^{\pm}$ further decays into $W^{\pm}\tilde{G}$, and the $\tilde{\chi}_1^0$ decays either into $\gamma\tilde{G}$ or $Z\tilde{G}$. The analysis therefore searches for events with 1 or 2 displaced or delayed γ, 3 or more jets and large $\not{E}_T$. The $\tilde{\chi}_1^0$ is excluded up to $\sim$ 220 GeV for a $c\tau_0(\tilde{\chi}_1^0)$ between 0.1 and 100 m, and up to $\sim$ 530 GeV for $c\tau_0(\tilde{\chi}_1^0) \approx 1$ m.

A.2 Searches for RPV Signatures

The MSSM potential allows for four terms violating the conservation of baryon and lepton numbers, and subsequently the conservation of R-parity (see Eqs. (1.59) and (1.60)). So far, only results on models forbidding such terms were presented. Since R-parity conservation is not enforced by any fundamental principle, RPV couplings can exist as long as they are small enough to be compatible with current observations. The impact of small RPV couplings on the production of SUSY particles can be neglected. However, RPV couplings allow the LSP to decay into

SM particles, for instance

$$\tilde{\chi}_1^0 \to \ell^\pm \ell^\mp \nu,\ \ell^\pm q\bar{q},\ \nu q\bar{q},\ qqq\,, \tag{A.1}$$

via virtual $\tilde{\ell}$ or $\tilde{q}$, and

$$\tilde{\chi}_1^0 \to \ell^\pm W^\mp,\ \nu Z,\ \nu h\,. \tag{A.2}$$

The LSP can then fully decay into detectable particles.[4] Therefore, unlike RPC SUSY searches, looking for final states with $\not{E}_T$ is not a key component of RPV SUSY searches. The focus will first be made on results assuming models with large enough RPV couplings for the SUSY particles to decay promptly.

A.2.1 Searches for Promptly Decaying Sparticles

A.2.1.1 Gluino Production

Searches for the pair production of gluinos decaying into q and the $\tilde{\chi}_1^0$ were conducted by ATLAS in the four three-body decay channels mentioned in the previous paragraph. The results are shown in Fig. A.10. The most stringent limit is obtained in the $\tilde{\chi}_1^0 \to \ell^\pm \ell^\mp \nu$ channel ($\ell = e, \mu$) by an analysis searching for final states with four or more electrically charged leptons (blue lines) [26]. Gluinos with a mass below $\sim$ 2.4 TeV are excluded for almost all the kinematically available $m_{\tilde{\chi}_1^0}$. The limit is decreased to around 1.8 TeV in the $\tilde{\chi}_1^0 \to \tau^\pm \tau^\mp \nu$ channel, due to the lower reconstruction efficiency for hadronic τ decays.

The two analyses probing the $\tilde{\chi}_1^0 \to \ell^\pm qq,\ \nu qq$ channels constrain $m_{\tilde{g}}$ to be higher than $\sim$ 2.2 TeV for a wide range of $m_{\tilde{\chi}_1^0}$. They are searching for final states with multiple jets and at least 1 isolated e or μ [27], and with multiple jets and 2 same-sign leptons (e or μ) or 3 or more leptons without any charge requirement [28].

The fully hadronic decay channel has been probed by a multijet analysis [29] excluding $m_{\tilde{g}}$ up to $\sim$ 2.2 TeV (khaki lines). The sensitivity decreases rapidly for high and low $m_{\tilde{\chi}_1^0}$ because the jets are too low in momentum or too collimated to be reconstructed.

The 1 lepton search aforementioned also probes the fully hadronic channel $\tilde{\chi}_1^0 \to tbs$ and excludes $m_{\tilde{g}}$ up to $\sim$ 2.3 TeV for a pure bino LSP (green lines). If the LSP is instead purely wino (higgsino), the limits around 150 (100) GeV weaker. If the LSP is instead purely wino (higgsino),in addition to $tt\tilde{\chi}_1^0$, the $\tilde{g}$ is also assumed to

---

[4] However, some SUSY models with non-zero RPV couplings can still provide a good DM candidate, for instance very light $\tilde{G}$ (whose lifetime is typically greater than the age of the Universe).

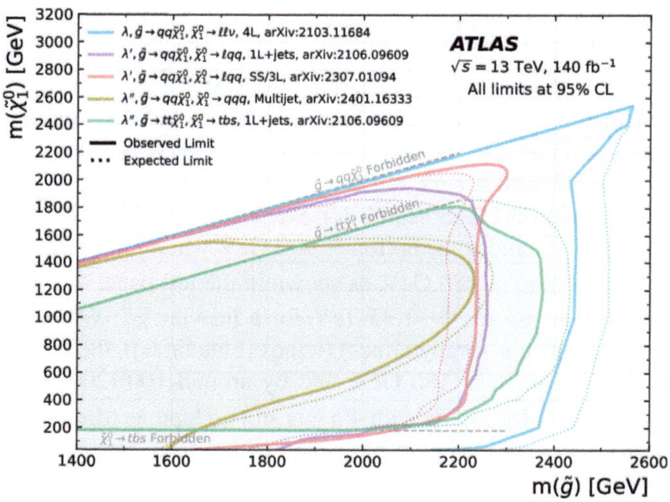

Fig. A.10 Summary of ATLAS constraints on models assuming a direct $\tilde{g}\tilde{g}$ production, with $\tilde{g} \to q\bar{q}\tilde{\chi}_1^0$, $t\bar{t}\tilde{\chi}_1^0$, and with the $\tilde{\chi}_1^0$ further decaying via RPV couplings into leptons and/or quarks, in the $m_{\tilde{\chi}_1^0}$ vs. $m_{\tilde{g}}$ plane, after LHC Run 2. Taken from [9]

decay into $bb\tilde{\chi}_1^0$ and $tb\tilde{\chi}_1^{\pm}$ ($tt\tilde{\chi}_2^0$ and $tb\tilde{\chi}_1^{\pm}$), with $\tilde{\chi}_{1,2}^0 \to tbs$ and $\tilde{\chi}_1^{\pm} \to bbs$, resulting in limits around 150 (100) GeV weaker.

On the CMS side, a search for high non-top jet multiplicity using 38.2 fb^{-1} of LHC Run 2 data [30] excludes $m_{\tilde{g}}$ up to $\sim$ 1.45 TeV when considering direct $\tilde{g}\tilde{g}$ production, with $\tilde{g} \to qq\widetilde{H}$ via an off-shell squark, where $\widetilde{H}$ is a higgsino with $m_{\widetilde{H}} = 0.6 \times m_{\tilde{g}}$. The $\widetilde{H}$ is assumed to further decay into qqq via an off-shell squark.

However, if the $\tilde{g}$ does not decay into the $\tilde{\chi}_1^0$, ATLAS excludes $m_{\tilde{g}}$ up to $\sim$ 1.75 TeV both for models where $\tilde{g} \to qqq$, qqb [29] and where $\tilde{g} \to t\tilde{t} \to tbs$, for almost all the kinematically allowed $m_{\tilde{t}}$ [27].

CMS excludes $m_{\tilde{g}}$ up to $\sim$ 1.75 TeV, $\sim$ 2.16 TeV and $\sim$ 1.75 TeV when considering direct $\tilde{g}\tilde{g}$ production with $\tilde{g} \to qqq$ [31], $\tilde{g} \to tbs$ and $\tilde{g} \to qq\bar{q}\bar{q} + e/\mu/\tau$ [32], respectively.

In addition, the multijet search using 128 fb^{-1} of LHC Run 2 data [31] excludes $m_{\tilde{g}}$ below $\sim$ 1.75 TeV by considering direct $\tilde{g}\tilde{g}$ production, with $\tilde{g} \to qqq$. Another search, looking for events with at least 2 jets and 2 isolated same-sign or 3 or more electrically charged leptons (e, μ and τ) [32], excludes $m_{\tilde{g}}$ below $\sim$ 2.16 TeV when considering direct $\tilde{g}\tilde{g}$ production, with $\tilde{g} \to tbs$. It also excludes $m_{\tilde{g}}$ up to $\sim$ 1.75 TeV for models where $\tilde{g} \to qq\bar{q}\bar{q}+e/\mu/\tau$, via the off-shell decay of a $\tilde{\chi}_1^0$.

A.2.1.2 Squark Production

As far as light-flavor squarks are concerned, the CMS search for high non-top jet multiplicity using 38.2 fb^{-1} of LHC Run 2 data [30] excludes $m_{\tilde{q}}$ up to ~ 700 GeV when considering direct $\tilde{q}\tilde{q}^*$ production, with $\tilde{q} \to q\widetilde{H}$, $m_{\widetilde{H}} = 0.75 \times m_{\tilde{q}}$ and $\widetilde{H} \to qqq$ via an off-shell squark.

Regarding stops, an ATLAS search for events with multiple jets and at least 1 isolated e or μ [27] excludes $m_{\tilde{t}}$ up to ~ 900 GeV when $\tilde{t} \to t\tilde{\chi}_1^0 \to ttbs$, with the $\tilde{\chi}_1^0$ being a pure bino of 180 GeV, as shown in the left panel of Fig. A.11 (green lines). The limit goes up to ~ 1.35 TeV for a heavier $\tilde{\chi}_1^0$. When considering $\tilde{t}$ decays into pure higgsinos (orange lines) (winos (blue lines)), the limit is decreased by ~ 100 GeV for $m_{\tilde{\chi}_1^0} \approx 180$ GeV, and by around 100 (200) for larger $m_{\tilde{\chi}_1^0}$. The analysis searching for events with 4 b jets and no leptons (dot-dot-dashed black line) excludes $\tilde{t}$ decaying into pure higgsinos with a mass up to ~ 900 GeV for $m_{\tilde{\chi}_1^\pm} \approx 800$ GeV.

The same analysis constrains models where $\tilde{t} \to b\tilde{\chi}_1^\pm \to bbbs$, with $\tilde{\chi}_1^\pm$ being a pure higgsino, as shown in the right panel of Fig. A.11. It excludes $m_{\tilde{t}}$ up to $\sim$ 780 GeV for $m_{\tilde{\chi}_1^\pm} \approx 120$ GeV, and up to ~ 930 GeV for larger $m_{\tilde{\chi}_1^\pm}$.

The preliminary result of the CMS search for events with t quarks, low $\not{E}_T$ and many additional quarks and gluons [34] constrains the direct $\tilde{t}\tilde{t}^*$ production, with

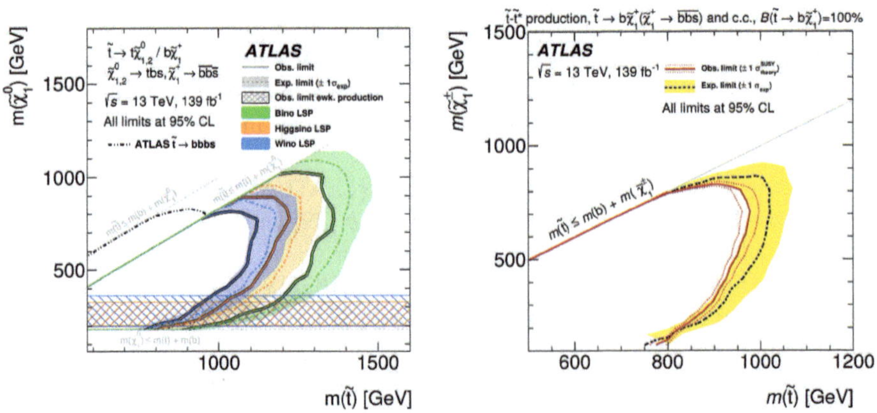

Fig. A.11 Right: ATLAS limits on the direct $\tilde{t}\tilde{t}^*$ production, in the $m_{\tilde{\chi}_1^0}$ vs. $m_{\tilde{t}}$ plane, assuming a purely bino, higgsino or wino $\tilde{\chi}_1^0$. If the $\tilde{\chi}_1^0$ is purely bino, $\tilde{t} \to t\tilde{\chi}_1^0$. If it is purely higgsino, $\tilde{t} \to t\tilde{\chi}_2^0$ and $\tilde{t} \to b\tilde{\chi}_1^\pm$. If it is instead purely wino, $\tilde{t} \to b\tilde{\chi}_1^\pm$. In every scenario, $\tilde{\chi}_{1,2}^0 \to tbs$ and $\tilde{\chi}_1^\pm \to bbs$. The analysis looks for events with multiple jets and at least 1 isolated e or μ. The dot-dot-dashed black line shows the limit obtained by another ATLAS search looking for 4 b jets, no leptons and assuming a higgsino LSP [33]. Taken from [27]. Right: ATLAS limits on the direct $\tilde{t}\tilde{t}^*$ production, with $\tilde{t} \to b\tilde{\chi}_1^\pm \to bbbs$, in the $m_{\tilde{\chi}_1^\pm}$ vs. $m_{\tilde{t}}$ plane, from the analysis searching for events with 4 b jets and no leptons [33]

$t\tilde{\chi}_1^0 \to tqqq$, where q is a light quark. It excludes $m_{\tilde{t}}$ up to $\sim$ 700 GeV for $m_{\tilde{\chi}_1^0} = 100$ GeV.

When considering direct $\tilde{t}$ decay into SM quarks, the ATLAS search for massive colored resonances excludes $m_{\tilde{t}}$ between 100 and 410 GeV for direct $\tilde{t}\tilde{t}^*$ production, with $\tilde{t} \to \bar{q}\bar{q}'$, using 36.7 fb^{-1} of LHC Run 2 data [35]. If one of the anti-quark is a $\bar{b}$ quark, $m_{\tilde{t}}$ is excluded between 100 and 470 GeV, and between 480 and 610 GeV.

The corresponding CMS search [36] uses 35.9 fb^{-1} of LHC Run 2 data to exclude $m_{\tilde{t}}$ between 80 and 520 GeV when both anti-quarks are light, and to exclude $m_{\tilde{t}}$ from the interval $[80, 270] \cup [285, 340] \cup [400, 525]$ GeV when one anti-quark is a $\bar{b}$ quark.

A.2.1.3 Slepton Production

The ATLAS search for events with 4 or more electrically charged leptons (e, μ and τ) [26] constrains the direct production of $\tilde{\ell}_L^+\tilde{\ell}_L^-$, $\tilde{\nu}\tilde{\nu}$ and $\tilde{\ell}_L^\pm\tilde{\nu}$, with $\tilde{\ell}_L^\pm \to \ell^\pm\tilde{\chi}_1^0$, $\tilde{\nu} \to \nu\tilde{\chi}_1^0$ and $\tilde{\chi}_1^0 \to \ell^+\ell^-\nu$. When $\ell = \{e, \mu\}$ ($\ell = \tau$), it excludes $m_{\tilde{\ell}_L^\pm/\tilde{\nu}}$ up to $\sim$ 1000 (700) GeV for nearly massless $\tilde{\chi}_1^0$, and up to $\sim$ 1180 (840) GeV for a heavier $\tilde{\chi}_1^0$.

Another ATLAS search [37] constrain the associated production of a $\tilde{\mu}_L$ and a t quark ($dg \to \tilde{\mu}_L t$ via a t quark mediator in the t-channel), with $\tilde{\mu}_L \to \mu\tilde{\chi}_1^0$ and $t \to e\nu b$. It excludes $m_{\tilde{\mu}_L}$ between approximately 100 and 640 GeV for a nearly massless $\tilde{\chi}_1^0$ and a $\tilde{\mu}_L dt$ coupling equal to 1, as shown by the red lines in the left plot of Fig. A.12. No model featuring a $\tilde{\chi}_1^0$ above 160 GeV is excluded. If the coupling

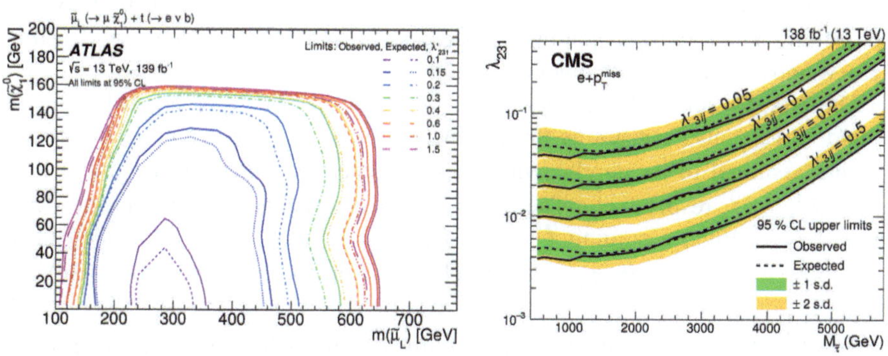

Fig. A.12 Right: ATLAS limits on the associated production of a $\tilde{mu}_L$ and a t quark, with $\tilde{\mu}_L \to \mu\tilde{\chi}_1^0$ and $t \to e\nu b$, in the $m_{\tilde{\chi}_1^0}$ vs. $m_{\tilde{\mu}_L}$ plane. Different values of λ'_{231}, the $dt\tilde{\mu}$ coupling are assumed. Taken from [37]. Right: CMS limits on the $q\bar{q}' \to \tilde{\tau} \to e\nu_\mu$ process, in the λ_{231} vs. $m_{\tilde{\tau}}$ plane. Different values of λ'_{3ij} $(i, j = 1, 2, 3)$ are assumed. The λ_{231} and λ'_{3ij} are the $\tilde{\tau}e\nu_\mu$ and $\tilde{\tau}q\bar{q}'$ couplings, respectively. Taken from [38]

is instead equal to 0.1, the $\tilde{\mu}_L$ is only excluded between approximately 230 and 360 GeV for a nearly massless $\tilde{\chi}_1^0$, and not excluded at all if $m_{\tilde{\chi}_1^0} \gtrsim 70$ GeV.

On the CMS side, a search for events with 1 electron or muon and $\not{E}_T$ corresponding to $q\bar{q}' \to \tilde{\tau} \to e\nu_\mu, \mu\nu_e$ set limits on $m_{\tilde{\tau}}$ and on the $\tilde{\tau}e\nu_\mu$ and $\tilde{\tau}\mu\nu_e$ couplings [38]. The constraints on the $\tilde{\tau}e\nu_\mu$ coupling are shown in the right plot of Fig. A.12. For $m_{\tilde{\tau}} = 1$ TeV, the $\tilde{\tau}e\nu_\mu$ coupling is constrained to be below $\sim 4 \times 10^{-3}$ (4×10^{-2}) when considering a $\tilde{\tau}q\bar{q}'$ coupling of 0.5 (0.05). The limits are weaker when $m_{\tilde{\tau}}$ increases. Limits on the $\tilde{\tau}\mu\nu_e$ coupling are almost identical.

A similar CMS analysis, searching for resonances decaying into $e\mu$ with 35.9 fb^{-1} of LHC Run 2 data [39], constrains the $q\bar{q} \to \tilde{\nu}_\tau \to e\mu$ process and excludes $m_{\tilde{\nu}_\tau}$ up to ~ 1.7 (3.7) TeV when the $\tilde{\nu}_\tau q\bar{q}$ and $\tilde{\nu}_\tau e\mu$ couplings are both equal to 0.01 (0.1).

A.2.1.4 Electroweak-ino Production

Turning now ourselves to the production of EW-inos, the ATLAS search for final states with either 2 same-sign leptons (e or μ), or 3 leptons [40] constrains the production of purely higgsino $\tilde{\chi}_2^0$, $\tilde{\chi}_1^\pm$ and $\tilde{\chi}_1^0$. It considers the direct production of nearly mass degenerate $\tilde{\chi}_1^\pm\tilde{\chi}_1^0$ and $\tilde{\chi}_2^0\tilde{\chi}_1^0$, with $\tilde{\chi}_1^\pm \to \nu W^\pm$ and $\tilde{\chi}_{1,2}^0 \to \ell^\pm W^\mp$. The three EW-inos are excluded up to ~ 440 GeV.

Another ATLAS search for events with 3 or more leptons (e, μ or τ) and an additional $W^\pm$, Z or h boson [41] constrains the associated production of purely wino, mass degenerate $\tilde{\chi}_1^\pm$ and $\tilde{\chi}_1^0$. The EW-ino decays are $\tilde{\chi}_1^\pm \to Z\ell^\pm, h\ell^\pm, W^\pm\nu$ and $\tilde{\chi}_1^0 \to Z\nu, h\nu, W^\pm\ell^\mp$. The left panel of Fig. A.13 shows the limits in the Z branching ratio vs. $m_{\tilde{\chi}_1^\pm/\tilde{\chi}_1^0}$ plane. The most stringent limit is obtained when only the Z channel is allowed, excluding $m_{\tilde{\chi}_1^\pm/\tilde{\chi}_1^0}$ up to ~ 970 GeV. The analysis is almost insensitive to decays in the other channels, and therefore does not constrain the EW-inos when the Z channel is forbidden.

The ATLAS search for events with 4 or more electrically charged leptons (e, μ and τ) [26] also constrains the production of $\tilde{\chi}_1^+\tilde{\chi}_1^-$ and $\tilde{\chi}_2^0\tilde{\chi}_1^\pm$. The $\tilde{\chi}_2^0$ and $\tilde{\chi}_1^\pm$ are assumed to be purely wino, mass degenerate, and to decay into $W^\pm\tilde{\chi}_1^0$ and $Z\tilde{\chi}_1^0$, $h\tilde{\chi}_1^0$, respectively. The $\tilde{\chi}_1^0$ is assumed to be purely bino and to decay into $\ell^+\ell^-\nu$. When assuming $\ell = \{e, \mu\}$ ($\ell = \tau$), $m_{\tilde{\chi}_2^0/\tilde{\chi}_1^\pm}$ is excluded up to $\sim 1500\,(940)$ GeV for $m_{\tilde{\chi}_1^0} = 100$ GeV and up to $\sim 1580\,(1100)$ GeV for larger $m_{\tilde{\chi}_1^0}$.

The ATLAS search for events with multiple jets and at least 1 isolated e or μ [27] constrains the associated production of purely wino, mass degenerate $\tilde{\chi}_1^\pm$ and $\tilde{\chi}_1^0$, decaying into $tb\,s$ and $bb\,s$, respectively. The $m_{\tilde{\chi}_1^\pm/\tilde{\chi}_1^0}$ is excluded between ~ 190 and 365 GeV. If instead the production of $\tilde{\chi}_2^0\tilde{\chi}_1^\pm$, $\tilde{\chi}_1^\pm\tilde{\chi}_1^0$ and $\tilde{\chi}_2^0\tilde{\chi}_1^0$ are considered, with the EW-inos being pure higgsinos, and $\tilde{\chi}_1^\pm \to bb\,s$ and $\tilde{\chi}_{1,2}^0 \to tb\,s$, EW-inos are excluded for masses between ~ 190 and 320 GeV.

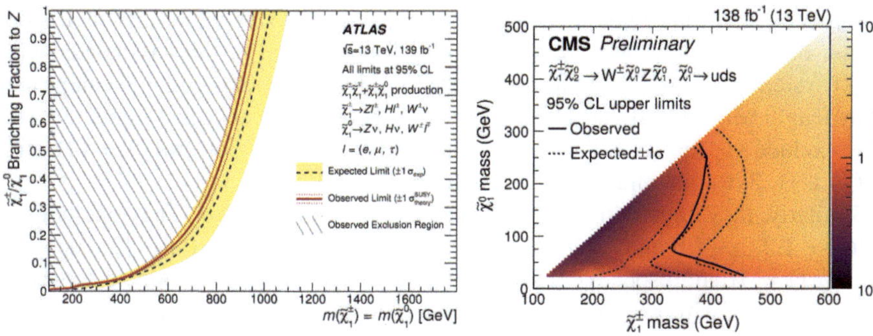

Fig. A.13 Right: ATLAS constraints on the direct production of $\tilde{\chi}_1^{\pm}\tilde{\chi}_1^{\mp}$ and $\tilde{\chi}_1^{\pm}\tilde{\chi}_1^{0}$, in the $\mathrm{BR}(\tilde{\chi}_1^{\pm}/\tilde{\chi}_1^{0} \to Zl^{\pm}/Z\nu)$ vs. $m_{\tilde{\chi}_1^{\pm}/\tilde{\chi}_1^{0}}$ plane. The $\tilde{\chi}_1^{\pm}$ and $\tilde{\chi}_1^{0}$ are assumed to be purely winos and to decay into ZX, hX and $W^{\pm}X$, with $X = \{e, \mu, \tau, \nu\}$ depending on the boson electric charge. The sum of the branching ratios to Z, W, and h is unity for each point, and the W and h branching ratios are equal. Taken from [41]. Right: Preliminary CMS constraints on the direct $\tilde{\chi}_2^{0}\tilde{\chi}_1^{\pm}$ production, in the $m_{\tilde{\chi}_1^{0}}$ vs. $m_{\tilde{\chi}_2^{0}/\tilde{\chi}_1^{\pm}}$ plane. The $\tilde{\chi}_2^{0}$ and $\tilde{\chi}_1^{\pm}$ are assumed to be purely winos and to decay into $Z\tilde{\chi}_1^{0}$ and $W^{\pm}\tilde{\chi}_1^{0}$, respectively. The $\tilde{\chi}_1^{0}$ is assumed to be purely bino and to decay into 3 light quarks. Taken from [42]

The CMS search for pair-produced multijet signatures using 128 fb^{-1} of LHC Run 2 data [31] also excludes higgsino masses between 75 and $\sim$ 78 GeV, and between $\sim$ 94 and $\sim$ 112 GeV when assuming the higgsinos to directly decay into 3 quarks.

Finally, the preliminary CMS search for multilepton events [42] constrains the associated production of purely wino, mass degenerate $\tilde{\chi}_2^{0}$ and $\tilde{\chi}_1^{0}$, decaying into $Z\tilde{\chi}_1^{0}$ and $W^{\pm}\tilde{\chi}_1^{0}$, respectively. The $\tilde{\chi}_1^{0}$ is assumed to be a pure bino and to decay into uds. The $m_{\tilde{\chi}_2^{0}/\tilde{\chi}_1^{\pm}}$ is excluded up to $\sim$ 450 GeV for $m_{\tilde{\chi}_1^{0}} = 25$ GeV, as shown in the right panel of Fig. A.13. The limit is reduced to $\sim$ 370 GeV for larger $m_{\tilde{\chi}_1^{0}}$. If instead $\tilde{\chi}_1^{0} \to udb$, the observed limit goes up to 600 GeV for $m_{\tilde{\chi}_1^{0}} = 20$ GeV, but goes down to $\sim$ 150 GeV for larger $m_{\tilde{\chi}_1^{0}}$.

A.2.2 Searches for Long-Lived Sparticles

The previous results were derived for RPV couplings important enough for the SUSY particles to decay promptly. However, if these couplings are too small, they might decay far away from the collision regions, leading to displaced vertices. Alternatively, displaced vertices can also arise with large RPV couplings if a produced SUSY particles is long-lived (for other reasons) but the SUSY particle it decays into has large RPV couplings.

Pair-produced long-lived gluinos, each decaying into tbs via an off-shell stop, are constrained by two CMS analyses looking for displaced vertices associated with a dijet system [43, 44]. The first one targets long-lived particles with a mean proper decay length between 0.1 and 100 mm, and the second one between 1 mm and 10 m. They exclude $m_{\tilde{g}}$ up to ~ 1.9, ~ 2.6 TeV and ~ 2.1 TeV for $c\tau_0(\tilde{g}) = 0.1$ mm, $c\tau_0(\tilde{g}) \in [1, 320]$ mm and $c\tau_0(\tilde{g}) = 10$ m, respectively.

Similarly, they constrain direct $\tilde{t}\tilde{t}^*$ production, with $\tilde{t} \to \bar{d}\bar{d}$. They exclude $m_{\tilde{t}}$ up to ~ 1.2, ~ 1.8 and ~ 1.3 TeV for $c\tau_0(\tilde{t}) = 0.1$ mm, $c\tau_0(\tilde{t}) \in [2, 100]$ mm and $c\tau_0(\tilde{t}) = 10$ m, respectively. In addition, [44] also constrains models assuming $\tilde{t} \to d\ell^+$, $b\ell^+$, with $\ell = \{e, \mu, \tau\}$. In these models, $m_{\tilde{t}}$ is excluded up to ~ 1.35, ~ 1.7 and ~ 1.1 TeV for $c\tau_0(\tilde{t}) = 1$ mm, $c\tau_0(\tilde{t}) \in [10, 55]$ mm and $c\tau_0(\tilde{t}) = 10$ m, respectively.

The corresponding ATLAS analysis [45] is searching for direct $\tilde{t}\tilde{t}^*$ production, with each $\tilde{t}$ forming an R-hadron decaying into μq, with $q = \{d, s, b\}$. It excludes $m_{\tilde{t}}$ up to ~ 1.75 GeV for $\tau_0(\tilde{t}) \approx 30$ mm. For a fixed $m_{\tilde{t}}$ of 1 TeV, $c\tau_0(\tilde{t})$ is excluded between 1.5 and 3 mm.

Long-lived charginos and neutralinos are constrained by ATLAS through a search for multiple energetic jets and a displaced vertex [46]. It constrains models where the lightest charginos and neutralinos are pure higgsinos, mass degenerate, and are directly produced, i.e. $pp \to \tilde{\chi}_2^0\tilde{\chi}_1^{\pm}$, $\tilde{\chi}_2^0\tilde{\chi}_1^0$, $\tilde{\chi}_1^+\tilde{\chi}_1^-$, $\tilde{\chi}_1^{\pm}\tilde{\chi}_1^0$, with $\widetilde{H} \to qqq$ ($\widetilde{H}$ stands for higgsinos). It excludes $m_{\widetilde{H}}$ up to ~ 1.6 TeV for $c\tau_0(\widetilde{H}) \approx 30$ mm. For a fixed $m_{\widetilde{H}}$ of 300 GeV, $c\tau_0(\widetilde{H})$ is excluded between 0.3 mm and 24 m.

It also constrains indirect $\tilde{\chi}_1^0$ production, through models where 2.4 TeV gluinos are pair produced, with $\tilde{g} \to q\tilde{\chi}_1^0$, and $\tilde{\chi}_1^0 \to qqq$. In such models, for a fixed $m_{\tilde{\chi}_1^0}$ of 100 GeV, $c\tau_0(\tilde{\chi}_1^0)$ is excluded between 1.2 and 60 mm. For a fixed $m_{\tilde{\chi}_1^0}$ of 2.4 TeV, $c\tau_0(\tilde{\chi}_1^0)$ is excluded between 6 mm and 1.8 m.

On the CMS side, one of the CMS analysis looking for displaced vertices associated with a dijet system [43] constrains direct $\tilde{\chi}_1^0\tilde{\chi}_1^0$ production, with $\tilde{\chi}_1^0 \to tbs$ via off-shell $\tilde{t}$. It excludes $m_{\tilde{\chi}_1^0}$ up to ~ 0.8, ~ 1.3 and ~ 1 TeV for a $c\tau_0(\tilde{\chi}_1^0)$ of 0.2, 10 and 100 mm, respectively.

Appendix B
Complete List of Experimental Results in the SModelS Database v3.0

This appendix adds additional information to the thesis. It lists all the analyses implemented in the SModelS database v3.0.

See Tables B.1, B.2, and B.3.

Table B.1 List of ATLAS Run 2 analyses and their types of results in the SModelS v2.3.0 database. The **New** column indicates the latest database version (from v2.2) that introduced significant changes for the analysis. Apart from the HSCP, disappearing tracks and displaced lepton searches, all analyses require $\not{E}_T$ in the final state (for conciseness omitted in the short descriptions). EW stands for electroweak (-ino or slepton) production. The last column specifies whether and how SRs are combined: "pyhf" means a HistFactory model is used through interface with Pyhf; "SLv1" ("SLv2") means that a covariance matrix (with a third moment) is used; and a blank means only the most sensitive SR is used. The EMs for ATLAS-SUSY-2019-08 changed between v2.3 and v3.0, but not between v2.1 and v2.3

ID	New	Short description	Lumi. [fb^{-1}]	UL_{obs}	UL_{exp}	EM	Comb.
ATLAS-EXOT-2018-06 [47]	v3.0	monojet	139.0	✓			
ATLAS-EXOT-2018-48 [48]	v3.0	di-top resonance	139.0	✓	✓		
ATLAS-EXOT-2019-03 [49]	v3.0	dijet resonance	139.0	✓	✓		
ATLAS-SUSY-2015-01 [50]		2 b-jets	3.2	✓			
ATLAS-SUSY-2015-02 [51]		1ℓ stop	3.2	✓		✓	
ATLAS-SUSY-2015-06 [52]		0ℓ + 2–6 jets	3.2			✓	
ATLAS-SUSY-2015-09 [53]		jets + 2 SS or $\geq 3\ell$	3.2	✓			
ATLAS-SUSY-2016-06 [54]		disappearing tracks	36.1			✓	
ATLAS-SUSY-2016-07 [55]		0ℓ + jets	36.1	✓		✓	
ATLAS-SUSY-2016-08 [56]		displaced vertices	32.8	✓			

(continued)

T. Pascal, *Beyond the Standard Model at the Large Hadron Collider: From Simplified Model Constraints to Artificial Proto-Modelling for Dispersed Signals*, Springer Theses, https://doi.org/10.1007/978-3-032-09391-2

Table B.1 (continued)

ID	New	Short description	Lumi. [fb^{-1}]	UL_{obs}	UL_{exp}	EM	Comb.
ATLAS-SUSY-2016-14 [57]		2 SS or 3 ℓ's + jets	36.1	✓			
ATLAS-SUSY-2016-15 [58]		0ℓ stop	36.1	✓			
ATLAS-SUSY-2016-16 [59]		1ℓ stop	36.1	✓		✓	
ATLAS-SUSY-2016-17 [60]		2 OS ℓ	36.1	✓			
ATLAS-SUSY-2016-19 [61]		2 b-jets + τ's	36.1	✓			
ATLAS-SUSY-2016-24 [62]		2–3 ℓ's, EW	36.1	✓		✓	
ATLAS-SUSY-2016-26 [63]		$\geq 2\,c$-jets	36.1	✓			
ATLAS-SUSY-2016-27 [64]		jets + γ	36.1	✓		✓	
ATLAS-SUSY-2016-28 [65]		2 b-jets	36.1	✓			
ATLAS-SUSY-2016-32 [66]		HSCP	36.1	✓	✓	✓	
ATLAS-SUSY-2016-33 [67]		2 SFOS ℓ's	36.1	✓			
ATLAS-SUSY-2017-01 [68]		$Wh(bb)$, EW	36.1	✓			
ATLAS-SUSY-2017-02 [69]		0ℓ + jets	36.1	✓	✓		
ATLAS-SUSY-2017-03 [70]		multi-ℓ EW	36.1	✓		✓	
ATLAS-SUSY-2018-04 [71]		2 hadronic taus	139.0	✓		✓	PYHF
ATLAS-SUSY-2018-05 [72]	v2.3	2ℓ + jets, EW	139.0	✓		✓	PYHF
ATLAS-SUSY-2018-05 [72]	v2.3	2ℓ + jets, strong	139.0			✓	
ATLAS-SUSY-2018-06 [73]		3ℓ, EW	139.0	✓	✓	✓	
ATLAS-SUSY-2018-08 [74]	v2.2	2 OS ℓ	139.0	✓		✓	
ATLAS-SUSY-2018-09 [75]	v3.0	2 SS ℓ	139.0	✓			
ATLAS-SUSY-2018-10 [76]		1ℓ + jets	139.0	✓		✓	
ATLAS-SUSY-2018-12 [77]		0ℓ + jets	139.0	✓	✓	✓	
ATLAS-SUSY-2018-13 [78]	v3.0	displaced jets	139.0			✓	
ATLAS-SUSY-2018-14 [79]		displaced vertices	139.0			✓	PYHF
ATLAS-SUSY-2018-16 [80]	v3.0	2 soft ℓ + jets, EW	139.0	✓	✓	✓	PYHF
ATLAS-SUSY-2018-16 [80]	v3.0	2 soft ℓ + jets, EW (hino)	139.0			✓	
ATLAS-SUSY-2018-22 [81]	v3.0	multi-jets	139.0	✓		✓	
ATLAS-SUSY-2018-23 [82]		$Wh(\gamma\gamma)$, EW	139.0	✓	✓		
ATLAS-SUSY-2018-31 [83]		$2b + 2h(bb)$	139.0	✓		✓	PYHF
ATLAS-SUSY-2018-32 [84]	v2.3	2 OS ℓ	139.0	✓		✓	PYHF
ATLAS-SUSY-2018-33 [85]	v3.0	displaced vertices	136.0			✓	
ATLAS-SUSY-2018-40 [86]	v2.2	$2b + 2h(\tau\tau)$	139.0	✓	✓	✓	
ATLAS-SUSY-2018-41 [87]	v2.3	hadr. EW search	139.0	✓	✓	✓	SLv1
ATLAS-SUSY-2018-42 [88]	v3.0	charged LLPs, dE/dx	139.0	✓	✓	✓	
ATLAS-SUSY-2019-02 [89]	v2.3	2 soft ℓ's, EW	139.0	✓		✓	SLv1
ATLAS-SUSY-2019-08 [90]	v3.0	$1\ell + h(bb)$, EW	139.0	✓		✓	PYHF
ATLAS-SUSY-2019-09 [91]	v2.2	3ℓ, EW	139.0	✓	✓	✓	PYHF

Table B.2 List of CMS Run 2 analyses and their types of results in the SMODELS v2.3.0 database. The **New** column indicates the latest database version (from v2.2) that introduced significant changes for the analysis. Apart from the HSCP, disappearing tracks and displaced lepton searches, all analyses require $\not{E}_T$ in the final state (for conciseness omitted in the short descriptions). EW stands for electroweak (-ino or slepton) production. The last column specifies whether and how SRs are combined: "pyhf" means a HISTFACTORY model is used through interface with PYHF; "SLv1" ("SLv2") means that a covariance matrix (with a third moment) is used; and a blank means only the most sensitive SR is used

ID	New	Short description	Lumi. [fb^{-1}]	UL_{obs}	UL_{exp}	EM	Comb.
CMS-EXO-19-001 [92]		non-prompt jets	137.0			✓	
CMS-EXO-19-010 [93]		disappearing tracks	101.0			✓	
CMS-EXO-19-012 [94]	v3.0	dijet resonance	137.0	✓			
CMS-EXO-20-004 [95]	v3.0	monojet	137.0			✓	SLv1
CMS-EXO-20-008 [96]	v3.0	di-bottom resonance	138.0	✓	✓		
CMS-PAS-EXO-16-036 [97]		HSCP	12.9	✓			
CMS-PAS-SUS-16-052 [98]		ISR jet + soft ℓ	35.9	✓		✓	SLv1
CMS-SUS-16-009 [99]		0ℓ + jets, top tag	2.3	✓	✓		
CMS-SUS-16-032 [100]		2 b- or 2 c-jets	35.9	✓			
CMS-SUS-16-033 [101]		0ℓ + jets	35.9	✓	✓	✓	
CMS-SUS-16-034 [102]		2 SFOS ℓ	35.9	✓			
CMS-SUS-16-035 [103]		2 SS ℓ	35.9	✓			
CMS-SUS-16-036 [104]		0ℓ + jets	35.9	✓	✓		
CMS-SUS-16-037 [105]		1ℓ + jets with MJ	35.9	✓			
CMS-SUS-16-039 [106]	v2.2	multi-ℓ, EW	35.9	✓		✓	SLv1
CMS-SUS-16-041 [107]		multi-ℓ + jets	35.9	✓			
CMS-SUS-16-042 [108]		1ℓ + jets	35.9	✓			
CMS-SUS-16-043 [109]		$Wh(bb)$, EW	35.9	✓			
CMS-SUS-16-045 [110]		$2b + 2h(\gamma\gamma)$	35.9	✓			
CMS-SUS-16-046 [111]		high-p_T γ	35.9	✓			
CMS-SUS-16-047 [112]		γ + jets, high H_T	35.9	✓			
CMS-SUS-16-048 [113]	v2.2	2 OS ℓ, soft	35.9			✓	SLv1
CMS-SUS-16-050 [114]	v2.2	0ℓ + top tag	35.9	✓	✓	✓	SLv1
CMS-SUS-16-051 [115]		1ℓ stop	35.9	✓	✓		
CMS-SUS-17-003 [116]		2 taus	35.9	✓			
CMS-SUS-17-004 [117]		EW combination	35.9	✓			
CMS-SUS-17-005 [118]		1ℓ + jets, top tag	35.9	✓	✓		
CMS-SUS-17-006 [119]		jets + boosted $h(bb)$	35.9	✓	✓		
CMS-SUS-17-009 [120]		SFOS ℓ	35.9	✓	✓		
CMS-SUS-17-010 [121]		2ℓ stop	35.9	✓	✓		
CMS-SUS-18-002 [122]		γ + (b-)jets, top tag	35.9	✓	✓		
CMS-SUS-18-004 [123]	v2.2	2–3 soft ℓ's	137.0	✓	✓		
CMS-SUS-18-007 [124]	v2.2	$2h(\gamma\gamma)$, EW	77.5	✓	✓		
CMS-SUS-19-006 [125]	v2.2	0ℓ + jets, MHT	137.0	✓	✓	✓	SLv1
CMS-SUS-19-008 [126]	v2.2	2–3ℓ + jets	137.0	✓	✓		
CMS-SUS-19-009 [127]	v2.3	1ℓ + jets, MHT	137.0	✓	✓		

(continued)

Table B.2 (continued)

ID	New	Short description	Lumi. [fb^{-1}]	UL_{obs}	UL_{exp}	EM	Comb.
CMS-SUS-19-010 [128]	v2.3	jets + top and W-tag	137.0	✓	✓		
CMS-SUS-19-011 [129]	v2.2	2ℓ stop	137.0	✓	✓		
CMS-SUS-19-013 [130]	v2.2	jets + boosted Z's	137.0	✓	✓		
CMS-SUS-20-001 [131]	v2.2	SFOS ℓ	137.0	✓	✓		
CMS-SUS-20-002 [132]	v2.2	stop combination	137.0	✓	✓		
CMS-SUS-20-004 [133]	v2.3	2 $h(bb)$, EW	137.0	✓	✓	✓	SLv2
CMS-SUS-21-002 [134]	v2.3	hadr. EW search	137.0	✓	✓	✓	SLv1
CMS-SUS-21-007 [135]	v3.0	1 ℓ + top and W-tag	138.0	✓			

Table B.3 List of 15 ATLAS and 18 CMS Run 1 analyses and their types of results in the SMODELS v3.0.0 database. The **New** column indicates the latest database version (from v2.2) that introduced significant changes for the analysis. Apart from the HSCP searches, all analyses require $\not{E}_T$ in the final state (for conciseness omitted in the short descriptions). EW stands for electroweak(-ino) production. The column **Comb.** from Tables B.1 and B.2 is omitted because none of the Run 1 analyses provides any information on background correlations

ID	New	Short description	Lumi. [fb^{-1}]	UL_{obs}	UL_{exp}	EM
ATLAS-EXOT-2013-11 [136]	v3.0	dijet resonance	20.3	✓		
ATLAS-SUSY-2013-02 [137]		0ℓ + 2–6 jets	20.3	✓		✓
ATLAS-SUSY-2013-04 [138]		0ℓ + 7–10 jets	20.3	✓		✓
ATLAS-SUSY-2013-05 [139]		0ℓ + $2b$-jets	20.1	✓		✓
ATLAS-SUSY-2013-08 [140]		$Z(\ell\ell)$ + b-jets	20.3	✓		
ATLAS-SUSY-2013-09 [141]		2 SS ℓ + 0–3 b-jets	20.3	✓		✓
ATLAS-SUSY-2013-11 [142]		2ℓ (e, μ), EW	20.3	✓		✓
ATLAS-SUSY-2013-12 [143]	v2.3	3ℓ (e, μ, τ), EW	20.3	✓		✓
ATLAS-SUSY-2013-15 [144]		1ℓ + 4 $(1b)$ jets	20.3	✓		✓
ATLAS-SUSY-2013-16 [145]		0ℓ + 6 $(2b)$ jets	20.1	✓		✓
ATLAS-SUSY-2013-18 [146]		jets + $\geq 3b$-jets	20.1	✓		✓
ATLAS-SUSY-2013-19 [147]		2 OS ℓ + $(b$-)jets	20.3	✓		
ATLAS-SUSY-2013-20 [148]		2ℓ (e, μ) + jets	20.3	✓		
ATLAS-SUSY-2013-21 [149]		monojet or c-jet	20.3			✓
ATLAS-SUSY-2013-23 [150]		1ℓ + 2 b-jets (or 2γ)	20.3	✓		
ATLAS-SUSY-2014-03 [151]		≥ 2 $(c$-)jets	20.3			✓
CMS-EXO-12-026 [152]		HSCP	18.8	✓		
CMS-EXO-12-059 [153]	v3.0	dijet resonance	19.7	✓	✓	
CMS-EXO-13-006 [154]		HSCP	18.8			✓
CMS-EXO-16-057 [155]	v3.0	di-bottom resonance	19.7	✓	✓	
CMS-PAS-SUS-13-015 [156]		≥ 5 $(1b)$ jets	19.4			✓

(continued)

Table B.3 (continued)

ID	New	Short description	Lumi. [fb^{-1}]	UL_{obs}	UL_{exp}	EM
CMS-PAS-SUS-13-016 [157]		$2\ell + \geq 4\ (2b)$ jets	19.7	✓		✓
CMS-PAS-SUS-13-018 [158]		1–2 b-jets, M_{CT}	19.4	✓		
CMS-PAS-SUS-13-023 [159]		0ℓ stop	18.9	✓		
CMS-SUS-12-024 [160]		$0\ell + \geq 3\ (1b)$ jets	19.4	✓		✓
CMS-SUS-12-028 [161]		multi (b-)jets, α_T	11.7	✓	✓	
CMS-SUS-13-002 [162]		$\geq 3\ell$ (+jets)	19.5	✓	✓	
CMS-SUS-13-004 [163]		$\geq 1b$-jet, Razor	19.3	✓		
CMS-SUS-13-006 [164]		multi-ℓ, EW	19.5	✓		
CMS-SUS-13-007 [165]		$1\ell + \geq 2b$-jets	19.3	✓		✓
CMS-SUS-13-011 [166]		$1\ell + \geq 4\ (1b)$jets	19.5	✓		✓
CMS-SUS-13-012 [167]		jets + $\not{H}_T$	19.5	✓	✓	✓
CMS-SUS-13-013 [168]		2 SS ℓ + (b-)jets	19.5	✓	✓	✓
CMS-SUS-13-019 [169]		≥ 2 jets, M_{T2}	19.5	✓		
CMS-SUS-14-010 [170]		b-jets + $4W$	19.5	✓	✓	
CMS-SUS-14-021 [171]		soft ℓ, low jet mult.	19.7	✓	✓	✓

Appendix C
Database Add-Ons

This appendix adds additional information to the thesis. It presents the database add-ons the user can specify when running SModelS.

When running SMODELS, the user has to specify which database to use. This is done in the `parameters.ini` file by giving the path to the `smodels-database` folder, the path to a pickle file or a URL path. Details are explained in the Using SMODELS section of the online manual. The available databases can be seen on the smodels-database-release page on Github.

Shorthand notations are available: `path=official` refers to the official database of the users' SMODELS version, while `path=latest` refers to the latest available database release. The '+' operator allows for extending the "official" or "latest" database with add-ons:

`+fastlim`: adds fastlim results (from early 8 TeV ATLAS analyses); from v2.1 onward

`+superseded`: adds results which were previously available but were superseded by newer ones; from v2.1 onward

`+nonaggregated`: adds analyses with non-aggregated SRs in addition to the aggregated results in CMS analyses; from v2.2 onward

`+full_llhds`: replaces simplified HISTFACTORY statistical models by full ones in ATLAS analyses; from v2.3 onward (careful, this increases a lot the runtime!)

Examples are "official+nonaggregated" or "official+nonaggregated+full_llhds". Note that order matters: results are replaced in the specified sequence, so "full_llhds+official" will fall back onto the official database with simplified HISTFACTORY statistical models. The add-ons can also be used alone, e.g. `path=full_llhds`, though this is of limited practical use. Finally, "debug" refers to a version of the database with extra information that is however not intended for usage by a regular user and only mentioned s here for completeness.

T. Pascal, *Beyond the Standard Model at the Large Hadron Collider: From Simplified Model Constraints to Artificial Proto-Modelling for Dispersed Signals*, Springer Theses, https://doi.org/10.1007/978-3-032-09391-2

References

1. Neutelings I (2016) Signatures of long-lived particles. https://tikz.net/bsm_longlived/. Inspired by J. Antonelli's presentation at ICHEP 2016: https://indico.cern.ch/event/432527/contributions/1072042/
2. Giudice GF, Romanino A (2004) Split supersymmetry. Nucl. Phys. B 699:65–89. [Erratum: Nucl.Phys.B 706, 487–487 (2005)]. https://doi.org/10.1016/j.nuclphysb.2004.08.001. arXiv:hep-ph/0406088
3. Hewett JL et al (2004) Signatures of long-lived gluinos in split supersymmetry. J High Energy Phys 09:070. https://doi.org/10.1088/1126-6708/2004/09/070. arXiv:hep-ph/0408248
4. Arkani-Hamed N et al (2005) Aspects of split supersymmetry. Nucl Phys B709:3–46. https://doi.org/10.1016/j.nuclphysb.2004.12.026. arXiv:hep-ph/0409232
5. Sirunyan AM et al (2019) Search for long-lived particles decaying into displaced jets in proton-proton collisions at $\sqrt{s} = 13$ TeV. Phys Rev D 99(3):032011. https://doi.org/10.1103/PhysRevD.99.032011. arXiv:1811.07991 [hep-ex]
6. Hayrapetyan A, CMS, et al (2025) Search for heavy long-lived charged particles with large ionization energy loss in proton-proton collisions at $\sqrt{s} = 13$ TeV. J High Energy Phys 4:109. https://doi.org/10.1007/JHEP04(2025)109
7. Sirunyan AM et al (2020) Searches for physics beyond the standard model with the M_{T2} variable in hadronic final states with and without disappearing tracks in proton-proton collisions at $\sqrt{s} = 13$ TeV. Eur Phys J C 80(1):3. https://doi.org/10.1140/epjc/s10052-019-7493-x. arXiv:1909.03460 [hep-ex]
8. Aad G et al (2022) Search for long-lived charginos based on a disappearing-track signature using 136 fb^{-1} of *pp* collisions at $\sqrt{s} = 13$ TeV with the ATLAS detector. Eur Phys J C 82(7):606. https://doi.org/10.1140/epjc/s10052-022-10489-5. arXiv:2201.02472 [hep-ex]
9. Aad G et al (2024) The quest to discover supersymmetry at the ATLAS experiment. arXiv:2403.02455 [hep-ex]
10. Carena M, Quiros M, Wagner CEM (1996) Opening the window for electroweak baryogenesis. Phys Lett B 380:81–91. https://doi.org/10.1016/0370-2693(96)00475-3. arXiv:hep-ph/9603420
11. Delepine D et al (1996) A Light stop and electroweak baryogenesis. Phys Lett B 386:183–188. https://doi.org/10.1016/0370-2693(96)00921-5. arXiv:hep-ph/9604440
12. Aaboud M et al (2019) Search for heavy charged long-lived particles in the ATLAS detector in 36.1 fb^{-1} of proton-proton collision data at $\sqrt{s} = 13$ TeV. Phys Rev D 99(9):092007. https://doi.org/10.1103/PhysRevD.99.092007. arXiv:1902.01636 [hep-ex]

T. Pascal, *Beyond the Standard Model at the Large Hadron Collider: From Simplified Model Constraints to Artificial Proto-Modelling for Dispersed Signals*, Springer Theses, https://doi.org/10.1007/978-3-032-09391-2

13. Hayrapetyan A et al (2024) Search for supersymmetry in final states with disappearing tracks in proton-proton collisions at s =13 TeV. Phys Rev D 109(7):072007. https://doi.org/10.1103/PhysRevD.109.072007. arXiv:2309.16823 [hep-ex]
14. Aad G et al (2021) Search for displaced leptons in $\sqrt{s} = 13$ TeV *pp* collisions with the ATLAS detector. Phys Rev Lett 127(5):051802. https://doi.org/10.1103/PhysRevLett.127.051802. arXiv:2011.07812 [hep-ex]
15. Tumasyan A et al (2022) Search for long-lived particles decaying to leptons with large impact parameter in proton–proton collisions at $\sqrt{s} = 13$ TeV. Eur Phys J C 82(2):153. https://doi.org/10.1140/epjc/s10052-022-10027-3. arXiv:2110.04809 [hep-ex]
16. Aad G et al (2023) Search for pairs of muons with small displacements in *pp* collisions at $\sqrt{s} = 13$ TeV with the ATLAS detector. Phys Lett B 846:138172. https://doi.org/10.1016/j.physletb.2023.138172. arXiv:2305.02005 [hep-ex]
17. Aad G et al (2023) Search for heavy, long-lived, charged particles with large ionisation energy loss in *pp* collisions at $\sqrt{s} = 13$ TeV using the ATLAS experiment and the full Run 2 dataset. J High Energy Phys 2306:158. https://doi.org/10.1007/JHEP06(2023)158. arXiv:2205.06013[hep-ex]
18. ATLAS (2023) Search for heavy, long lived charged particles with large specific ionisation and low-beta in 140 fb-1 of p-p collisions at sqrt(s) = 13 TeV using the ATLAS experiment. ATLAS-CONF-2023-044. CERN, Geneva. https://cds.cern.ch/record/2870112. All figures including auxiliary figures are available at https://atlas.web.cern.ch/Atlas/GROUPS/PHYSICS/CONFNOTES/ATLAS-CONF-2023-044
19. Ibe M, Matsumoto S, Sato R (2013) Mass splitting between charged and neutral winos at two-loop level. Phys Lett B 721:252–260. https://doi.org/10.1016/j.physletb.2013.03.015. arXiv:1212.5989 [hep-ph]
20. Sirunyan AM et al (2020) Search for disappearing tracks in proton-proton collisions at $\sqrt{s} = 13$ TeV. Phys Lett B 806:135502. https://doi.org/10.1016/j.physletb.2020.135502. arXiv:2004.05153 [hep-ex]
21. Aad G et al. (2023) Search in diphoton and dielectron final states for displaced production of Higgs or Z bosons with the ATLAS detector in s =13 TeV *pp* collisions. Phys Rev D 108(1):012012. https://doi.org/10.1103/PhysRevD.108.012012. arXiv:2304.12885 [hep-ex]
22. Allanach BC et al (2002) The snowmass points and slopes: benchmarks for SUSY searches. Eur Phys J C 25:113–123. Ed. by Norman Graf. https://doi.org/10.1007/s10052-002-0949-3. arXiv:hep-ph/0202233
23. Aad G et al (2014) Search for nonpointing and delayed photons in the diphoton and missing transverse momentum final state in 8 TeV *pp* collisions at the LHC using the ATLAS detector. Phys Rev D 90(11):112005. https://doi.org/10.1103/PhysRevD.90.112005. arXiv:1409.5542 [hep-ex]
24. Chatrchyan S et al (2013) Search for long-lived particles decaying to photons and missing energy in proton-proton collisions at $\sqrt{s} = 7$ TeV. Phys Lett B 722:273–294. https://doi.org/10.1016/j.physletb.2013.04.027. arXiv:1212.1838 [hep-ex]
25. Sirunyan AM et al (2019) Search for long-lived particles using delayed photons in proton-proton collisions at $\sqrt{s} = 13$ TeV. Phys Rev D 100(11):112003. https://doi.org/10.1103/PhysRevD.100.112003. arXiv:1909.06166 [hep-ex]
26. Aad G et al (2021) Search for supersymmetry in events with four or more charged leptons in 139 fb^{-1} of $\sqrt{s} = 13$ TeV *pp* collisions with the ATLAS detector. J High Energy Phys 07:167. https://doi.org/10.1007/JHEP07(2021)167. arXiv:2103.11684 [hep-ex]
27. Aad G et al (2021) Search for R-parity-violating supersymmetry in a final state containing leptons and many jets with the ATLAS experiment using $\sqrt{s} = 13TeV$ proton-proton collision data. Eur Phys J C 81(11):1023. https://doi.org/10.1140/epjc/s10052-021-09761-x. arXiv:2106.09609 [hep-ex]
28. Aad G et al (2024) Search for pair production of squarks or gluinos decaying via sleptons or weak bosons in final states with two same-sign or three leptons with the ATLAS detector. J High Energy Phys 02:107. https://doi.org/10.1007/JHEP02(2024)107. arXiv:2307.01094 [hep-ex]

29. Aad G et al (2024) A search for R-parity-violating supersymmetry in final states containing many jets in *pp* collisions at $\sqrt{s} = 13$ TeV with the ATLAS detector. J High Energy Phys 05:003. https://doi.org/10.1007/JHEP05(2024)003. arXiv:2401.16333 [hep-ex]
30. Sirunyan AM et al (2018) Search for pair-produced resonances each decaying into at least four quarks in proton-proton collisions at $\sqrt{s} = 13$ TeV. Phys Rev Lett 121(14):141802. https://doi.org/10.1103/PhysRevLett.121.141802. arXiv:1806.01058 [hep-ex]
31. Hayrapetyan A et al (2024) Searches for pair-produced multijet resonances using data scouting in proton-proton collisions at $\sqrt{s} = 13$ TeV. arXiv:2404.02992 [hep-ex]
32. Sirunyan AM et al (2020) Search for physics beyond the standard model in events with jets and two same-sign or at least three charged leptons in proton-proton collisions at $\sqrt{s} = 13$ TeV. Eur. Phys. J. C 80(8):752. https://doi.org/10.1140/epjc/s10052-020-8168-3. arXiv:2001.10086 [hep-ex]
33. Aad G et al (2021) Search for phenomena beyond the standard model in events with large b-jet multiplicity using the ATLAS detector at the LHC. Eur Phys J C 81(1):11. [Erratum: Eur Phys J C 81:249 (2021)]. https://doi.org/10.1140/epjc/s10052-020-08730-0. arXiv:2010.01015 [hep-ex]
34. Chekhovsky V, CMS, et al (2025) Search for top squarks in final states with many light-flavor jets and 0, 1, or 2 charged leptons in proton-proton collisions at $\sqrt{s} = 13$ TeV. J High Energy Phys 10:236. https://doi.org/10.1007/JHEP10(2025)236
35. Aaboud M et al (2018) A search for pair-produced resonances in four-jet final states at $\sqrt{s} = 13$ TeV with the ATLAS detector. Eur Phys J. C 78(3):250. https://doi.org/10.1140/epjc/s10052-018-5693-4. arXiv:1710.07171 [hep-ex]
36. Sirunyan AM et al (2018) Search for pair-produced resonances decaying to quark pairs in proton-proton collisions at $\sqrt{s} = 13$ TeV. Phys Rev D 98(11):112014. https://doi.org/10.1103/PhysRevD.98.112014. arXiv:1808.03124 [hep-ex]
37. Aad G et al (2022) A search for an unexpected asymmetry in the production of e $+ \mu$-and e $- \mu+$ pairs in proton-proton collisions recorded by the ATLAS detector at s =13 TeV. Phys Lett B 830:137106. https://doi.org/10.1016/j.physletb.2022.137106. arXiv:2112.08090 [hep-ex]
38. Tumasyan A et al (2022) Search for new physics in the lepton plus missing transverse momentum final state in proton-proton collisions at $\sqrt{s} = 13$ TeV. J High Energy Phys 07:067. https://doi.org/10.1007/JHEP07(2022)067. arXiv:2202.06075 [hep-ex]
39. Sirunyan AM et al (2018) Search for lepton-flavor violating decays of heavy resonances and quantum black holes to eμfinal states in proton-proton collisions at $\sqrt{s} = 13$ TeV. J High Energy Phys 04:073. https://doi.org/10.1007/JHEP04(2018)073. arXiv:1802.01122 [hep-ex]
40. Aad G et al (2023) Search for direct production of winos and higgsinos in events with two same-charge leptons or three leptons in *pp* collision data at $\sqrt{s} = 13$ TeV with the ATLAS detector. J High Energy Phys 11:150. https://doi.org/10.1007/JHEP11(2023)150. arXiv:2305.09322[hep-ex]
41. Aad G et al (2021) Search for trilepton resonances from chargino and neutralino pair production in $\sqrt{s} = 13$ TeV *pp* collisions with the ATLAS detector. Phys Rev D 103(11):112003. https://doi.org/10.1103/PhysRevD.103.112003. arXiv:2011.10543 [hep-ex]
42. CMS (2024) Search for hadronic R-parity violating decays of electroweak superpartners using jet scaling patterns in multilepton events at $\sqrt{s} = 13$ TeV. Report Number CMS-PAS-SUS-23-015
43. Sirunyan AM et al (2021) Search for long-lived particles decaying to jets with displaced vertices in proton-proton collisions at $\sqrt{s} = 13$ TeV. Phys Rev D 104(5):052011. https://doi.org/10.1103/PhysRevD.104.052011. arXiv:2104.13474 [hep-ex]
44. Sirunyan AM et al (2021) Search for long-lived particles using displaced jets in proton-proton collisions at $\sqrt{s} = 13$ TeV. Phys Rev D 104(1):012015. https://doi.org/10.1103/PhysRevD.104.012015. arXiv:2012.01581 [hep-ex]

45. Aad G et al (2020) Search for long-lived, massive particles in events with a displaced vertex and a muon with large impact parameter in *pp* collisions at $\sqrt{s} = 13$ TeV with the ATLAS detector. Phys Rev D 102(3):032006. https://doi.org/10.1103/PhysRevD.102.032006. arXiv:2003.11956 [hep-ex]
46. Aad G et al (2023) Search for long-lived, massive particles in events with displaced vertices and multiple jets in *pp* collisions at $\sqrt{s} = 13$ TeV with the ATLAS detector. J High Energy Phys 2306:200. https://doi.org/10.1007/JHEP06(2023)200. arXiv:2301.13866 [hep-ex]

References for Appendix B

47. Aad G et al (2021) Search for new phenomena in events with an energetic jet and missing transverse momentum in *pp* collisions at $\sqrt{s} = 13$ TeV with the ATLAS detector. Phys Rev D 103(11):112006. https://doi.org/10.1103/PhysRevD.103.112006. arXiv:2102.10874[hep-ex]
48. Aad G et al (2020) Search for *tt* resonances in fully hadronic final states in *pp* collisions at $\sqrt{s} = 13$ TeV with the ATLAS detector. J High Energy Phys 10:061. https://doi.org/10.1007/JHEP10(2020)061. arXiv:2005.05138 [hep-ex]
49. Aad G et al (2020) Search for new resonances in mass distributions of jet pairs using 139 fb^{-1} of *pp* collisions at $\sqrt{s} = 13$ TeV with the ATLAS detector. J High Energy Phys 03:145. https://doi.org/10.1007/JHEP03(2020)145. arXiv:1910.08447 [hep-ex]
50. Aaboud M et al (2016) Search for bottom squark pair production in proton-proton collisions at $\sqrt{s} = 13$ TeV with the ATLAS detector. Eur Phys J C76(10):547. https://doi.org/10.1140/epjc/s10052-016-4382-4. arXiv:1606.08772 [hep-ex]
51. Aaboud M et al (2016) Search for top squarks in final states with one isolated lepton, jets, and missing transverse momentum in $\sqrt{s} = 13$ TeV *pp* collisions with the ATLAS detector. Phys Rev D94(5):052009. https://doi.org/10.1103/PhysRevD.94.052009. arXiv:1606.03903 [hep-ex]
52. Aaboud M et al (2016) Search for squarks and gluinos in final states with jets and missing transverse momentum at $\sqrt{s} = 13$ TeV with the ATLAS detector. Eur Phys J C76(7):392. https://doi.org/10.1140/epjc/s10052-016-4184-8. arXiv:1605.03814 [hep-ex]
53. Aad G et al (2016) Search for supersymmetry at $\sqrt{s} = 13$ TeV in final states with jets and two same-sign leptons or three leptons with the ATLAS detector. Eur Phys J C76(5):259. https://doi.org/10.1140/epjc/s10052-016-4095-8. arXiv: 1602.09058[hep-ex]
54. Aaboud M et al (2018) Search for long-lived charginos based on a disappearing-track signature in *pp* collisions at $\sqrt{s} = 13$ TeV with the ATLAS detector. J High Energy Phys 06:022. https://doi.org/10.1007/JHEP06(2018)022. arXiv:1712.02118 [hep-ex]
55. Aaboud M et al (2018) Search for squarks and gluinos in final states with jets and missing transverse momentum using 36 fb^{-1} of $\sqrt{s} = 13$ TeV *pp* collision data with the ATLAS detector. Phys Rev D97(11):112001. https://doi.org/10.1103/PhysRevD.97.112001. arXiv:1712.02332 [hep-ex]
56. Aaboud M et al (2018) Search for long-lived, massive particles in events with displaced vertices and missing transverse momentum in $\sqrt{s} = 13$ TeV *pp* collisions with the ATLAS detector. Phys Rev D97(5):052012. https://doi.org/10.1103/PhysRevD.97.052012. arXiv:1710.04901 [hep-ex]
57. Aaboud M et al (2017) Search for supersymmetry in final states with two same-sign or three leptons and jets using 36 fb^{-1} of $\sqrt{s} = 13$ TeV *pp* collision data with the ATLAS detector. J High Energy Phys 09:084 https://doi.org/10.1007/JHEP09(2017)084. [Erratum: JHEP08,121(2019)]. https://doi.org/10.1007/JHEP08(2019)121. arXiv:1706.03731 [hep-ex]
58. Aaboud M et al (2017) Search for a scalar partner of the top quark in the jets plus missing transverse momentum final state at $\sqrt{s} = 13$ TeV with the ATLAS detector. J High Energy Phys 12:085. https://doi.org/10.1007/JHEP12(2017)085. arXiv:1709.04183 [hep-ex]

59. Aaboud M et al (2018) Search for top-squark pair production in final states with one lepton, jets, and missing transverse momentum using 36 fb^{-1} of $\sqrt{s} = 13$ TeV *pp* collision data with the ATLAS detector. J High Energy Phys 06:108. https://doi.org/10.1007/JHEP06(2018)108. arXiv:1711.11520 [hep-ex]
60. Aaboud M et al (2017) Search for direct top squark pair production in final states with two leptons in $\sqrt{s} = 13$ TeV *pp* collisions with the ATLAS detector. Eur Phys J C77(12):898. https://doi.org/10.1140/epjc/s10052-017-5445-x. arXiv:1708.03247 [hep-ex]
61. Aaboud M et al (2018) Search for top squarks decaying to tau sleptons in *pp* collisions at $\sqrt{s} = 13$ TeV with the ATLAS detector. Phys Rev D98(3):032008. https://doi.org/10.1103/PhysRevD.98.032008. arXiv:1803.10178 [hep-ex]
62. Aaboud M et al (2018) Search for electroweak production of supersymmetric particles in final states with two or three leptons at $\sqrt{s} = 13$ TeV with the ATLAS detector. Eur Phys J C78(12):995. https://doi.org/10.1140/epjc/s10052-018-6423-7. arXiv:1803.02762 [hep-ex]
63. Aaboud M et al (2018) Search for supersymmetry in final states with charm jets and missing transverse momentum in 13 TeV *pp* collisions with the ATLAS detector. J High Energy Phys 09:050. https://doi.org/10.1007/JHEP09(2018)050. arXiv:1805.01649 [hep-ex]
64. Aaboud M et al (2018) Search for photonic signatures of gauge-mediated supersymmetry in 13 TeV *pp* collisions with the ATLAS detector. Phys Rev D97(9):092006. https://doi.org/10.1103/PhysRevD.97.092006. arXiv:1802.03158 [hep-ex]
65. Aaboud M et al (2017) Search for supersymmetry in events with b-tagged jets and missing transverse momentum in *pp* collisions at $\sqrt{s} = 13$ TeV with the ATLAS detector. J High Energy Phys 11:195. https://doi.org/10.1007/JHEP11(2017)195. arXiv:1708.09266 [hep-ex]
66. Aaboud M et al (2019) Search for heavy charged long-lived particles in the ATLAS detector in 36.1 fb^{-1} of proton-proton collision data at $\sqrt{s} = 13$ TeV. Phys Rev D99(9):092007. https://doi.org/10.1103/PhysRevD.99.092007. arXiv:1902.01636 [hep-ex]
67. Aaboud M et al (2018) Search for new phenomena using the invariant mass distribution of same-flavour opposite-sign dilepton pairs in events with missing transverse momentum in $\sqrt{s} = 13$ TeV *pp* collisions with the ATLAS detector. Eur Phys J C78(8):625. https://doi.org/10.1140/epjc/s10052-018-6081-9. arXiv:1805.11381 [hep-ex]
68. Aaboud M et al (2019) Search for chargino and neutralino production in final states with a Higgs boson and missing transverse momentum at $\sqrt{s} = 13$ TeV with the ATLAS detector. Phys Rev D100(1):012006. https://doi.org/10.1103/PhysRevD.100.012006. arXiv:1812.09432 [hep-ex]
69. Aaboud M et al (2018) Search for pair production of higgsinos in final states with at least three b-tagged jets in $\sqrt{s} = 13$ TeV *pp* collisions using the ATLAS detector. Phys Rev D98(9):092002. https://doi.org/10.1103/PhysRevD.98.092002. arXiv:1806.04030 [hep-ex]
70. Aaboud M et al (2018) Search for chargino-neutralino production using recursive jigsaw reconstruction in final states with two or three charged leptons in proton-proton collisions at $\sqrt{s} = 13$ TeV with the ATLAS detector. Phys Rev D98(9):092012. https://doi.org/10.1103/PhysRevD.98.092012. arXiv:1806.02293 [hep-ex]
71. Aad G et al (2020) Search for direct stau production in events with two hadronic τ-leptons in $\sqrt{s} = 13$ TeV *pp* collisions with the ATLAS detector. Phys Rev D101(3):032009. https://doi.org/10.1103/PhysRevD.101.032009. arXiv:1911.06660 [hep-ex]
72. Aad G et al (2023) Searches for new phenomena in events with two leptons, jets, and missing transverse momentum in 139 fb^{-1} of $\sqrt{s} = 13$ TeV *pp* collisions with the ATLAS detector. Eur Phys J C 83(6):515. https://doi.org/10.1140/epjc/s10052-023-11434-w. arXiv:2204.13072 [hep-ex]
73. Aad G et al (2020) Search for chargino-neutralino production with mass splittings near the electroweak scale in three-lepton final states in $\sqrt{s} = 13$ TeV *pp* collisions with the ATLAS detector. Phys Rev D101(7):072001. https://doi.org/10.1103/PhysRevD.101.072001. arXiv:1912.08479 [hep-ex]

74. Aad G et al (2021) Search for new phenomena in events with two opposite-charge leptons, jets and missing transverse momentum in pp collisions at $\sqrt{s} = 13$ TeV with the ATLAS detector. J High Energy Phys 04:165. https://doi.org/10.1007/JHEP04(2021)165. arXiv:2102.01444 [hep-ex]
75. Aad G et al (2020) Search for squarks and gluinos in final states with same-sign leptons and jets using 139 fb^{-1} of data collected with the ATLAS detector. J High Energy Phys 06:046. https://doi.org/10.1007/JHEP06(2020)046. arXiv:1909.08457 [hep-ex]
76. Aad G et al (2021) Search for squarks and gluinos in final states with one isolated lepton, jets, and missing transverse momentum at $\sqrt{s} = 13$ with the ATLAS detector. Eur Phys J C 81(7):600. https://doi.org/10.1140/epjc/s10052-021-09344-w. arXiv:2101.01629 [hep-ex]
77. Aad G et al (2020) Search for a scalar partner of the top quark in the all-hadronic $t\bar{t}$ plus missing transverse momentum final state at $\sqrt{s} = 13$ TeV with the ATLAS detector. Eur Phys J C 80(8):737. https://doi.org/10.1140/epjc/s10052-020-8102-8. arXiv:2004.14060 [hep-ex]
78. Aad G et al (2023) Search for long-lived, massive particles in events with displaced vertices and multiple jets in pp collisions at $\sqrt{s} = 13$ TeV with the ATLAS detector. J High Energy Phys 2306:200. https://doi.org/10.1007/JHEP06(2023)200. arXiv:2301.13866 [hep-ex]
79. Aad G et al (2021) Search for displaced leptons in $\sqrt{s} = 13$ TeV pp collisions with the ATLAS detector. Phys Rev Lett 127(5):051802. https://doi.org/10.1103/PhysRevLett.127.051802. arXiv:2011.07812 [hep-ex]
80. Aad G et al (2020) Searches for electroweak production of supersymmetric particles with compressed mass spectra in $\sqrt{s} = 13$ TeV pp collisions with the ATLAS detector. Phys Rev D 101(5):052005. https://doi.org/10.1103/PhysRevD.101.052005. arXiv:1911.12606 [hep-ex]
81. Aad G et al (2021) Search for squarks and gluinos in final states with jets and missing transverse momentum using 139 fb^{-1} of $\sqrt{s} = 13$ TeV pp collision data with the ATLAS detector. J High Energy Phys 02:143. https://doi.org/10.1007/JHEP02(2021)143. arXiv:2010.14293 [hep-ex]
82. Aad G et al (2020) Search for direct production of electroweakinos in final states with missing transverse momentum and a Higgs boson decaying into photons in pp collisions at $\sqrt{s} = 13$ TeV with the ATLAS detector. J High Energy Phys 10:005. https://doi.org/10.1007/JHEP10(2020)005. arXiv:2004.10894 [hep-ex]
83. Aad G et al (2019) Search for bottom-squark pair production with the ATLAS detector in final states containing Higgs bosons, b-jets and missing transverse momentum. J High Energy Phys 12:060. https://doi.org/10.1007/JHEP12(2019)060. arXiv:1908.03122 [hep-ex]
84. Aad G et al (2020) Search for electroweak production of charginos and sleptons decaying into final states with two leptons and missing transverse momentum in $\sqrt{s} = 13$ TeV pp collisions using the ATLAS detector. Eur Phys J C80(2):123. https://doi.org/10.1140/epjc/s10052-019-7594-6. arXiv:1908.08215 [hep-ex]
85. Aad G et al (2020) Search for long-lived, massive particles in events with a displaced vertex and a muon with large impact parameter in pp collisions at $\sqrt{s} = 13$ TeV with the ATLAS detector. Phys Rev D 102(3):032006. https://doi.org/10.1103/PhysRevD.102.032006. arXiv:2003.11956 [hep-ex]
86. Aad G et al (2021) Search for bottom-squark pair production in pp collision events at $\sqrt{s} = 13$ TeV with hadronically decaying τ-leptons, b-jets and missing transverse momentum using the ATLAS detector. Phys Rev D 104(3):032014. https://doi.org/10.1103/PhysRevD.104.032014. arXiv:2103.08189 [hep-ex]
87. Aad G et al (2021) Search for charginos and neutralinos in final states with two boosted hadronically decaying bosons and missing transverse momentum in pp collisions at $\sqrt{s} = 13$ TeV with the ATLAS detector. Phys Rev D 104(11):112010. https://doi.org/10.1103/PhysRevD.104.112010. arXiv:2108.07586 [hep-ex]
88. Aad G et al (2023) Search for heavy, long-lived, charged particles with large ionisation energy loss in pp collisions at $\sqrt{s} = 13$ TeV using the ATLAS experiment and the full Run 2 dataset. J High Energy Phys 2306:158. https://doi.org/10.1007/JHEP06(2023)158. arXiv:2205.06013 [hep-ex]

89. Aad G et al (2023) Search for direct pair production of sleptons and charginos decaying to two leptons and neutralinos with mass splittings near the W-boson mass in $\sqrt{s} = 13$ TeV *pp* collisions with the ATLAS detector. J High Energy Phys 06:031. https://doi.org/10.1007/JHEP06(2023)031. arXiv:2209.13935 [hep-ex]
90. Aad G et al (2020) Search for direct production of electroweakinos in final states with one lepton, missing transverse momentum and a Higgs boson decaying into two *b*-jets in *pp* collisions at $\sqrt{s} = 13$ TeV with the ATLAS detector. Eur Phys J C80(8):691. https://doi.org/10.1140/epjc/s10052-020-8050-3. arXiv:1909.09226 [hep-ex]
91. Aad G et al (2021) Search for chargino-neutralino pair production in final states with three leptons and missing transverse momentum in $\sqrt{s} = 13$ TeV *pp* collisions with the ATLAS detector. Eur Phys J C 81(12):1118. https://doi.org/10.1140/epjc/s10052-021-09749-7. arXiv:2106.01676 [hep-ex]
92. Sirunyan AM et al (2019) Search for long-lived particles using nonprompt jets and missing transverse momentum with proton-proton collisions at $\sqrt{s} = 13$ TeV. Phys Lett B797:134876. https://doi.org/10.1016/j.physletb.2019.134876. arXiv:1906.06441 [hep-ex]
93. Sirunyan AM et al (2020) Search for disappearing tracks in proton-proton collisions at $\sqrt{s} = 13$ TeV. Phys Lett B 806:135502. https://doi.org/10.1016/j.physletb.2020.135502. arXiv:2004.05153 [hep-ex]
94. Sirunyan AM et al (2020) Search for high mass dijet resonances with a new background prediction method in proton-proton collisions at $\sqrt{s} = 13$ TeV. J High Energy Phys 05:033. https://doi.org/10.1007/JHEP05(2020)033. arXiv:1911.03947 [hep-ex]
95. Tumasyan A et al (2021) Search for new particles in events with energetic jets and large missing transverse momentum in proton-proton collisions at $\sqrt{s} = 13$ TeV. J High Energy Phys 11:153. https://doi.org/10.1007/JHEP11(2021)153. arXiv:2107.13021 [hep-ex]
96. Tumasyan A et al (2023) Search for narrow resonances in the b-tagged dijet mass spectrum in proton-proton collisions at s =13 TeV. Phys Rev D 108(1):012009. https://doi.org/10.1103/PhysRevD.108.012009. arXiv:2205.01835 [hep-ex]
97. CMS Collaboration (2016) Search for heavy stable charged particles with 12.9 fb^{-1} of 2016 data. Tech. Rep. CMS-PAS-EXO-16-036. CERN, Geneva
98. CMS Collaboration (2017) Search for supersymmetry in events with at least one soft lepton, low jet multiplicity, and missing transverse momentum in proton-proton collisions at $\sqrt{s} = 13$ TeV. Tech. Rep. CMS-PAS-SUS-16-052. CERN, Geneva
99. Khachatryan V et al (2017) Search for supersymmetry in the all-hadronic final state using top quark tagging in *pp* collisions at $\sqrt{s} = 13$ TeV. Phys Rev D96(1):012004. https://doi.org/10.1103/PhysRevD.96.012004. arXiv:1701.01954 [hep-ex]
100. Sirunyan AM et al (2018) Search for the pair production of third-generation squarks with two-body decays to a bottom or charm quark and a neutralino in proton-proton collisions at $\sqrt{s} = 13$ TeV. Phys Lett B778:263–291. https://doi.org/10.1016/j.physletb.2018.01.012. arXiv:1707.07274 [hep-ex]
101. Sirunyan AM et al (2017) Search for supersymmetry in multijet events with missing transverse momentum in proton-proton collisions at 13 TeV. Phys Rev D96(3):032003. https://doi.org/10.1103/PhysRevD.96.032003. arXiv:1704.07781 [hep-ex]
102. Sirunyan AM et al (2018) Search for new phenomena in final states with two opposite-charge, same-flavor leptons, jets, and missing transverse momentum in *pp* collisions at $\sqrt{s} = 13$ TeV. J High Energy Phys 03:076. https://doi.org/10.1007/s13130-018-7845-2,10.1007/JHEP03(2018)076. arXiv:1709.08908 [hep-ex]
103. Sirunyan AM et al (2017) Search for physics beyond the standard model in events with two leptons of same sign, missing transverse momentum, and jets in proton-proton collisions at $\sqrt{s} = 13$ TeV. Eur Phys J C77(9):578. https://doi.org/10.1140/epjc/s10052-017-5079-z. arXiv:1704.07323 [hep-ex]
104. Sirunyan AM et al (2017) Search for new phenomena with the M_{T2} variable in the all hadronic final state produced in proton-proton collisions at $\sqrt{s} = 13$ TeV. Eur Phys J C77(10):710. https://doi.org/10.1140/epjc/s10052-017-5267-x. arXiv:1705.04650 [hep-ex]

105. Sirunyan AM et al (017) Search for supersymmetry in *pp* collisions at $\sqrt{s} = 13$ TeV in the single-lepton final state using the sum of masses of large-radius jets. Phys Rev Lett 119(15):151802. https://doi.org/10.1103/PhysRevLett.119.151802. arXiv:1705.04673 [hep-ex]
106. Sirunyan AM et al (2018) Search for electroweak production of charginos and neutralinos in multilepton final states in proton-proton collisions at $\sqrt{s} = 13$ TeV. J High Energy Phys 03:166. https://doi.org/10.1007/JHEP03(2018)166. arXiv:1709.05406 [hep-ex]
107. Sirunyan AM et al (2018) Search for supersymmetry in events with at least three electrons or muons, jets, and missing transverse momentum in proton-proton collisions at $\sqrt{s} = 13$ TeV. J High Energy Phys 02:067. https://doi.org/10.1007/JHEP02(2018)067. arXiv:1710.09154 [hep-ex]
108. Sirunyan AM et al (2018) Search for supersymmetry in events with one lepton and multiple jets exploiting the angular correlation between the lepton and the missing transverse momentum in proton-proton collisions at $\sqrt{s} = 13$ TeV. Phys Lett B780:384–409. https://doi.org/10.1016/j.physletb.2018.03.028. arXiv:1709.09814 [hep-ex]
109. Sirunyan AM et al (2017) Search for electroweak production of charginos and neutralinos in WH events in proton-proton collisions at $\sqrt{s} = 13$ TeV. J High Energy Phys 11:029. https://doi.org/10.1007/JHEP11(2017)029. arXiv:1706.09933 [hep-ex]
110. Sirunyan AM et al (2018) Search for supersymmetry with Higgs boson to diphoton decays using the razor variables at $\sqrt{s} = 13$ TeV. Phys Lett B779:166–190. https://doi.org/10.1016/j.physletb.2017.12.069. arXiv:1709.00384 [hep-ex]
111. Sirunyan AM et al (2018) Search for gauge-mediated supersymmetry in events with at least one photon and missing transverse momentum in pp collisions at $\sqrt{s} = 13$ TeV. Phys Lett B780:118–143. https://doi.org/10.1016/j.physletb.2018.02.045. arXiv:1711.08008 [hep-ex]
112. Sirunyan AM et al (2017) Search for supersymmetry in events with at least one photon, missing transverse momentum, and large transverse event activity in proton-proton collisions at $\sqrt{s} = 13$ TeV. J High Energy Phys 12:142. https://doi.org/10.1007/JHEP12(2017)142. arXiv:1707.06193 [hep-ex]
113. Sirunyan AM et al (2018) Search for new physics in events with two soft oppositely charged leptons and missing transverse momentum in proton-proton collisions at $\sqrt{s} = 13$ TeV. Phys Lett B 782:440–467. https://doi.org/10.1016/j.physletb.2018.05.062. arXiv:1801.01846 [hep-ex]
114. Sirunyan AM et al (2018) Search for supersymmetry in proton-proton collisions at 13 TeV using identified top quarks. Phys Rev D97(1):012007. https://doi.org/10.1103/PhysRevD.97.012007. arXiv:1710.11188 [hep-ex]
115. Sirunyan AM et al (2017) Search for top squark pair production in *pp* collisions at $\sqrt{s} = 13$ TeV using single lepton events. J High Energy Phys 10:019. https://doi.org/10.1007/JHEP10(2017)019. arXiv:1706.04402 [hep-ex]
116. Sirunyan AM et al (2018) Search for supersymmetry in events with a τ lepton pair and missing transverse momentum in proton-proton collisions at $\sqrt{s} = 13$ TeV. J High Energy Phys 11:151. https://doi.org/10.1007/JHEP11(2018)151. arXiv:1807.02048 [hep-ex]
117. Sirunyan AM et al (2018) Combined search for electroweak production of charginos and neutralinos in proton-proton collisions at $\sqrt{s} = 13$ TeV. J High Energy Phys 03:160. https://doi.org/10.1007/JHEP03(2018)160. arXiv:1801.03957 [hep-ex]
118. Sirunyan AM et al (2018) Search for top squarks decaying via four-body or chargino-mediated modes in single-lepton final states in proton-proton collisions at $\sqrt{s} = 13$ TeV. J High Energy Phys 09:065. https://doi.org/10.1007/JHEP09(2018)065. arXiv:1805.05784 [hep-ex]
119. Sirunyan AM et al (2018) Search for physics beyond the standard model in events with high-momentum higgs bosons and missing transverse momentum in proton-proton collisions at 13 TeV. Phys Rev Lett 120(24):241801. https://doi.org/10.1103/PhysRevLett.120.241801. arXiv:1712.08501 [hep-ex]

120. Sirunyan AM et al (2019) Search for supersymmetric partners of electrons and muons in proton-proton collisions at $\sqrt{s} = 13$ TeV. Phys Lett B790:140–166. https://doi.org/10.1016/j.physletb.2019.01.005. arXiv:1806.05264 [hep-ex]
121. Sirunyan AM et al (2018) Searches for pair production of charginos and top squarks in final states with two oppositely charged leptons in proton-proton collisions at $\sqrt{s} = 13$ TeV. J High Energy Phys 11:079. https://doi.org/10.1007/JHEP11(2018)079. arXiv:1807.07799 [hep-ex]
122. Sirunyan AM et al (2019) Search for supersymmetry in events with a photon, jets, b-jets, and missing transverse momentum in proton–proton collisions at 13 TeV. Eur Phys J C79(5):444. https://doi.org/10.1140/epjc/s10052-019-6926-x. arXiv:1901.06726 [hep-ex]
123. Tumasyan A et al (2022) Search for supersymmetry in final states with two or three soft leptons and missing transverse momentum in proton-proton collisions at $\sqrt{s} = 13$ TeV. J High Energy Phys 04:091. https://doi.org/10.1007/JHEP04(2022)091. arXiv:2111.06296 [hep-ex]
124. Sirunyan AM et al (2019) Search for supersymmetry using Higgs boson to diphoton decays at $\sqrt{s} = 13$ TeV. J High Energy Phys 11:109. https://doi.org/10.1007/JHEP11(2019)109. arXiv:1908.08500 [hep-ex]
125. Sirunyan AM et al (2019) Search for supersymmetry in proton-proton collisions at 13 TeV in final states with jets and missing transverse momentum. J High Energy Phys 10:244. https://doi.org/10.1007/JHEP10(2019)244. arXiv:1908.04722 [hep-ex]
126. Sirunyan AM et al (2020) Search for physics beyond the standard model in events with jets and two same-sign or at least three charged leptons in proton-proton collisions at $\sqrt{s} = 13$ TeV. Eur Phys J C 80(8):752. https://doi.org/10.1140/epjc/s10052-020-8168-3. arXiv:2001.10086 [hep-ex]
127. Sirunyan AM et al (2020) Search for direct top squark pair production in events with one lepton, jets, and missing transverse momentum at 13 TeV with the CMS experiment. J High Energy Phys 05:032. https://doi.org/10.1007/JHEP05(2020)032. arXiv:1912.08887 [hep-ex]
128. Sirunyan AM et al (2021) Search for top squark production in fully-hadronic final states in proton-proton collisions at $\sqrt{s} = 13$ TeV. Phys Rev D 104(5):052001. https://doi.org/10.1103/PhysRevD.104.052001. arXiv:2103.01290 [hep-ex]
129. Sirunyan AM et al (2021) Search for top squark pair production using dilepton final states in pp collision data collected at $\sqrt{s} = 13$ TeV. Eur Phys J C 81(1):3. https://doi.org/10.1140/epjc/s10052-020-08701-5. arXiv:2008.05936 [hep-ex]
130. Sirunyan AM et al (2020) Search for supersymmetry in proton-proton collisions at $\sqrt{s} = 13$ TeV in events with high-momentum Z bosons and missing transverse momentum. J High Energy Phys 09:149. https://doi.org/10.1007/JHEP09(2020)149. arXiv:2008.04422 [hep-ex]
131. Sirunyan AM et al (2021) Search for supersymmetry in final states with two oppositely charged same-flavor leptons and missing transverse momentum in proton-proton collisions at $\sqrt{s} = 13$ TeV. J High Energy Phys 04:123. https://doi.org/10.1007/JHEP04(2021)123. arXiv:2012.08600 [hep-ex]
132. Tumasyan A et al (2021) Combined searches for the production of supersymmetric top quark partners in proton–proton collisions at $\sqrt{s} = 13$ TeV. Eur Phys J C 81(11):970. https://doi.org/10.1140/epjc/s10052-021-09721-5. arXiv:2107.10892 [hep-ex]
133. Tumasyan A et al (2022) Search for higgsinos decaying to two Higgs bosons and missing transverse momentum in proton-proton collisions at $\sqrt{s} = 13$ TeV. J High Energy Phys 05:014. https://doi.org/10.1007/JHEP05(2022)014. arXiv:2201.04206 [hep-ex]
134. Tumasyan A et al (2023) Search for electroweak production of charginos and neutralinos at s =13TeV in final states containing hadronic decays of WW, WZ, or WH and missing transverse momentum. Phys Lett B 842:137460. https://doi.org/10.1016/j.physletb.2022.137460. arXiv:2205.09597 [hep-ex]
135. Tumasyan A et al (2023) Search for supersymmetry in final states with a single electron or muon using angular correlations and heavy-object identification in proton-proton collisions at $\sqrt{s} = 13$ TeV. J High Energy Phys 09:149. https://doi.org/10.1007/JHEP09(2023)149. arXiv:2211.08476 [hep-ex]

136. Aad G et al (2015) Search for new phenomena in the dijet mass distribution using $p - p$ collision data at $\sqrt{s} = 8$ TeV with the ATLAS detector. Phys Rev D 91(5):052007. https://doi.org/10.1103/PhysRevD.91.052007. arXiv:1407.1376 [hep-ex]
137. Aad G et al (2014) Search for squarks and gluinos with the ATLAS detector in final states with jets and missing transverse momentum using $\sqrt{s} = 8$ TeV proton–proton collision data. J High Energy Phys 1409:176. https://doi.org/10.1007/JHEP09(2014)176. arXiv:1405.7875 [hep-ex]
138. Aad G et al (2013) Search for new phenomena in final states with large jet multiplicities and missing transverse momentum at $\sqrt{s} = 8$ TeV proton-proton collisions using the ATLAS experiment. J High Energy Phys 10:130 https://doi.org/10.1007/JHEP10(2013)130. [Erratum: JHEP01,109(2014)]. https://doi.org/10.1007/JHEP01(2014)109. arXiv:1308.1841 [hep-ex]
139. Aad G et al (2013) Search for direct third-generation squark pair production in final states with missing transverse momentum and two b-jets in $\sqrt{s} = 8$ TeV pp collisions with the ATLAS detector. J High Energy Phys 1310:189. https://doi.org/10.1007/JHEP10(2013)189. arXiv:1308.2631 [hep-ex]
140. Aad G et al (2014) Search for direct top squark pair production in events with a Z boson, b-jets and missing transverse momentum in $\sqrt{s} = 8$ TeV pp collisions with the ATLAS detector. Eur Phys J C74(6):2883. https://doi.org/10.1140/epjc/s10052-014-2883-6. arXiv:1403.5222 [hep-ex]
141. Aad G et al (2014) Search for supersymmetry at $\sqrt{s} = 8$ TeV in final states with jets and two same-sign leptons or three leptons with the ATLAS detector. J High Energy Phys 06:035. https://doi.org/10.1007/JHEP06(2014)035. arXiv:1404.2500 [hep-ex]
142. Aad G et al (2014) Search for direct production of charginos, neutralinos and sleptons in final states with two leptons and missing transverse momentum in pp collisions at $\sqrt{s} = 8$ TeV with the ATLAS detector. J High Energy Phys 05:071. https://doi.org/10.1007/JHEP05(2014)071. arXiv:1403.5294 [hep-ex]
143. Aad G et al (2014) Search for direct production of charginos and neutralinos in events with three leptons and missing transverse momentum in $\sqrt{s} = 8$TeV pp collisions with the ATLAS detector. J High Energy Phys 1404:169. https://doi.org/10.1007/JHEP04(2014)169. arXiv:1402.7029 [hep-ex]
144. Aad G et al (2014) Search for top squark pair production in final states with one isolated lepton, jets, and missing transverse momentum in $\sqrt{s} = 8$ TeV pp collisions with the ATLAS detector. arXiv:1407.0583 [hep-ex]
145. Aad G et al (2014) Search for direct pair production of the top squark in all-hadronic final states in proton-proton collisions at $\sqrt{s} = 8$ TeV with the ATLAS detector. J High Energy Phys 09:015. https://doi.org/10.1007/JHEP09(2014)015. arXiv: 1406.1122 [hep-ex]
146. Aad G et al (2014) Search for strong production of supersymmetric particles in final states with missing transverse momentum and at least three b-jets at $s = 8$ TeV proton-proton collisions with the ATLAS detector. J High Energy Phys 10:024. https://doi.org/10.1007/JHEP10(2014)024. arXiv:1407.0600 [hep-ex]
147. Aad G et al (2014) Search for direct top-squark pair production in final states with two leptons in pp collisions at $\sqrt{s} = 8$ TeV with the ATLAS detector. J High Energy Phys 1406:124. https://doi.org/10.1007/JHEP06(2014)124. arXiv:1403.4853 [hep-ex]
148. Aad G et al (2015) Search for squarks and gluinos in events with isolated leptons, jets and missing transverse momentum at $\sqrt{s} = 8$ TeV with the ATLAS detector. J High Energy Phys 04:116. https://doi.org/10.1007/JHEP04(2015)116. arXiv:1501.03555 [hep-ex]
149. Aad G et al (2014) Search for pair-produced third-generation squarks decaying via charm quarks or in compressed supersymmetric scenarios in pp collisions at $\sqrt{s} = 8$ TeV with the ATLAS detector. Phys Rev D90(5):052008. https://doi.org/10.1103/PhysRevD.90.052008. arXiv:1407.0608 [hep-ex]
150. Aad G et al (2015) Search for direct pair production of a chargino and a neutralino decaying to the 125 GeV Higgs boson in $\sqrt{s} = 8$ TeV pp collisions with the ATLAS detector. Eur Phys J C75(5):208. https://doi.org/10.1140/epjc/s10052-015-3408-7. arXiv:1501.07110 [hep-ex]

151. Aad G et al (2015) Search for scalar charm quark pair production in *pp* collisions at $\sqrt{s}$ = 8 TeV with the ATLAS detector. Phys Rev Lett 114(16):161801. https://doi.org/10.1103/PhysRevLett.114.161801. arXiv:1501.01325 [hep-ex]
152. Chatrchyan S et al (2013) Searches for long-lived charged particles in pp collisions at $\sqrt{s}$ = 7 and 8 TeV. J High Energy Phys 07:122. https://doi.org/10.1007/JHEP07(2013)122. arXiv:1305.0491 [hep-ex]
153. Khachatryan V et al (2015) Search for resonances and quantum black holes using dijet mass spectra in proton-proton collisions at $\sqrt{s}$ = 8 TeV. Phys Rev D 91(5):052009. https://doi.org/10.1103/PhysRevD.91.052009. arXiv:1501.04198 [hep-ex]
154. Khachatryan V et al (2015) Constraints on the pMSSM, AMSB model and on other models from the search for long-lived charged particles in proton-proton collisions at $\sqrt{s}$ = 8 TeV. Eur Phys J C75(7):325. https://doi.org/10.1140/epjc/s10052-015-3533-3. arXiv:1502.02522 [hep-ex]
155. Sirunyan AM et al (2018) Search for narrow resonances in the b-tagged dijet mass spectrum in proton-proton collisions at $\sqrt{s}$ = 8 TeV. Phys Rev Lett 120(20):201801. https://doi.org/10.1103/PhysRevLett.120.201801. arXiv:1802.06149 [hep-ex]
156. CMS Collaboration (2013) Search for top squarks in multijet events with large missing momentum in proton-proton collisions at 8 TeV. Tech. Rep. CMS-PAS-SUS-13-015. CERN, Geneva
157. CMS Collaboration (2013) Search for supersymmetry in pp collisions at $\sqrt{s}$ = 8 TeV in events with two opposite sign leptons, large number of jets, b-tagged jets, and large missing transverse energy. Tech. Rep. CMS-PAS-SUS-13-016. CERN, Geneva
158. CMS Collaboration (2014) Search for direct production of bottom squark pairs. Tech. Rep. CMS- PAS-SUS-13-018. CERN, Geneva
159. CMS Collaboration (2015) A search for scalar top quark production and decay to all hadronic final states in pp collisions at $\sqrt{s}$ = 8 TeV. Tech. Rep. CMS-PAS-SUS-13-023. CERN, Geneva
160. Chatrchyan S et al (2013) Search for gluino mediated bottom- and top-squark production in multijet final states in *pp* collisions at 8 TeV. Phys. Lett. B725:243–270. https://doi.org/10.1016/j.physletb.2013.06.058. arXiv: 1305.2390 [hep-ex]
161. Chatrchyan S et al (2013) Search for supersymmetry in hadronic final states with missing transverse energy using the variables α_T and b-quark multiplicity in *pp* collisions at $\sqrt{s}$ = 8 TeV. Eur Phys J C73(9):2568. https://doi.org/10.1140/epjc/s10052-013-2568-6. arXiv:1303.2985 [hep-ex]
162. Chatrchyan S et al (2014) Search for anomalous production of events with three or more leptons in *pp* collisions at $\sqrt{s}$ = 8 TeV. Phys Rev D90:032006. https://doi.org/10.1103/PhysRevD.90.032006. arXiv:1404.5801 [hep-ex]
163. Khachatryan V et al (2015) Search for supersymmetry using razor variables in events with *b*-tagged jets in *pp* collisions at $\sqrt{s}$ = 8 TeV. Phys Rev D91:052018. https://doi.org/10.1103/PhysRevD.91.052018. arXiv:1502.00300 [hep-ex]
164. Khachatryan V et al (2014) Searches for electroweak production of charginos, neutralinos, and sleptons decaying to leptons and W, Z, and Higgs bosons in *pp* collisions at 8 TeV. Eur Phys J C74(9):3036. https://doi.org/10.1140/epjc/s10052-014-3036-7. arXiv:1405.7570 [hep-ex].
165. Chatrchyan S et al (2014) Search for supersymmetry in *pp* collisions at $\sqrt{s}$ = 8 TeV in events with a single lepton, large jet multiplicity, and multiple b jets. Phys. Lett. B733:328–353. https://doi.org/10.1016/j.physletb.2014.04.023. arXiv:1311.4937 [hep-ex]
166. Chatrchyan S et al (2013) Search for top-squark pair production in the single-lepton final state in *pp* collisions at $\sqrt{s}$ = 8 TeV. Eur Phys J C73(12):2677. https://doi.org/10.1140/epjc/s10052-013-2677-2. arXiv:1308.1586 [hep-ex]
167. Chatrchyan S et al (2014) Search for new physics in the multijet and missing transverse momentum final state in proton-proton collisions at $\sqrt{s}$ = 8 TeV. J High Energy Phys 06:055. https://doi.org/10.1007/JHEP06(2014)055. arXiv:1402.4770 [hep-ex]

168. Chatrchyan S et al (2014) Search for new physics in events with same-sign dileptons and jets in *pp* collisions at $\sqrt{s} = 8$ TeV. J High Energy Phys 1401:163. https://doi.org/10.1007/JHEP01(2014)163. arXiv:1311.6736
169. Khachatryan V et al (2015) Searches for supersymmetry using the M_{T2} variable in hadronic events produced in *pp* collisions at 8 TeV. J High Energy Phys 05:078. https://doi.org/10.1007/JHEP05(2015)078. arXiv: 1502.04358 [hep-ex]
170. Khachatryan V et al (2015) Searches for supersymmetry based on events with b jets and four W bosons in *pp* collisions at 8 TeV. Phys Lett B745:5–28. https://doi.org/10.1016/j.physletb.2015.04.002. arXiv:1412.4109 [hep-ex]
171. Khachatryan V et al (2016) Search for supersymmetry in events with soft leptons, low jet multiplicity, and missing transverse energy in proton–proton collisions at $\sqrt{s} = 8$ TeV. Phys Lett B759:9–35. https://doi.org/10.1016/j.physletb.2016.05.033. arXiv:1512.08002 [hep-ex]

MIX
Papier aus verantwortungsvollen Quellen
Paper from responsible sources
FSC® C105338
FSC
www.fsc.org

If you have any concerns about our products,
you can contact us on
ProductSafety@springernature.com

In case Publisher is established outside the EU,
the EU authorized representative is:
Springer Nature Customer Service Center GmbH
Europaplatz 3, 69115 Heidelberg, Germany

Printed by Libri Plureos GmbH
in Hamburg, Germany